STP 1263

Thermomechanical Fatigue Behavior of Materials: Second Volume

Michael J. Verrilli and Michael G. Castelli, Editors

ASTM Publication Number (PCN):
04-012630-30

ASTM
100 Barr Harbor Drive
West Conshohocken, PA 19428-2959
Printed in the U.S.A.

Library of Congress Cataloging-in-Publication Data

Thermomechanical fatigue behavior of materials. Second volume /
 Michael J. Verrilli and Michael G. Castelli, editors.
 (STP : 1263)
 Contains papers presented at the Second Symposium on
Thermomechanical Fatigue Behavior of materials held 14–15 November
1994 in Phoenix, AZ"—Foreword.
 "ASTM publication code number (PCN) 04-012630-30."
 Includes indexes.
 ISBN 0-8031-2001-X
 1. Alloys—Fatigue. 2. Alloys—Thermomechanical properties.
3. Composite materials—Thermomechanical properties. 4. Fracture
mechanics. I. Verrilli, Michael J. II. Castelli, Michael G.
TA483.T48 1996
620.1'617—dc20   96-19174
                                                            CIP

Photocopy Rights

Authorization to photocopy items for internal, personal, or educational classroom use, or the internal, personal, or educational classroom use of specific clients, is granted by the American Society for Testing and Materials (ASTM) provided that the appropriate fee is paid to the Copyright
Clearance Center, 222 Rosewood Drive, Danvers, MA 01923, Tel: 508-750-8400 online: http://
www.copyright.com/.

Peer Review Policy

Each paper published in this volume was evaluated by three peer reviewers. The authors
addressed all of the reviewers' comments to the satisfaction of both the technical editor(s) and the
ASTM Committee on Publications.

To make technical information available as quickly as possible, the peer-reviewed papers in this
publication were prepared "camera-ready" as submitted by the authors.

The quality of the papers in this publication reflects not only the obvious efforts of the authors
and the technical editor(s), but also the work of these peer reviewers. The ASTM Committee on
Publications acknowledges with appreciation their dedication and contribution of time and effort on
behalf of ASTM.

Printed in Ann Arbor, MI
1996

**Books are to be returned on or before
the last date below.**

LIBREX —

Foreword

This publication, *Thermomechanical Fatigue Behavior of Materials: Second Volume,* contains papers presented at the Second Symposium on Thermomechanical Fatigue Behavior of Materials held 14–15 November 1994 in Phoenix, AZ. The symposium was sponsored by ASTM Committee E-8 on Fatigue and Fracture. Michael J. Verrilli, of the NASA Lewis Research Center in Cleveland, and Michael G. Castelli, with NYMA, Inc., NASA LeRC Group in Brook Park, OH, presided as symposium chairmen and are editors of the resulting publication.

Contents

Overview

Background

Virtually all high-temperature components experience service cycles that include simultaneous temperature and load cycling, or thermomechanical fatigue (TMF). Materials testing and characterization are required to capture the often unique synergistic effects of combined thermal and mechanical loading. This information can make possible the proper formulation of models used for component lifetime prediction and design, and can guide materials development.

The paper included in this volume were written in conjunction with a symposium organized to disseminate current research in the area of TMF behavior of materials. ASTM, through the members of Committee E-8 on Fatigue and Fracture, has traditionally had a keen interest in thermal and thermomechanical fatigue, as evidenced by the numerous STPs which discuss the issue. In 1968, the first ASTM paper on TMF appeared in STP 459, *Fatigue at High Temperature.* Carden and Slade discussed the behavior of Hastelloy X under strain-controlled isothermal and TMF conditions. *The Handbook of Fatigue Testing* (STP 566, published in 1974) described a technique for thermal fatigue testing of coupon specimens as well as the structural TMF test system for the airframe of the Concorde. STP 612, *Thermal Fatigue of Materials and Components* (1975) is the proceedings of the first comprehensive ASTM symposium on thermal and thermomechanical fatigue. Paper topics included TMF test techniques, life prediction methods, and TMF behavior of advanced materials such as ceramics and directionally-solidifed superalloys. A symposium entitled "Low Cycle Fatigue" (STP 942) held in 1988 contained five papers on thermal and thermomechanical fatigue. TMF test technqiues, deformation behavior and modeling, and observation of microstructural damage were presented. The first ASTM STP devoted to TMF of materials (and the predecessor to this volume) was the proceedings of the 1991 symposium on *TMF Behavior of Materials* (STP 1186). Several papers discussed the role of environmental attack on performance and life modeling of high-temperature alloys subjected to TMF loadings. In addition, this STP contains two papers which discuss TMF of metal matrix composites, an indication of the emerging interest in this class of materials for high-temperature applications.

ASTM is also actively pursuing development of a standard practice for TMF testing. Numerous standard practices for isothermal low-cycle fatigue testing exist (including ASTM E606 for strain-controlled testing and E466 for load-controlled testing), but none exist for TMF. However, the first standard for strain-controlled TMF testing of metallic materials is under development by an ISO working group in conjunction with ASTM Committee E-8 on Fatigue and Fracture. We expect that the resulting international standard will be the foundation of an ASTM standard.

Summary of the Papers

High-Temperature Structural Alloys

Most papers in this section discuss high-temperature alloys used for gas turbine engines, such as Ni-base superalloys and titanium alloys. Steels which are subjected to TMF conditions in power generation applications are discussed as well. The topics of the papers in this section on TMF behavior of high-temperature alloys include crack initiation and growth, novel experimental techniques, deformation modeling, and the role of coatings on life and microcracking.

Chataigner and Remy studied the TMF behavior of a chromium-aluminum coated [001] single crystal using a diamond-shaped strain-temperature cycle. They found no difference between the lives of coated and bare specimens. A life prediction model based on microcrack propagation due to fatigue and oxidation damage is evaluated.

Kraft and Mughrabi examined the crack evolution and microstructural changes of a single-crystal superalloy subjected to in-phase, out-of-phase, and diamond TMF cycle types. The morphology of the γ' structure after TMF cycling was found to be dependent on cycle type. The maximum tensile stress response of the [001] oriented specimens governed life for all the cycle types.

Meyer-Olbersleben et al. performed thermal fatigue (TF) experiments on blade-shaped, wedge specimens made of single-crystal superalloys. They proposed an "integrated" approach where the temperature-strain history measured during TF experiments is used as the basic cycle for a TMF investigation. This method is suggested as an alternative to finite element calculations to deduce the stress history of wedge specimens.

Bressers, Martínez-Esnaola, Timm, and co-workers contributed three papers examining the role of a coating on the TMF behavior of single-crystal Ni-base superalloys. In the first contribution, Bressers et al. studied the effect of TMF cycle type on the lives of a coated and uncoated single-crystal superalloy. This study reports the various modes of crack initiation, crack growth, and the stress and inelastic strain response due to in-phase cycle and $-135°$ lag cycle. For uncoated specimens, the cycle type significantly affected the mode of crack initiation. Also, life debits due to the presence of the coating varied as a function of strain range and cycle type. In the second paper by this group, Martínez-Esnaola et al. investigated cracking of the coating on the Ni-base single crystals subjected to the $-135°$ lag TMF cycle. The mode of coating crack initiation depended on the applied mechanical strain range, while crack initiation of bare specimens occurred via a single mode. A fracture mechanics model was applied to examine the effects of parameters such as coating thickness and temperature on the coating toughness, strain to cracking, and crack density. In the third contribution, Bressers et al. used a crack shielding model in an effort to explain the experimentally-observed debit in TMF life due to the presence of the coating on the single-crystal specimens. Higher crack-growth rates of the main crack were observed in coated specimens relative to the uncoated material. The crack shielding model was used in a parametric study to stimulate the growth of interacting, parallel cracks. The results of the analysis indicated that crack shielding effects due to the presence of the coating did not play a primary role in the life difference, and that other factors should be investigated as the potential cause, such as presence of residual stresses or thermal expansion mismatch of the coating and substrate.

Two papers discussed TMF of stainless steels. Zamrick and his co-workers compared the TMF and high-temperature LCF behavior of type 316 stainless steel. Yamauchi et al. conducted structural thermal fatigue tests on tubes of 304 stainless steel to simulate the service conditions. A FEM stress analysis revealed the stress state and temperature-strain phasing for the inner and outer surfaces of the pipe which experienced through-thickness gradients during the tests. The analysis, combined with uniaxial specimen tests, explained the experimentally-observed difference of crack initiation life between the inner and outer surfaces.

Arnold et al. present their recent developments in viscoplastic deformation modeling. The model utilizes an evolutionary law that has nonlinear kinematic hardening and both thermal and strain-induced recovery mechanisms. One tensorial internal state variable is employed. A unique aspect of the present model is the inclusion of nonlinear hardening in the evoluation law for the back stress. Verification of the proposed model is shown using non-standard isothermal and thermomechanical deformation tests on a titanium alloy commonly used as the matrix in SiC fiber-reinforced composites.

A novel test method to assess the role of temperature in determining the operative fracture mode and crack growth rates in superalloy single crystals is presented in the paper by Cunningham and DeLuca. The technique involves varying temperature with crack length according to a user-supplied function and was shown to work with several specimen geometries. Applications of the test method for screening of temperature-dependent crack growth behavior and model verification are discussed.

Gao et al. describe a unique thermal fatigue test rig fitted with a chamber that enables testing under various environments, including flowing hydrogen. The performance of the rig and the associated test procedures were evaluated through experimental testing of a γ TiAl alloy.

Dai et al. discuss thermal mechanical fatigue crack growth (TMFCG) results obtained for two titanium alloys. Tests were conducted using several strain-temperature phasings, and the ability of several fracture mechanics parameters to correlate the data was evaluated. Also, a model to predict TMFCG rates is presented and its application to estimate lives of engine components is discussed.

Titanium Matrix Composites

Over the past several years, silicon-carbon-fiber-reinforced titanium matrix composites (TMCs) have received considerable attention in the aeronautics and aerospace research communities for potential use in advanced high-temperature airframe and propulsion system applications. The obvious attractions of TMCs are the high stiffness and strength-to-weight ratios achievable at elevated temperatures, relative to current generation structural alloys. The papers included in the TMC section of this publication discuss many of the complex phenomenological behaviors and analytical modeling issues which arise under TMF loading conditions.

Coker et al. present a deformation analysis of a [0/90] TMC. A micromechanics approach is taken which treats the crossply as a three-constituent material consisting of a linear-elastic [0] fiber, a viscoplastic matrix in the [0] ply, and a viscoplastic [90] ply with damage to simulate fiber/matrix (f/m) interface separation, The authors clearly show the importance of treating the TMC as a thermoviscoplastic medium and the need to account for f/m separation when assessing [0/90] crossply macroscopic response. The contribution by Roberston and Mall features a modified Method of Cells micromechanics approach coupled with a unique f/m interface failure scheme based upon a probabalistic failure criterion. The proposed methodology incorporates the effects of both normal and shear f/m interface failures. Verification of the analysis is conducted under TMF loadings where the model appears to capture the progression of the interfacial damage with cycles.

Johnson et al. present a detailed experimental evaluation of the fatigue behavior of a [0/90] TMC subjected to a generic hypersonic flight profile. Material response under isolated segments of the flight profile are also examined to help identify critical combinations of load and time at temperature. Results indicate that sustained load at temperature had a more deleterious effect on fatigue life than that of a combined nonisothermal temperature profile and mechanical loading. Significant strain accumulations and eventual failure of the composite under sustained load conditions were found to result primarily from [90] f/m interface separation and sustained load crack growth, rather than more traditional creep mechanisms such as viscoplastic deformation of the matrix. Aksoy et al. also examine the fatigue performance of a TMC subjected to a mission cycle, but here the cycle was designed to simulate

the stress-temperature-time profile in a TMC ring reinforced impeller of a turboshaft engine. Results indicate that although the 14-minute mission cycle life was found to be significantly less than that revealed under isothermal conditions at a much faster loading rate (as expected), the failure mechanisms appeared to be very similar.

The paper contributed by Neu extends the concept of mechanistic maps to TMCs and presents unique TMF damage mechanisms maps for unidirectional laminates loaded in the fiber direction. Extensive experimental data and observations are weighted to guide the use of adopted and derived life prediction models and specify mechanistic regions of the maps. Combined life and damage mechanism maps are then constructed over a wide range of stress and temperature using the characterized prediction models. Ball presents experimental results on both [0] and [0/90] TMCs, along with a continuum damage-mechanics-based lifing approach. Damage is incorporated into the material constitutive equations at the ply level prior to the use of classical lamination theory to obtain the laminate response. Three types of damage are considered, including fiber breakage, f/m debonding, and matrix microcracking.

Nicholas and Johnson present a systematic study of the potential interactions between cyclic fatigue and creep (superimposed hold times) in [0] and [0/90] TMCs. Cyclic conditions involving low-frequency cycling and/or hold times at relatively high temperatures were found to result in failures dominated by time-dependent mechanisms with little or no contribution from fatigue-induced failure mechanisms. This observation was elucidated through a linear damage summation model which treats cycle- and time-dependent mechanisms separately. Blatt et al. also employ a linear summation model, but here in the context of understanding and predicting fatigue crack growth (FCG) rates. A unique study is presented examining the FCG behavior of a unidirectional TMC under TMF conditions. Results indicate that the amount of cycle time spent at or near T_{max} conditions was a key factor influencing the FCG rate. The proposed model appeared to be successful at predicting the FCG rate of a proof test involving a continually changing temperature and load range to produce a constant FCG rate.

Concluding Remarks

We feel that the work presented here is an outstanding reflection of the latest research in this demanding field and a noteworthy contribution to the literature. The contributions from both U.S. and international authors give a global perspective of the concerns and approaches. Finally, we would like to express our gratitude to the authors, reviewers, and ASTM staff for their hard work and resulting contributions to this STP.

Michael J. Verrilli

Symposium co-chairman and co-editor;
    NASA Lewis Research Center, Cleveland,
    Ohio.

Michael G. Castelli

Symposium co-chairman and co-editor;
    NYMA, Inc., NASA LeRC Group, Brook
    Park, Ohio.

High-Temperature Structural Alloys

Eric Chataigner and Luc Remy *

THERMOMECHANICAL FATIGUE BEHAVIOUR OF COATED AND BARE
NICKEL-BASED SUPERALLOY SINGLE CRYSTALS

REFERENCE: Chataigner, E. and Remy, L., **"Thermomechanical Fatigue Behaviour
of Coated and Bare Nickel-Based Superalloy Single Crystals,"** Thermomechanical
Fatigue Behavior of Materials: Second Volume, ASTM STP 1263, Michael J. Verrilli and
Michael G. Castelli, Eds., American Society for Testing and Materials, 1996.

ABSTRACT: The thermal-mechanical fatigue behaviour of chromium-aluminium
coated [001] single crystals of AM1, a nickel-base superalloy for turbine blades, is
studied using a "diamond" shape cycle from 600° to 1100°C. Comparison with bare
specimens does not show any significant difference in thermal-mechanical fatigue nor in
isothermal low cycle fatigue at high temperature. Metallographic observations on fracture
surfaces and longitudinal sections of specimens tested to fatigue life or to a definite
fraction of expected life have shown that the major crack tends to initiate from casting
micropores in the sub-surface area very early in bare and coated specimens, under low
cycle fatigue or thermal-mechanical fatigue. But the interaction between oxidation and
fatigue cracking seems to play a major role. A simple model proposed by Reuchet and
Rémy has been identified for this single crystal superalloy. Its application to the life
prediction under low cycle fatigue and thermal-mechanical fatigue for bare and coated
single crystals with different orientations is shown.

KEYWORDS: thermomechanical fatigue, nickel-based superalloy, single crystals,
coatings, lifetime prediction.

INTRODUCTION

Cooled turbine blades in jet engines must have a good resistance to thermal-mechanical
cyclic loadings due to the superposition of centrifugal loads and thermal stresses

* Centre des Matériaux P.M. FOURT E.N.S.M.P. URA 866 - B.P. 87 91003
EVRY Cedex (FRANCE)

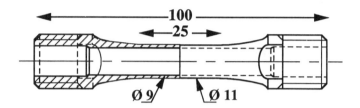

Figure 1 : TMF specimen geometry.

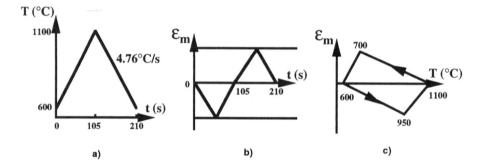

Figure 2 : Thermal-mechanical fatigue cycle:
a) Temperature versus time diagram,
b) Mechanical strain versus time diagram,
c) Mechanical strain versus temperature diagram.

the constrained thermal expansion, due to temperature gradients across the components section. The TMF tests were performed using a specific cycle presented on Fig. 2 which simulates thermal loading conditions experienced in service at the leading edge of a blade in a jet engine and was used in a previous investigation in the bare condition [19]. A mechanical strain-temperature loop was used from 600° to 1100°C (873 to 1373K) with peak strains at intermediate temperature : 950°C (1223K) in compression on heating and 700°C (973K) in tension on cooling.

Our own TMF test facilities used a micro-computer to generate two synchronous temperature and mechanical strain signals and a lamp furnace to heat the sample, as described in earlier publications [2, 3]. Thermal cycling time is 210 s. During the test, temperature is also measured by a coaxial thermocouple located and attached on the cylindrical part.

Smooth specimen testing is especially appropriate to investigate the life to engineering crack initiation. Crack growth was monitored using the d.c. potential drop technique in all the specimens and the plastic replication technique in some specimens. This second procedure which necessitates test interruptions, enabled cracks as small as 10μm in surface length to be detected by scanning electron microscopy (SEM). Specimens were sectioned and broken in order to get an experimental calibration curve between surface crack length "a_s" and crack depth "a_p". Surface cracks observed in different TMF specimens with a depth in the range 0.05 to about 0.8mm have a semi-elliptical shape. Experimental a_s and a_p data can be fitted by the equation $a_p = 0.38\ a_s$ using a least square method, as previous LCF data on the same alloy [18] as well as on other single crystal alloys [20]. For coated specimens, tests were conducted too up to different fractions of expected life and the crack distribution was observed on longitudinal sections by SEM.

Tests were stopped when the major crack grows through wall thickness or slightly before 1mm depth (this corresponds to our definition of Nf). Previous work on solid LCF specimens [10, 11] has shown that using a potential drop technique, a conventional fatigue life can be defined to 0.3mm crack depth, referred to as Ni. Consistent fatigue lives are thus given by solid and hollow specimens under LCF and Ni data under TMF and LCF can be reliably compared. In order to describe more closely the initiation phase and to trigger differences between coated and bare specimens, data to 0.1mm (referred to as Ne) were also estimated whenever possible.

RESULTS

Cyclic Stress-Strain Behaviour under TMF

Stepwise-increasing strain TMF tests were carried out on coated [001] specimens. Fig. 3 shows the variation of stress as a function of mechanical strain. On this kind of material, with a symmetrical strain range imposed, the hysteresis loops are stabilized after a few cycles. The stress-mechanical strain response in non-isothermal conditions is not usual and necessitates some comments. Because the material behaviour is different at the temperature of each peak strain, the inelastic strain is mostly created during the heating phase in compression. Though strain cycling was fully reversed, the stress cycle is unbalanced with a positive tensile mean stress. The particular shape of the TMF hysteresis loops is the result of the combined variations of elastic modulus E, monotonic yield strength σ_y and imposed mechanical strain ε_m with temperature T. A crystallographic model has been recently proposed which uses viscoplastic constitutive equations at the level of the slip system considering both cube and octahedral slip planes

[21]. This model describes the stress-strain loop as well as the active slip systems of AM1 single crystals under isothermal conditions [21] and was shown to predict quite accurately the present stress-strain loops under TMF [22].

TMF and LCF Life of Bare and Coated Specimens with a [001] Crystallographic Orientation

Test results--TMF life results for 1 mm crack depth (Nf) of the [001] specimens are plotted versus the total mechanical strain range in the bare condition (Fig. 4). There is no significant difference between the different batches of material. The endurance of coated specimens is almost the same as that of bare specimens for the TMF cycle used (Fig. 5).

The life to engineering crack initiation in the bare condition is mostly spent in micro-crack growth as shown by the variation of crack depth, which has been deduced from the observations of plastic replicas, versus the fraction of total life (Fig.6). Cracks a few tens μm in depth can be actually detected within about 5 pct of total life. Whatever the mechanical strain range, cracks propagated in stage II mode, i.e. mode I opening, from the initiation site to 0.4mm about. But when cracks are longer they deviated and tend to follow a crystallographic path, with a higher crack growth rate. Contrarily to LCF data [18], short TMF lives do not seem to fit the straight line for longer lives as a function of mechanical stain range.

The observations of interrupted tests after some fraction of expected life has shown that a major crack initiates quite early in the TMF life of coated specimens (see for instance Fig.6). The endurance of chromaluminized (C1A coated) specimens tested in LCF at 950°c at a frequency of 0.05 Hz is compared with that of bare samples under the same test conditions using a total mechanical strain range in Fig.7. There is no noticeable difference within the experimental scatter.

The life of chromaluminized specimens under TMF is compared with that under LCF at 950 and at 1100 °C versus total mechanical strain range and versus stress range in Figs. 8a and 8b respectively. The TMF life versus stress range is in pretty good agreement with LCF life at 950°C. With the total mechanical strain range, TMF life is well described by LCF life at 1100°C.

Metallographic observations--The damage mechanisms of AM1 single crystals under TMF cycling of different orientations were recently reported in the bare condition [19] and can be summarised as follows : at high strain ranges, the major crack nucleates from a large casting micropore at the surface or in the subsurface area; at low strain ranges, the major crack nucleates from oxidised areas at the surface.

The major crack seems always to initiate at a subsurface casting micropore beneath the coating layer from the observation of fracture surfaces of coated [001] specimens tested in LCF at 950°C. Fig. 9 shows that the main crack looks as a surface initiated crack despite this crack initiation mechanism. Numerous cracks actually form in the coating but they stop at the interface between coating and substrate and do not grow into the substrate. At the lower strain ranges, the crack initiation mechanism seems unaltered but the cracks are much more oxidised (Fig. 10); localised oxidation occurs at the interface between coating and substrate.

Under TMF cycling, the fracture surfaces of coated [001] specimens tested at high strain ranges, look like those tested in LCF at 950°C, Fig.11 and the major crack initiates at a subsurface casting micropore. At low strain ranges, the initiation mechanisms of the

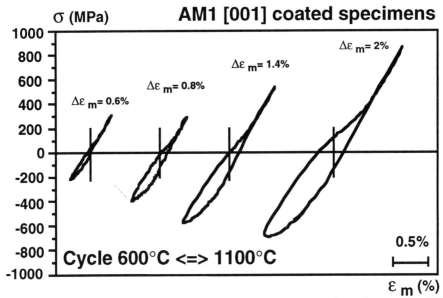

FIG. 3--Stress-mechanical strain hysteresis loops for a coated specimen with a [001] crystallographic orientation.

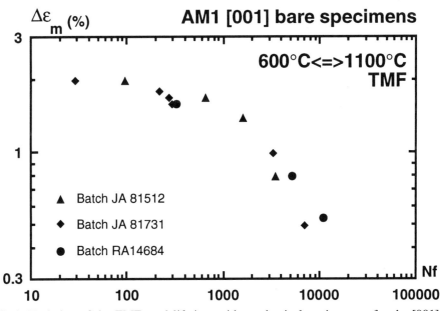

FIG. 4--Variation of the TMF total lifetime with mechanical strain range for the [001] orientation in the bare condition : comparison between the different batches of material.

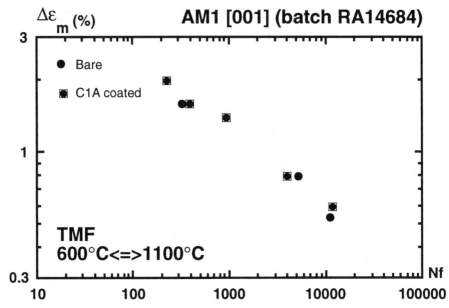

FIG. 5--Variation of the TMF total lifetime with mechanical strain range for the [001] orientation : comparison between bare and coated specimens.

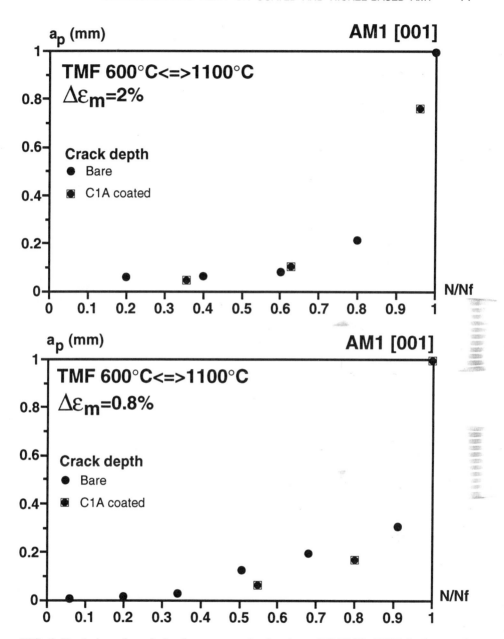

FIG. 6--Evolution of crack depth a_p versus the fraction of TMF life N/Nf for bare and coated specimens with the [001] orientation submitted to two different strain ranges: $\Delta\varepsilon_m$=2% and $\Delta\varepsilon_m$=0.8%.

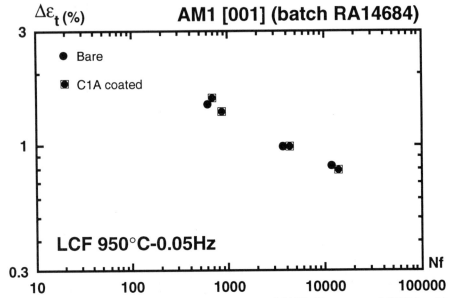

FIG. 7--Variation of the LCF total lifetime at 950°C (frequency 0.05Hz) with mechanical strain range for the [001] orientation : comparison between bare and coated specimens.

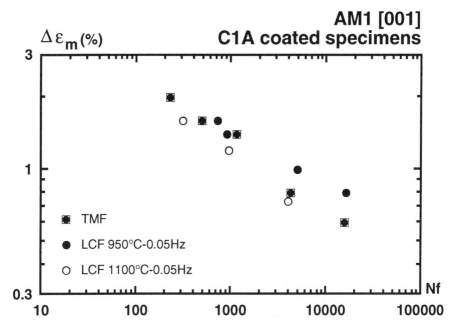

FIG. 8a--Comparison between the TMF and the LCF total lifetime with mechanical strain range for the [001] orientation.

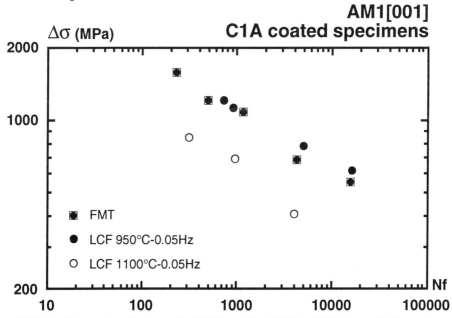

FIG. 8b--Comparison between the TMF and the LCF total lifetime with the stress for the [001] orientation.

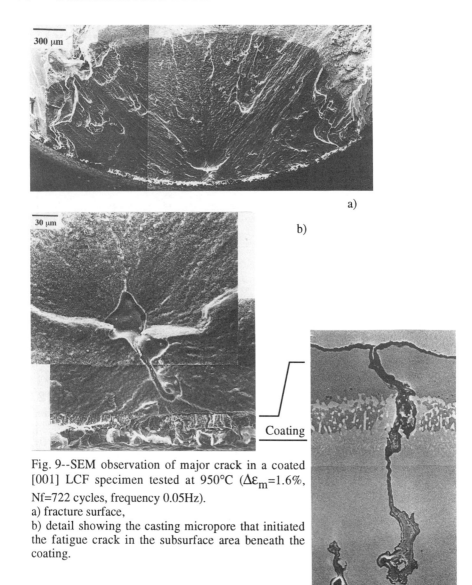

a)

b)

Coating

Fig. 9--SEM observation of major crack in a coated [001] LCF specimen tested at 950°C ($\Delta\varepsilon_m$=1.6%, Nf=722 cycles, frequency 0.05Hz).
a) fracture surface,
b) detail showing the casting micropore that initiated the fatigue crack in the subsurface area beneath the coating.

Fig. 10--SEM micrograph of a longitudinal section in a coated [001] LCF specimen tested at 950°C ($\Delta\varepsilon_m$=0.8%, Nf=16040 cycles, frequency 0.05Hz): an oxidised crack has grown through the coating and linked an internal casting micropore with the external surface.

major crack are less clear due to the heavy oxidation of the fracture surfaces. Surface-like cracks might initiate at subsurface micropores (Fig.12). The coating displays large strains, is oxidised and numerous cracks form in it. Strong localised oxidation occurs at the coating-substrate interface that can give rise to local delamination of the coating, as previously reported for aluminised IN100 specimens tested under TMF [23].

DISCUSSION

This investigation of the endurance of [001] single crystals of superalloy AM1 in TMF and LCF at high temperatures has shown no major difference between bare and C1A coated specimens, for the investigated test conditions. The initiation of the major crack takes place in most cases at casting micropores which are located in the sub-surface area. Cracks seem to behave as surface cracks very early and the initiation period is at most a small fraction of total life. In addition, this study and the previous results on the same alloy [18, 19] have pointed out the importance of oxidation.

Therefore models that have been developed for conventionally cast superalloys to account for the interaction between fatigue and oxidation and describe fatigue damage as a micro-crack process, should be applicable to the superalloy single crystals [12, 13, 14]. For sake of simplicity, we decide to use the simple model proposed by Reuchet and Rémy [12]. The physical basis of this model is thus first recalled and its application to the TMF and LCF lives of AM1 single crystals is then shown.

Oxidation-Fatigue Damage Equation in Fatigue at Elevated Temperature

As proposed by Reuchet and Rémy [12] a simple way to account for the interaction between fatigue and oxidation is to superimpose both kinds of damage. These authors consider that LCF damage in conventionally cast superalloys is mostly the growth of a dominant micro-crack and that the initiation period can be neglected for practical purposes [12, 14]. Thus, the damage equations were derived assuming that the elementary crack advance results from an advance due to crack opening under fatigue and from an additional contribution due to oxidation at the crack tip. The damage equation can then be written as follows :

$$\frac{da}{dN} = \left(\frac{da}{dN}\right)_{fat} + \left(\frac{da}{dN}\right)_{ox} \tag{1}$$

The fatigue contribution to the crack advance is estimated through the crack opening displacement model proposed by Tomkins [24] assuming that the crack is opened only by a tensile stress :

$$\left(\frac{da}{dN}\right)_{fat} = B \cdot a \tag{2}$$

where
$$B = \Delta\varepsilon_{in} \left(\frac{1}{\cos\left(\dfrac{\pi}{2}\dfrac{\sigma_{max}}{\sigma_u}\right)} - 1 \right) \tag{3}$$

$$a = \Delta l_{ox} \cdot \frac{(\exp B.N - 1)}{B} \qquad (7)$$

In this equation the crack length is expressed as a linear function of a term which depends on Tomkins's coefficient and on the number of elapsed cycles. It is worth noting that the slope of this function is exactly the oxidation contribution, Δl_{ox}. Thus for these interrupted tests we can deduce the oxide length formed at the crack tip at every cycle.

As it has been already demonstrated in other superalloys [10, 14, 27], we assume that the depth of interdendritic oxide spikes varies as follows in the absence of any loading :

$$\Delta l_{ox}^{o} = \alpha_0 \cdot \Delta t^{1/4} \qquad (8)$$

where α_0 is the oxidation constant of the interdendritic spaces at a given temperature and Δt is the cycle period. Under stress, the same law applies provided that the maximum stress is below a threshold σ_0. Above this threshold the depth of oxide formed at every cycle increases with maximum stress according to a power law :

$$\Delta l_{ox} = \alpha(\sigma_{max}).\Delta t^{1/4} \qquad (9)$$

where $\alpha(\sigma_{max}) = \alpha_0 (\sigma_{max}/\sigma_0)^n$ when $\sigma_{max} > \sigma_0$

$\alpha(\sigma_{mox}) = \alpha_0$ when $\sigma_{max} \leq \sigma_0$

where the constant n is 4.5 at both 950 and 1100°C. Fig. 13 shows the comparison between the fitted curve and experimental crack growth data for two different strain ranges on [001] bare specimens tested in LCF at 950°C. A pretty good correlation is actually observed if one remembers that there are only two fitting parameters at a given temperature in the model. These parameters were thus identified from such experimental crack growth data using only [001] bare specimens tested in LCF at 650, 950 and 1100°C at a frequency of 0.05 Hz.

The application of the model to LCF life is shown for total life at 1100°C and for the life to 0.3 mm crack depth at 950°C in Figs.14 and 15 in the case of [001] bare specimens. As information from some of the specimens was used to identify the model coefficients, this merely reflects the degree of correlation provided by the model.

Fig. 16 compares the prediction of LCF life to 0.3 mm crack depth at 950°C at a frequency of 0.05 Hz for different orientations [001], [111], [101], [213] and the transverse orientation [010] and experimental data [18]. The influence of orientation is well predicted. Further the model is able to predict the life under TMF. In this case the oxidation term is obtained by integration of the oxidation constant over the whole temperature-time cycle and the average oxidation constant, $\overline{\alpha}(\sigma_{max})$, is given by :

$$\overline{\alpha}^4(\sigma_{max}) = \int_0^{\Delta t} \alpha^4 [\sigma_{max}(t), T(t)] \cdot dt / \Delta t \qquad (10)$$

The fatigue contribution is deduced from Tomkins's model applied to the maximum stress of the whole stress-strain loop at the temperature of the peak tensile stress.

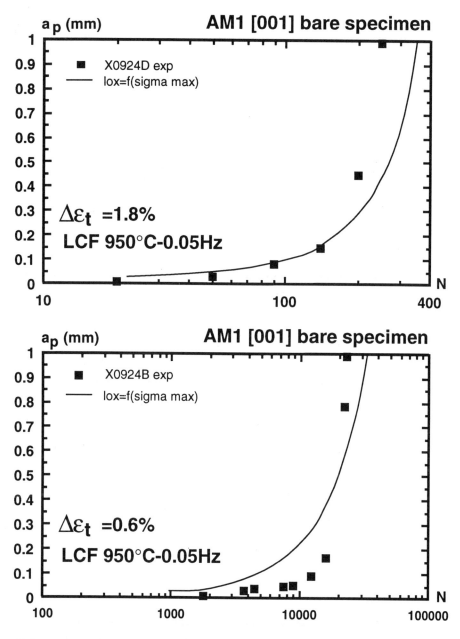

FIG. 13--Evolution of crack depth ap versus the number of cycles in bare [001] specimens tested in LCF at 950°C : comparison between the curve computed by the model and experimental data (solid symbols) for two different strain ranges.

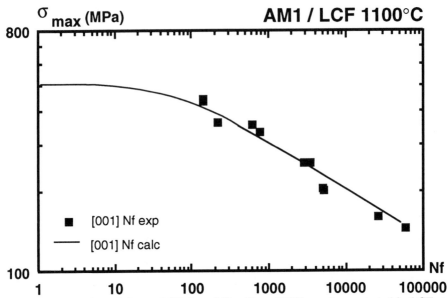

FIG. 14--Variation of the total lifetime (Nf) of bare [001] specimens tested in LCF at 1100°C with the peak tensile stress (σmax) : comparison between model (solid curve) and experiment (symbols).

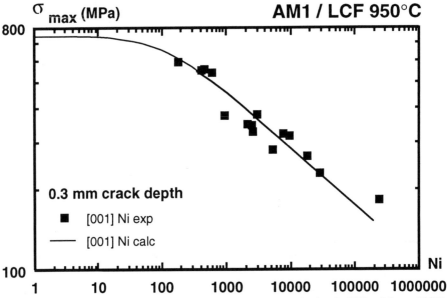

FIG. 15--Variation of the total lifetime to 0.3mm crack depth (Ni) of bare [001] specimens tested in LCF at 950°C with the peak tensile stress (σmax) : comparison between model (solid curve) and experiment (symbols).

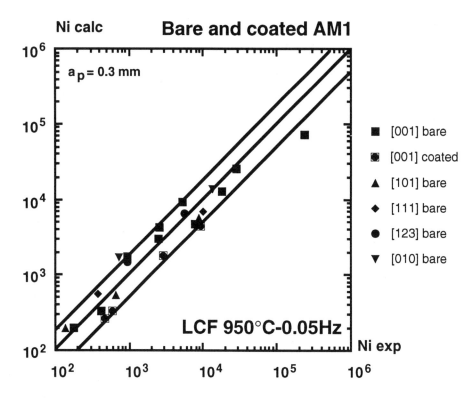

FIG. 16--Comparison between predicted and experimental lifetime to 0.3 mm crack depth (Ni) of bare [001] specimens tested in LCF at 950°C with different orientation.

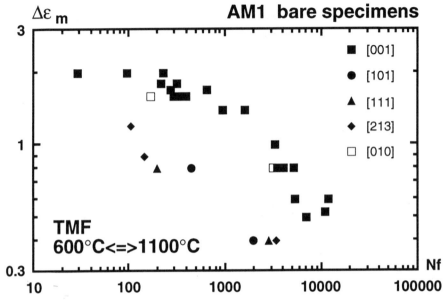

FIG. 17--Variation of the total lifetime (Nf) of bare AM1 specimens of different orientations tested in TMF with mechanical strain range ($\Delta\varepsilon_m$).

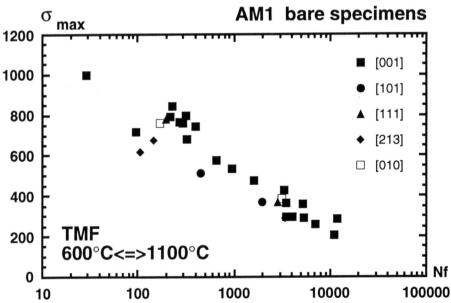

FIG. 18--Variation of the total lifetime (Nf) of bare AM1 specimens of different orientations tested in TMF with the peak tensile stress (σ_{max}).

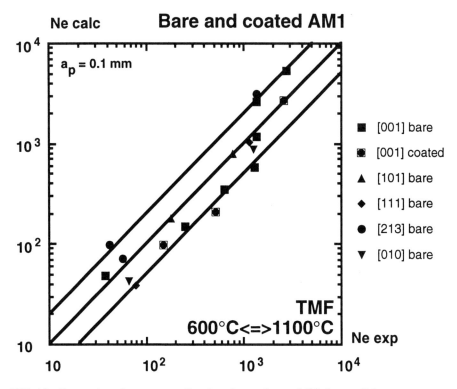

FIG. 19--Comparison between predicted and experimental lifetime to 0.1 mm crack depth (Ne) of bare specimens with different orientations and C1A coated [001] specimens tested in TMF.

[18] Fleury E. and Remy L., Materials Science and Engineering, vol.A167, 1992, pp.23-30.

[19] Fleury E. and Remy L., Metallurgical Transactions, vol.14 A, 1994, pp.99-109.

[20] Chieragatti R. and Remy L., Materials Science and Engineering, vol. A 141, 1991, pp. 1-9 and 11-22.

[21] Hanriot F., Cailletaud G. and Remy L., in High Temperature Constitutive Modeling. Theory and Applications, Freed A.D. and Walker K.P. Eds, American Society of Mechanical Engineers, New-York, Vol.26 and Vol.121, 1991, pp.139-150.

[22] Hanriot F., Thesis, Ecole des Mines de Paris, 1993.

[23] Bernard H. and Remy L., in Advanced Materials and Processes, Proc. Conf., EUROMAT 89, Exner H.E. and Schumacher V., Eds., Aachen (FRG), 1990, vol.1, pp.529-534.

[24] Remy L., F.Rezai-Aria, R.Danzer, and W.Hoffelner, in Low Cycle Fatigue, ASTM STP 942, Solomon H.D., Halford G.R., Kaisand L.R., and Leis B.N., Eds., American Society for Testing and Materials, Philadelphia, 1988, pp.1115-1132.

[25] Tomkins B., Philosophical Magazine, vol.18, 1968, pp.1041-1066.

[26] M.François and Remy L., Metallurgical Transactions A, vol.21, 1990, pp.949-958.

[27] Vasseur E., Thesis, Ecole des Mines de Paris, 1993.

[28] Reger M. and Remy L., Metallurgical Transactions A, vol. 19, 1988, pp. 2259-2268.

[29] Gabb T.P., Gayda J. and Miner R.V., Metallurgical Transactions A, vol. 17, 1986, pp. 497-505.

Stephan A. Kraft[1], Haël Mughrabi[1]

THERMO-MECHANICAL FATIGUE OF THE MONOCRYSTALLINE NICKEL-BASE SUPERALLOY CMSX-6

REFERENCE: Kraft, S. A. and Mughrabi, H., **"Thermo-Mechanical Fatigue of the Monocrystalline Nickel-Base Superalloy CMSX-6,"** Thermomechanical Fatigue Behavior of Materials: Second Volume, ASTM STP 1263, Michael J. Verrilli and Michael G. Castelli, Eds., American Society for Testing and Materials, 1996.

ABSTRACT: Total strain-controlled thermo-mechanical fatigue (TMF) tests were performed between 600 and 1100°C on as-grown near to [001]-orientated specimens of the γ' precipitate-strengthened monocrystalline nickel-base superalloy CMSX-6. The strain-temperature cycle (in-phase, out-of-phase, diamond) has a strong influence on the mechanical and microstructural events occurring during TMF. For given temperature intervals, the fatigue lives were shortest for out-of-phase tests and longest for in-phase tests, respectively. For all cycle shapes applied, the maximum tensile stress level was concluded to be the lifetime-limiting factor. The failure mode of the investigated alloy depends also on the conducted strain-temperature history. While strongly localized crystallographic shearing along {111} planes was dominant for in- and out-of-phase cycling, creep-induced damage occurred in diamond cycle tests. The evolution and coarsening of the microstructure were also studied. The latter is compared with the formation of stress-, strain- and diffusion-induced coarsened raft-like structures found in creep testing of similar materials.

KEYWORDS: single crystal, cyclic softening/hardening behaviour, nickel-base superalloy, low-cycle fatigue, thermo-mechanical fatigue, microstructure, rafting, fatigue life, fatigue damage.

[1] Research Associate and Professor, respectively, Institut für Werkstoffwissenschaften, Lehrstuhl I, Universität Erlangen-Nürnberg, Martensstrasse 5, D-91058 Erlangen, Federal Republic of Germany

INTRODUCTION

Purpose Of Thermo-Mechanical Fatigue Testing

The aim of thermo-mechanical fatigue tests (TMF) is to simulate the loading conditions of turbine blades during engine operation in gas turbines. During start-up and shut-down of a turbine engine, temperature gradients are built up, e.g. between the socket and the top or the leading and trailing edge of the turbine blade, leading to stress or strain gradients and therefore microstructural changes in the material [1,2]. TMF-tests [3-9] are able to simulate these conditions in the laboratory and are thus more realistic than monotonic creep or fatigue tests.

Monocrystalline γ'-strengthened nickel-based superalloys with a high volume content of the ordered coherent γ'-phase are in use for turbine blades due to their excellent high temperature creep [10-15], low-cycle fatigue [16-19] and hot corrosion [20] resistance. While the named properties have been studied extensively, the thermo-mechanical fatigue behaviour in the low-cycle range

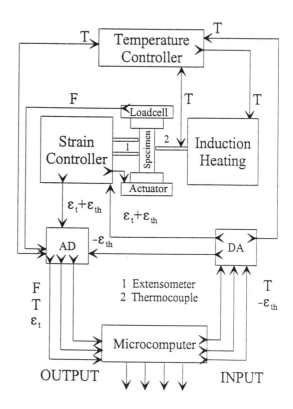

FIG. 1--Schematic diagram of the control cycles for thermo-mechanical fatigue tests.

has not been investigated in similar detail [3-9].The microstructural changes occurring under service conditions in the turbine blade are usually considered to result from creep deformation under the influence of diffusional processes at high temperatures [10-14]. The application of thermal and mechanical loads simultaneously approximates more closely the conditions prevailing in service [6]. The comparison of the mechanical and microstructural behaviour observed in laboratory tests with that found in the turbine blades can help to understand the real deformation and failure mechanisms and may lead to constructive improvements of components and therefore to a higher level of integrity in the future.

EXPERIMENTAL PROCEDURE

Testing Method

The TMF tests were performed using a servohydraulic test system (MTS 880), equipped appropriately, as shown schematically in fig.1. The testing equipment is the same as that described in [5,21]. Three digital input signals (temperature T, thermal expansion - ε_{th} and the sum of total strain ε_t and thermal strain: $\varepsilon_t + \varepsilon_{th}$) are generated in the microcomputer and fed into the digital analog converter (DA). For symmetrical push-pull thermo-mechanical fatigue tests with a constant total strain amplitude $\Delta\varepsilon_t/2$ ($\Delta\varepsilon_t$: total strain range) and a constant total strain rate $\dot\varepsilon_t$, the thermal expansion strain ε_{th} of the material, calculated via the temperature-dependent thermal expansion coefficient $\alpha(T)$ has to be combined with ε_t to generate the analog command signal $\varepsilon_t + \varepsilon_{th}$. This signal is used as the closed-loop feedback signal for the test system.

The total strain ε_t is calculated afterwards by addition of the negative value of the thermal expansion $-\varepsilon_{th}$ in the analog-digital (AD) converter. The triangular temperature signal T is created simultaneously to control the induction heating device necessary for rapid heating and cooling of the specimen.

The incoming signals of force F (measured by the load cell), temperature T and total strain ε_t are recorded by the microcomputer as peak-valley data and hysteresis data. The stress values $\sigma = F/A$ (A: cross section of the specimen) and the values of plastic strain ε_{pl} are calculated afterwards.

Measurement and Test Parameters

For the measurement and the control of the total strain ε_t and the temperature T a water-cooled high-temperature axial extensometer with ceramic rods and a Pt-10PtRh thermocouple are attached directly at the surface of the gauge length of the solid specimens. Heating was performed by a 200 kHz induction furnace.

All tests were conducted in air with total strain amplitudes ($\Delta\varepsilon_t/2$) of 5×10^{-3} and 6×10^{-3}, employing a triangular waveform with a constant cycle time (t_c) of 300 s leading to total strain rates of $\dot\varepsilon_t = 6.67\times10^{-5}$ s^{-1} and 8×10^{-5} s^{-1}, respectively. The temperature interval used in the tests extended from 600 to 1100°C with lower temperatures T_l of 600, 700 and 800°C and upper temperatures T_u of 900, 1000 and 1100°C. The cooling rates dT/dt is limited to 200°C/min by the heat transfer of the specimen to the hydraulic grips and the environment during unforced cooling. The temperature gradient

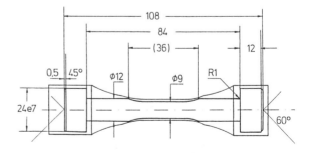

FIG. 2--Specimen geometry. Dimensions in millimeters.

over the gauge length was maintained constant within 5 K as measured on the surface. Due to the comparably low frequency of the induction furnace (200 kHz) and the cycle time t_c of 300 s the temperature gradient to the inner area of the cross section of tubular specimens was negligible [21]. This was concluded to be also true for compact specimens, since no difference in mechanical response could be found compared to tubular specimens.

Specimen Geometry

The geometry of the near-net-shaped cast[1] (outer profile) and subsequently machined (inner profile) specimens is shown in fig. 2.

The overall length of 108 mm with cylindrical heads was chosen to avoid overheating and destruction of the hydraulic grips. The gauge length of 36 mm allows the direct attachment of the high-temperature axial extensometer (length: 12 mm) and the thermocouple combined with the surrounding induction coil (length $\approx$ 50 mm) avoiding axial and radial temperature gradients. The diameter of 9 mm of the cross section reduces the risk of inhomogeneous deformation due to buckling of the specimen during compression.

Thermo-Mechanical Cycling

Four different temperature-total strain cycles with fully compensated thermal expansion and symmetrical total strain amplitudes were conducted in air (fig. 3). Two general types of testing can be distinguished: in-phase (IP) and out-of-phase (OP) testing, with the total strain at the highest temperature being largest in tension (IP) or in compression (OP), respectively, and clockwise-diamond (CD) and counter-clockwise-diamond (CCD), the intermediate temperatures being reached at largest tensile and

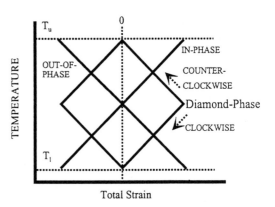

FIG. 3--Temperature-total strain cycles for the different TMF-tests. From [9].

compressive strains, respectively, and the highest and lowest temperatures at zero total strain. For comparison with loading under service conditions, the CCD cycle shape has been used preferentially [4,6].

Microstructural And Fractographic Observations

The microstructural changes which occurred during TMF were investigated on sections cut parallel to {001} planes by standard metallographic techniques, in particular by transmission electron microscopy (TEM), using a Philips EM 400T, equipped with energy-dispersive analysis of X-rays (EDX). Observations of the γ/γ'-morphology were

---

[1] Thyssen Guss AG, Bochum, Germany

performed by scanning electron microscopy (SEM), using a Jeol JSM 6400 MK2 microscope. The fracture surfaces were studied by SEM.

MATERIAL

The alloy used in this investigation is the monocrystalline nickel-based superalloy CMSX-6[1] in the as-grown, near [001]-orientation. The composition of this alloy (wt.%) is as follows: C 0.02, Cr 10.00, Co 5.00, Mo 3.00, Ti 4.75, Hf 0.10, Al 4.85, B 0.03, Zr 0.08, Ta 2.00, Si 0.02, Ni balance. The heat treatment (slightly different from that used by Cannon-Muskegon) is divided into two parts, the solutioning process in vacuum (1205°C/1h/ 1240°C/3h/ 1270°C/3h/ 1274°C/3h/inert gas quenching) and the precipitation ageing (1080°C/4h/ 870°C/16h/air quenching). The axial orientation of the specimens was found to be off-oriented within 10° away from [001]. All specimens were tested in an uncoated condition. For the TMF-tests, the surface of the specimens was polished electrochemically. In fig. 4a the morphology of the γ-γ'-structure (SEM) exhibits a uniform distribution of coherent and nearly cuboidal particles with an average edge length of 460 nm. The volume fraction of the γ'-phase was determined as 0.72 by EDX measurements [22]. The constrained misfit parameter: $\delta = 2(a_0^\gamma - a_0^{\gamma'})/(a_0^\gamma + a_0^{\gamma'})$, where a_0^γ and $a_0^{\gamma'}$ are the lattice parameters of the γ and γ' phase, respectively, was measured by X-ray diffractometry [23] and found to be negative with $|\delta| < 10^{-3}$. The thermal expansion coefficient α was measured to be 10 to 21×10^{-6} K^{-1} between room temperature and 1200°C. Figure 4b shows the undeformed microstructure of the alloy in the TEM. The greatest fraction of the volume is dislocation-free, only the interdendritic region (cf. fig. 4d) contains some disloca-

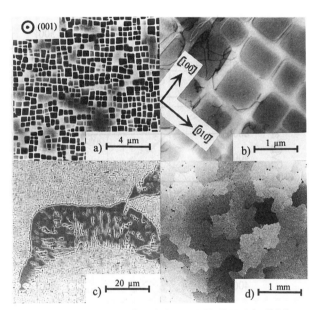

FIG. 4--Microstructure of undeformed CMSX-6 in (001)-sections, a) precipitation morphology (γ'-etched, SEM), b) dislocation distribution (TEM), c) eutectic γ' in the interdendritic area (SEM), d) dendritic structure and micropores (SEM).

---

[1] TM Cannon-Muskegon, USA

plastic yielding. In general, the stress level is comparatively low, when the total strain is near zero (CD- and CCD-tests at $\varepsilon_t=0$) or the temperature is high (IP- and OP-tests at T_u). The IP- (fig. 5a, b) and the OP-curves (fig. 5c, d) (and the CD- (fig. 5e, f) and CCD-curves (fig. 5g, h), respectively) show a point-symmetrical behaviour with respect to the origin of the axes. The stress levels are highest for IP-tests in compression, for OP-tests in tension. For CD- and CCD-tests, intermediate and comparatively symmetric peak stresses are found in tension and compression, since similar temperatures are reached at these points of the hysteresis loops. The plastic strain amplitudes $\Delta\varepsilon_{pl}/2$ (half the hysteresis loop width at $\sigma=0$) are very small for IP- and OP-tests, the greatest values for $\Delta\varepsilon_{pl}/2$ are

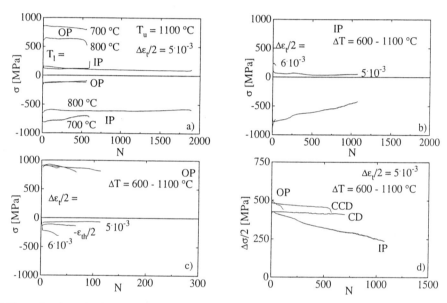

FIG. 7--Cyclic softening/hardening curves for the different tests. a) Comparison of IP- and OP-testing with different lower temperatures T_l of 700 and 800°C, b) OP-testing with to-tal strain amplitudes of 5×10^{-3}, 6×10^{-3} and $\Delta\varepsilon_t/2=-\Delta\varepsilon_{th}/2$. The latter case corresponds to fully constrained thermal cycling. c) IP-testing with total strain amplitudes of 5×10^{-3} and 6×10^{-3} d) influence of the cycle shape on the cyclic softening/hardening behaviour with otherwise similar test conditions.

reached in the CD- and CCD-tests. The influence of different temperature ranges on the (stabilized) hysteresis loop shapes for IP- and OP-testing (fig. 6) shows an increase of the amount of plastic deformation during one cycle for a higher T_l of 700°C (fig. 6a, d) at a constant $\Delta\varepsilon_t/2$ of 5×10^{-3} and $T_u=1100$°C compared to the 600-1100°C cycle (fig. 6b, e). The reduction of T_u to 1000°C leads to a very small value of $\Delta\varepsilon_{pl}/2$ (fig. 6c, f).

Cyclic Softening/Hardening Behaviour--Figure 7 shows the influence of different test parameters on the cyclic softening/hardening behaviour under TMF-conditions. In general, cyclic softening occurs in all tests after a small increase of the stress amplitude in the first ten cycles.

Comparing IP- and OP-tests with constant $\Delta\varepsilon_t/2$ of 5×10^{-3} and different temperature intervals ΔT of 700-1100°C and 800-1100°C (fig. 7a), the stress levels reached at T_u (compression for OP, tension for IP) are not influenced by the lower temperature T_l, whereas the stress levels reached at T_l are higher for lower T_L.

The influence of $\Delta\varepsilon_t/2$ on the maximum and minimum stress values for OP- (fig. 7b) and IP- (fig.7c) tests shows no remarkable dependence of the cyclic softening/hardening behaviour, only the stress amplitudes are comparatively higher for higher $\Delta\varepsilon_t/2$ values. For a given temperature interval ΔT and a given total strain amplitude $\Delta\varepsilon_t/2$, the stress amplitude $\Delta\sigma/2$ is dependent on the cycle shape. OP-testing leads to the highest $\Delta\sigma/2$, IP-testing to the lowest $\Delta\sigma/2$, CD- and CCD-testing are intermediate concerning the stress amplitude (fig. 7d).

Lifetime--The endurance of this material under TMF conditions is strongly dependent on the temperature interval ΔT, the total strain amplitude $\Delta\varepsilon_t/2$ and the

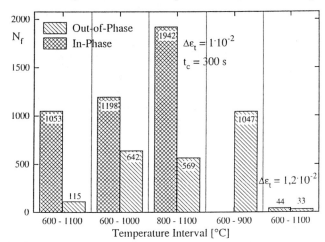

FIG. 8--Comparison of number of cycles till fracture for IP- and OP-testing with different temperature intervals and similar $\Delta\varepsilon_t/2$ of 5×10^{-3} (first seven columns from left). For the two outermost columns on the right $\Delta\varepsilon_t/2 = 6\times10^{-3}$. From [9].

cycle shape (fig. 7). In general, the stress amplitude reached in the different test cycles under otherwise similar conditions seems to be the life-limiting factor (cf. fig. 7d). As shown in fig. 7b and 7c, higher $\Delta\varepsilon_t/2$ lead to smaller numbers of cycles to failure, as does a increase of the temperature interval (fig. 7a, 8). The comparison of IP- and OP-testing under varying test parameters with respect to the lifetime illustrates that OP-testing is always connected with lower numbers of cycles till fracture (fig. 8).

Failure Mechanisms

Macroscopic Cracking--Under TMF conditions, the appearance of the main crack varies strongly. Dependent on the cycle shape and the other testing conditions, the fracture surfaces indicate a creep-like damage in a few and crystallographic shear on one or two slip systems in all other cases (fig. 9). The CCD-test (fig. 9a, ΔT=600-1100°C, $\Delta\varepsilon_t/2$=5×10^{-3}) leads to a more creep-like fracture, IP-testing between 600 and 1100°C and

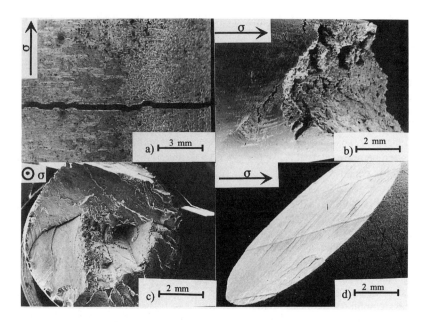

FIG. 9--Macroscopic cracks of different specimens. a) CCD, $\Delta T=600\text{-}1100°C$, $\Delta\varepsilon_t/2=5\times10^{-3}$, b) IP, $\Delta T=600\text{-}1100°C$, $\Delta\varepsilon_t/2=6\times10^{-3}$, c) OP, $\Delta T=700\text{-}1100°C$, $\Delta\varepsilon_t/2=5\times10^{-3}$, d) IP, $\Delta T=600\text{-}1000°C$, $\Delta\varepsilon_t/2=5\times10^{-3}$.

at a higher strain amplitude of $\Delta\varepsilon_t/2=6\times10^{-3}$ (fig. 9b) to a mixed character between creep and crystallographic shear, whereas for OP-testing ($\Delta T=700\text{-}1100°C$, $\Delta\varepsilon_t/2=5\times10^{-3}$) and IP-testing ($\Delta T=600\text{-}1000°C$, $\Delta\varepsilon_t/2=5\times10^{-3}$) (fig. 9c, d) pure crystallographic cracking along {111}-planes occurs.

Crack Initiation at the Surface--Figure 10 shows the development of a short crack initiated at a surface pore. The oxidation of the surface leads to a diffusional flow of the oxide-forming elements Cr and Al to the surface and to a depletion of γ' under the surface. As a consequence of the pore the depleted zone under it is bigger than at the other parts of the surface of the specimen (fig. 10a). Short cracks can grow through this zone and oxidation will keep them open (fig. 10b). This process can occur in every cycle, when tensile stresses act on the microcrack, so that the crack can grow and branch (fig 10c).

Coarsening Of The Precipitate Morphology

The initially cuboidal γ' structure of this type of alloys changes drastically under the influence of stress, strain and temperature. During TMF, several different appearances of the structure will develop, figure 11 shows the main results. IP-testing (fig. 11a), where the highest temperatures are reached in tension, leads to a directionally coarsened precipitate structure perpendicular to the stress axis (which lies near the [001]-orientation). For OP-testing (fig. 11b), the direction of coarsening is parallel to the stress axis, the so called "rafts" or "plates" are arranged along the two <001>-directions lying perpendicular to the stress axis. The same directional coarsening occurs during CCD- (fig. 11c) and CD-cycling (fig. 11d). In CCD-testing, also a plate-like structure develops perpendicular to the loading direction, although the highest temperature is reached in compression. CD-testing leads to an OP-like structure with the plates orientated parallel to the stress axis, the highest temperatures are reached in tension.

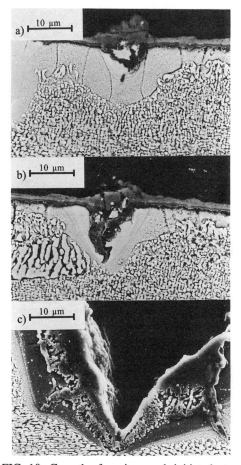

DISCUSSION

The discussion of the TMF-behaviour of CMSX-6 single crystals must describe and relate the macroscopic mechanical behaviour, the fracture mechanisms and the microscopical changes.

The hysteresis loop shapes for the different applied total strain-temperature cycles are in good agreement with other studies [3,4,6] and show the typical behaviour of this type of material with narrow hysteresis loops for IP- and OP-testing and more widely opened hysteresis loops for CD- and CCD-testing. The strong decrease of the peak stress of the first cycle compared to that of the second cycle observed in the IP- and OP-tests (compare to hysteresis-loops in fig. 5 a-d) can be explained by the increase of the initially very low dislocation density and must be compared with the

FIG. 10--Growth of a microcrack initiated at a surface pore, see text for details. Stress axis lies horizontal.

minimum of strain rate normally measured in creep tests [10-14] which corresponds to the maximum deformation resistance. The cyclic softening behaviour after the increase of the stress level in the first ten cycles is due to changes in the dislocation and precipitation structure. The deformed structure becomes weaker and weaker. The interfacial interaction between γ and γ' must play an important role due to the coherency strains and stresses and their superposition with the externally applied stresses (and strains). Under conditions of zero net total strain, it is likely that the generated dislocation networks serve mainly to reduce the coherency strains [5], cf. also [16]. The morphology of the γ'-precipitates undergoes time-, temperature-, stress- and strain-dependent changes. It was found in creep tests [11,12] and also during

LCF [18] and TMF [4,24] that the direction of the formation of plate- or raft-like structures in IP- and OP-tests depends on the direction of stress at high temperatures. In CCD- and CD-tests, the high temperatures are reached in tension (CD) and compression (CCD), respectively. Nonetheless, in our work the rafts develop parallel to the stress axis in CD-tests and perpendicular in CCD-tests. This suggests that rafting is governed not only by the external stress but also by the internal stresses and by the built-up plastic tensile and compressive strains [9]. These locally acting internal stresses result from the superposition of the coherency stresses (misfit) between γ and γ' and the internal stresses induced during deforma-

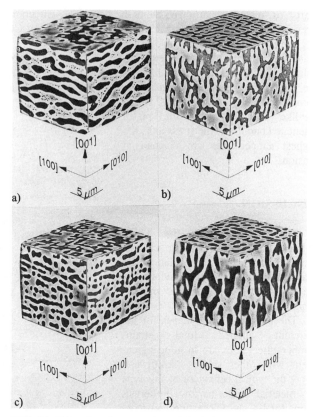

FIG. 11--Directional coarsening of the initially cuboidal γ'-particles, $\Delta T = 600\text{-}1100°C$, $\Delta\varepsilon_t/2 = 5\times10^{-3}$. a) IP, b) OP, c) CCD, d) CD. From [9].

tion. Compared to former studies on the same material with a different heat treatment [5], which did not exhibit similar rafting, it must be suggested that the initial microstructure has a strong influence on the coarsening behaviour.

Fracture starts always at the surface of the specimens, normally under the influence of structural defects, especially micropores at the surface which are preferentially oxidized [3,7]. The fracture mechanism is strongly dependent on the cycle shape. In CCD- and CD-tests, a more "creep-like" fracture occurs due to the intermediate stresses reached at the strain reversal points, whereas the IP- and OP-testing lead to a more "fatigue-like" fracture because of the higher stress levels reached at the minimum temperature. Crystallographic shearing is mainly observed under IP-conditions [5] which leads to the assumption that the high compressive stresses at low temperatures during IP-cycling cause inhomogeneous glide on localized {111}-planes and the formation of slip bands.

Regarding the lifetime of CMSX-6 under TMF-conditions, OP-testing is found to be the most severe loading condition due to the high tensile stresses reached at low temperature [5].

CONCLUSIONS

- Controlled thermo-mechanical fatigue testing provides valuable insight into the material behaviour under conditions of simultaneously varying strains/stresses and temperature.

- For a given temperature interval ΔT and a total strain amplitude $\Delta\varepsilon_t/2$, the mechanical data differ strongly, dependent on the cycle shape applied.

- The magnitude of the tensile peak stress seems to be the major life-limiting factor, in-phase testing always leads to a higher number of cycles to fracture than out-of-phase testing under otherwise similar conditions.

- Increasing lower temperatures, decreasing temperature intervals, decreasing upper temperatures and decreasing total strain amplitudes enhance the lifetime of the specimen for a given type of cycling.

- The morphology changes of the precipitate structure are concluded to be connected not only to the externally measured stresses, but also to the superposition of internal stresses and external load.

- TMF fracture starts at structural defects at the surface of the specimen, especially at micropores which are preferentially oxidized.

- In general deformation occurs inhomogeneously. The fracture mode is mainly fatigue-induced. Diamond TMF cycling leads to cracks with a higher amount of creep-induced failure, out-of-phase testing and, in particular, in-phase testing promote mainly a shear fracture mode with pronounced crystallographic cracking.

REFERENCES

[1]   H.E. Miller and W.L. Chambers, in Superalloys II C.T. Sims, N.S. Stoloff and W.C. Hagel, Eds., John Wiley & Sons, New York, 1987, pp. 27-57.

[2]   S. Draper, D. Hull and R. Dreshfield, Metallurgical Transactions A, Vol. 20, 1989, pp. 683-688.

[3]   F. Meyer-Olbersleben, D. Goldschmidt and F. Rézai-Aria, in Superalloys 1992, S.D. Antolovich, R.W. Stusrud, R.A. MacKay, D.L. Anton, T. Khan, R.D. Kissinger, D.L. Klarstrom, Eds., The Minerals, Metals & Materials Society, Warrendale, 1992, pp. 785-794.

[4]   J.Y. Guédou and Y. Honnorat, in Thermo-Mechanical Fatigue Behaviour of Materials, H. Sehitoglu, Ed., ASTM STP 1186, American Society for Testing and Materials , Philadelphia, 1993, pp. 157-171.

[5]   S. Kraft, R. Zauter and H. Mughrabi, Fatigue & Fracture of Engineering Materials & Structures, Vol. 16, 1993, pp. 237-253.

[6]   D.A. Boismier and H. Sehitoglu, Transactions of the ASME, Vol. 112, 1990, pp. 68-79.

[7]   E. Fleury and L. Rémy, in Advances in Fracture Research, ICF 7, Pergamon Press, Houston, TX, 1989, pp. 1133-1142.

[8]   J.W. Holmes, F.A. McClintock, K.S. O'Hara and M.E. Conners, in Low Cycle Fatigue, ASTM STP 942, H.D. Solomon, G.R. Halford, L.R. Kaisand, and B.N. Leis, Eds., American Society for Testing and Materials, Philadelphia, 1988, pp. 672-691.

[9]   S. Kraft, I. Altenberger and H. Mughrabi, Scripta Metallurgica et Materialia, Vol. 32, 1994, pp.411-416.

[10]   R.A. MacKay and M.V. Nathal, Acta Metallurgica et Materialia, Vol. 38, 1990. pp. 993-1005.

[11]   S.H. Ai, V. Lupinc and G. Onofrio, Scripta Metallurgica et Materialia, Vol. 29, 1993, pp. 1385-1390.

[12]   A. Pineau, Acta Metallurgica, Vol. 24, 1976, pp. 559-564.

[13]   M. Ignat, J.-Y. Buffière and J.M. Chaix, Acta Metallurgica et Materialia, Vol. 41, 1993, pp. 855-862.

[14]   T.M. Pollock and A.S. Argon, Acta Metallurgica et Materialia, Vol. 40, 1992, pp. 1-30.

[15]   C. Carry and J.L. Strudel, Acta Metallurgica, Vol. 25, 1977, pp. 767-777.

[16]   U. Glatzel and M. Feller-Kniepmeier, Scripta Metallurgica et Materialia, Vol. 25, 1991, pp.1845-1850.

[17]   T.P. Gabb, R.V. Miner and J. Gayda, Scripta Metallurgica, Vol. 20, 1986, pp. 513-518.

[18]   P.D. Portella, C. Kirimtay and K. Naseband, in Proc. of 13.Vortragsveranstaltung der AG Hochtemperaturwerkstoffe, VDEh, Düsseldorf, 1990, pp. 145-152.

[19]   T.P. Gabb, J. Gayda and R.V. Miner, Metallurgical Transactions A, Vol. 17, 1986, pp. 497-505.

[20]   F. S. Pettit and S. Giggins, in Superalloys II, C.T. Sims, N.S. Stoloff and W.C. Hagel, Eds., John Wiley & Sons, New York, 1987, pp. 327-358.

[21]   R. Zauter, H.-J. Christ and H. Mughrabi, Metallurgical And Materials Transactions A, Vol. 25, 1994, pp. 401-406.

[22]   D. Blavette and A. Bostel, Acta Metallurgica, Vol. 32, 1984, pp. 811-816.

[23]   H. Biermann, private communications, 1994.The X-ray technique is described in detail in: H.-A. Kuhn, H. Biermann, T. Ungár and H. Mughrabi, Acta Metallurgica et Materialia, Vol. 39, 1991, pp. 2783-2794.

[24]   N. Marchand, G.L. Espérance and R.M. Pelloux, in Low Cycle Fatigue, ASTM STP 942, H.D. Solomon, G.R. Halford, L.R. Kaisand and B.N. Leis, Eds., American Society for Testing and Materials, Philadelphia, 1988, pp. 638-656.

Frank Meyer-Olbersleben,[1] Carlos C. Engler-Pinto Jr.,[1] and Farhad Rézaï-Aria[1]

ON THERMAL FATIGUE OF NICKEL-BASED SUPERALLOYS

REFERENCE: Meyer-Olbersleben, F., Engler-Pinto Jr., C. C., and Rézaï-Aria, F., **"On Thermal Fatigue of Nickel-Based Superalloys,"** Thermomechanical Fatigue Behavior of Materials: Second Volume, ASTM STP 1263, Michael J. Verrilli and Michael G. Castelli, Eds., American Society for Testing and Materials, 1996.

ABSTRACT: The thermal fatigue (TF) behaviour of two single crystal nickel-based super-alloys, SRR99 and CMSX-4, is reported. Single edge wedge specimens are rapidly heated by induction heating of the wedge tip to a maximum temperature between 1000°C and 1175°C and cooled to 200°C by forced air. A constant cycle period is employed for all experiments. The strain distribution along the edge of the TF specimens is measured. Changing the induction frequency leads to different strain ranges. CMSX-4 shows crack initiation always on cast microporosities. SRR99 presents an additional oxidation/spalling/re-oxidation mechanism under low strain loading. An "integrated approach" combining TF and TMF (thermo-mechanical fatigue) is proposed. The applied TMF temperature-strain cycle is deduced from the measured TF-cycle. Under this new temperature-strain cycle the crack initiation life and the total life in TMF for SRR99 are compared with the TF results.

KEYWORDS: thermal fatigue, thermo-mechanical fatigue, high temperature low cycle fatigue, superalloy, single crystal, induction heating, strain measurement

INTRODUCTION

Blades and vanes are key components in gas turbine engines. Thermal gradients in these components, in particular during transient regimes of start-up and shut-down operations, produce a complex thermal and mechanical fatigue loading which limits the life of these components. Additionally, centrifugal and gas bending forces lead to creep and the aggressive environment produces oxidation and corrosion.

Testing of blades and vanes under service conditions is prohibitively expensive [1]. Accurate and reliable laboratory assessment methods are therefore required. High temperature isothermal low cycle fatigue (HTLCF) and non-isothermal thermal fatigue (TF) as well as thermo-mechanical fatigue (TMF) tests are the main experiments used.

Specimens with blade-shaped sections such as Glenny's tapered discs [2] and single or double-edge wedge specimens [3-4] are generally used in TF experiments. No external constraint is imposed to the specimens during rapid heating and cooling. The transient thermal gradients generated in the specimens yield mechanical strains. In this kind of experiment, stresses and strains must be calculated by a suitable numerical thermal and mechanical analysis.

[1]Postdoc, PhD student and research associate, respectively, Laboratory of Mechanical Metallurgy, Swiss Federal Institute of Technology, CH - 1015 Lausanne, Switzerland.

Cylindrical LCF specimens are usually employed in TMF investigations. A strain-time cycle is superposed onto a temperature-time cycle. The external constraint replaces the internal constraint that occurs in TF. In general, the TMF temperature-strain histories are simplified cycles deduced from calculations of critical component parts.

The aim of this contribution is to report the recent progress which has been achieved in the Laboratory of Mechanical Metallurgy, Swiss Federal Institute of Technology, on TF and TMF assessment of nickel-based superalloys. The TF behaviours of two single crystal (SC) alloys, SRR99 and CMSX-4, are reported. Blade-shaped single edge wedge specimens with the crystallographic orientation [001] parallel to the main loading axis are used. The temperature-strain loops are measured both at the central zone and over the total length along the edge of the TF specimens. The influence of the maximum temperature of the thermal cycle as well as of the mechanical strain range on the TF crack initiation life are reported. A new temperature-strain cycle deduced from TF strain measurements is proposed for TMF investigations.

EXPERIMENTAL PROCEDURES

Alloys

The chemical compositions of both SC nickel-based superalloys, SRR99 and CMSX-4, are given in Table 1. The densities are $8.56 \, kg/dm^3$ for SRR99 and $8.70 \, kg/dm^3$ for CMSX-4.

TABLE 1--Chemical composition of SRR99 and CMSX-4 in weight%.

Alloy	Ni	Co	Al	Ti	Ta	Re	W	Mo	Cr	Hf	C
SRR99	65.4	5.0	5.6	2.2	2.9	...	9.5	0.03	8.5	...	0.020
CMSX-4	61.2	9.5	5.5	0.9	6.3	3.0	6.3	0.6	6.4	0.08	0.007

The alloys were supplied by MTU Munich GmbH, Germany, in the non-hipped condition. The heat treatments were as follows: SRR99 1280°C/1h, 1290°C/2h, 1300°C/0.5h, 1305°C/0.5h, 1080°C/4h, 870°C/16h, furnace cooling and CMSX-4 1290°C/3h, 1315°C/6h, 1140°C/4h, 870°C/16h, furnace cooling.

Both superalloys are constituted of a γ-matrix and cuboidal γ'-precipitates of about 70% volume fraction. In CMSX-4, only slight differences in the γ'-shape and the γ'-size throughout the dendritic microstructure were found, while SRR99 presented coarser γ'-precipitates in interdendritic regions. The cast microporosity was measured by an image analyser Kontron IBAS 200 AT. In both alloys the mean diameter of porosity was about 10 μm with a maximum value of 60 μm and a volume fraction of less than 0.2%.

Thermal fatigue test procedures

The TF tests were carried out on blade-shaped single edge wedge specimens, Fig.1a. They were cut from large single crystal cast slabs by electroerosion machining, then ground parallel to the edge and polished (1 μm diamond paste). The [001] crystallographic orientation of each specimen was determined by the Laue method. It was within 14° in the direction of the thin edge for all specimens.

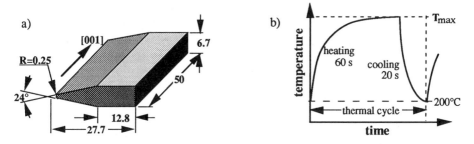

FIG. 1--a) Thermal fatigue specimen (dimensions in mm),
b) Thermal fatigue cycle.

Two similar thermal fatigue rigs for externally unconstrained rapid heating and cooling were designed, Fig. 2. Each TF rig consists of a high frequency oscillator for heating and a nozzle for forced air cooling. Two 6.0 kW Hüttinger solid state induction generators with 200 kHz and 3000 kHz were employed [4]. A Chromel-Alumel (type K) thermocouple (TC) spot-welded at 0.5 mm from the edge was used to monitor the test. The temperature was controlled by means of programmable Eurotherm 818P controllers.

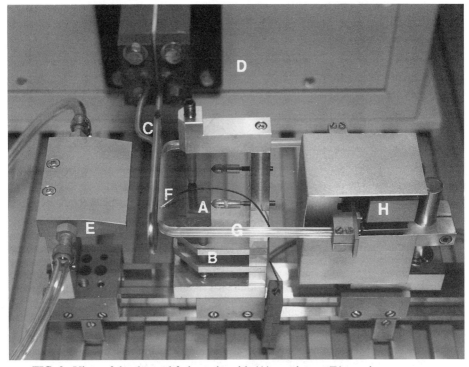

FIG. 2--View of the thermal fatigue rig with (A) specimen, (B) specimen support,
(C) induction coil, (D) HF-generator, (E) air nozzle, (F) thermocouple
and (G, H) extensometer device.

The TF tests were performed under temperature-time cycles of 60 s heating and 20 s cooling periods, Fig. 1b. Different maximum cycle temperatures, T_{max}, between 1000°C and 1175°C at the wedge tip were used, while the minimum cycle temperature, T_{min}, was maintained at 200°C. The tests were regularly interrupted to investigate crack initiation and propagation by means of scanning electron microscopy (SEM) and optical microscopy. The experiments were stopped when the crack growth rate became very low (corresponding to about 5 mm crack length). The specimens were cut and cross sections parallel and perpendicular to the edge were polished and examined by SEM.

The accurate measurement of temperature during thermal cycling was the main challenge in the test set-up. In fact, spot-welded TC wires act as heat extractors and reduce locally the surface temperature. It was found that the smaller the TC wires diameter, the more accurate the temperature could be measured, Fig. 3 [4]. Verifications by pyrometer measurements (bi-colour Quotienten-Pyrometer Maurer) for temperatures above 800°C revealed that even a TC with 0.1 mm wires diameter locally decreases the temperature, but negligibly. Therefore, type K thermocouples with separated 0.1 mm spot-welded wires were used. A data acquisition system was employed to collect time, temperature and deformation data. Ten points per second per channel were measured.

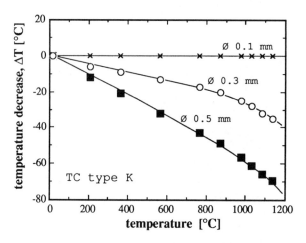

FIG. 3--Localized temperature decrease, ΔT, as a function of TC diameter (compared to the temperature measured with TC Ø=0.1 mm).

The thermal gradients along the edge length (50 mm) were determined. Ten thermocouples were spot-welded near the edge radius of a dummy specimen. The maximum thermal gradient along the edge during thermal cycling was less than 20°C over 10 mm from both free sides while it was about 5°C within the central zone of 30 mm for both test rigs. In addition, transient thermal gradients were measured during stabilized thermal cycles by means of a total of 30 thermocouples, sequentially welded at different points of the chord. At each sequence 10 thermocouples were welded in a plane perpendicular to the edge, Fig. 4. The maximum gradients in depth on the first millimeter are produced during the heating and cooling shocks, respectively 35°C/mm and 55°C/mm. The transient time-temperature cycles measured at the surface were used as boundary conditions in a thermal and mechanical analysis by a finite element method (FEM) to calculate stresses and strains [5].

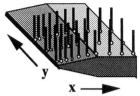

FIG. 4--Schematic presentation of the thermocouple locations.

Modifications of the specimen geometry (leading edge radius, wedge angle, width, thickness, etc.) [2-3,6] and of the temperature-time characteristics (T_{min}, T_{max}, heating or cooling rate, etc.) [3,7-9] are generally used to alter the TF test conditions. In fact, these modifications change the thermal gradients inside the TF specimens. The induction heating frequencies used (200 kHz and 3000 kHz) enable changing the transient thermal gradients while maintaining a unique temperature-time cycle in the wedge tip region, Fig. 1b. The different skin depths lead to two different temperature distributions and thermal gradients, Fig. 5. Therefore, two different temperature-strain histories are obtained.

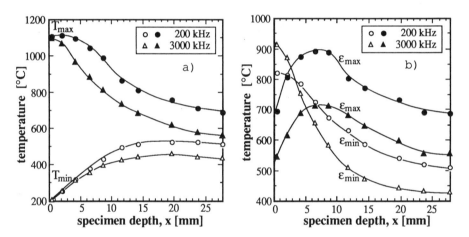

FIG. 5--Surface temperature distribution in depth at the middle plane of the TF specimen for 200 kHz and 3000 kHz at the instant of:
a) T_{min} and T_{max}, b) ε_{min} and ε_{max} (see Fig. 8 for definition).

Thermal fatigue temperature-strain cycles

The measurement of thermal and mechanical strains of blades in operation is difficult [10]. The temperature-strain histories must then be calculated. To overcome this difficulty, an axial MTS extensometer was modified and mounted on the test rig (see Fig. 2) with a total freedom of movement in the x-y plane of the TF specimen. The strain could then be measured as a function of time and temperature at the edge of the TF specimen. The TF strain measurement is a useful technique to investigate the behaviour of anisotropic directionally solidified (DS) and single crystal (SC) alloys, for which advanced FE design procedures are required.

The strain measurements on blade-shaped TF specimens reported for the first time in the literature were performed over 50 mm on the 200 kHz TF rig [9]. Actually, these measurements represent a global mean strain since the strain distribution along the wedge tip was neglected. The extensometer was thus readapted to measure the strain over 20 mm within the central part of the specimen.

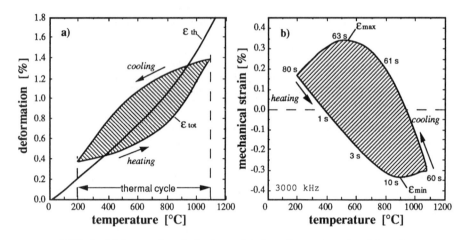

FIG. 6--a) Measured strain in thermal fatigue (gauge length = 20 mm),
b) mechanical strain $\varepsilon_{mec} = \varepsilon_{tot} - \varepsilon_{th}$.

The total strain versus temperature loop, $\varepsilon_{tot}(T)$, measured at the wedge tip of the TF specimens differs from the pure thermal expansion the material would exhibit under unconstrained conditions, $\varepsilon_{th}(T)$, as depicted in Fig. 6a. The thermal expansion is calculated by the well-known relationship $\varepsilon_{th}(T) = \alpha(T) \times \Delta T$, where $\alpha(T)$ is the coefficient of thermal expansion from dilatometry and ΔT is the difference between the instantaneous temperature and a reference temperature (normally 20°C). Therefore, the thermally induced mechanical strain, ε_{mec} (shown in Fig. 6b for the wedge tip), is the difference between the measured total strain $\varepsilon_{tot}(T)$ and the unconstrained thermal strain:

$$\varepsilon_{mec}(T) = \varepsilon_{tot}(T) - \varepsilon_{th}(T) \tag{1}$$

The temperature-mechanical strain loops at the edge over 50 mm and 20 mm gauge lengths are compared in Fig. 7 for the 3000 kHz rig at $T_{max} = 1100°C$. The central part of the wedge tip is subjected to a wider, hence more damaging temperature-strain loop. Therefore, cracks initiate mostly in the middle region of the TF specimens [4].

The temperature-mechanical strain loops for both test rigs are compared in Fig. 8 ($T_{max} = 1100°C$). The mechanical strain range $\Delta\varepsilon_{mec} = \varepsilon_{max} - \varepsilon_{min}$, as well as the mean mechanical strain $\varepsilon_{mean} = (\varepsilon_{max} + \varepsilon_{min})/2$ and the strain ratio $R_\varepsilon = \varepsilon_{min}/\varepsilon_{max}$, are different for each TF rig. The 3000 kHz TF rig with higher thermal gradients induces a wider temperature-mechanical strain loop. The hysteresis loops of 3000 kHz and 200 kHz TF test-rigs are thus named "high strain" and "low strain" cycles, respectively.

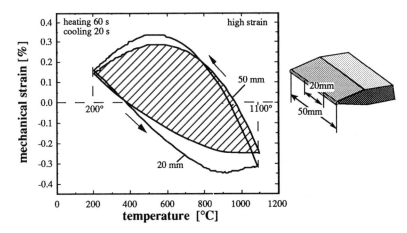

FIG. 7--Temperature-mechanical strain loops measured over the total length (50 mm) and the centre part (20 mm) of the edge.

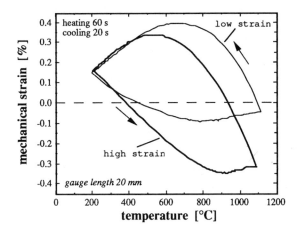

FIG. 8--Temperature-mechanical strain curves of the high strain cycle and the low strain cycle at T_{max}=1100°C, 20 mm gauge length.

Symmetric or asymmetric diamond-shaped temperature-mechanical strain cycles employed in thermo-mechanical fatigue (TMF) assessments [11-12] are in general based on FE-analysis of temperature-stress-strain fields of real components. In these cycles, maximum and minimum mechanical strains occur at intermediate temperatures while zero mechanical strain takes place at T_{min} and T_{max}. The measured cycles, however, show residual mechanical strains at T_{min} and T_{max} because thermal gradients are still present.

RESULTS

Crack initiation and propagation

In SRR99 and CMSX-4, residual cast microporosities on the wedge tip lead to stress concentration. For both alloys under high strain loading, cracks always initiate on these porosities at the specimen surface, Fig. 9a+b.

Under low strain loading the number of thermal cycles (N) to crack initiation is much higher. Nevertheless, the same mechanism of crack initiation on microporosities was observed for CMSX-4, Fig. 9d. For SRR99, a complex mechanism of oxidation/ spalling/ reoxidation combined with the effect of the residual cast microporosity was identified as the crack initiation mechanism, Fig. 9c.

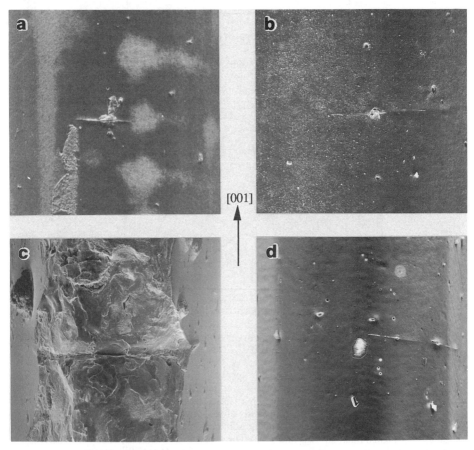

FIG. 9--View on the wedge tip, SEM-micrographs $\frac{200\ \mu m}{}$:
a) SRR99, T_{max}=1150°C, $\Delta\varepsilon_{mec}$=0.75%, N=   680,
b) CMSX-4, T_{max}=1150°C, $\Delta\varepsilon_{mec}$=0.75%, N= 1080,
c) SRR99, T_{max}=1150°C, $\Delta\varepsilon_{mec}$=0.53%, N= 5200,
d) CMSX-4, T_{max}=1100°C, $\Delta\varepsilon_{mec}$=0.48%, N=28000.

Wedge tips of CMSX-4 specimens remained almost intact during low strain loading even after very high numbers of thermal cycles (Fig. 9d), showing its high resistance to oxidation, while wedge tips of SRR99 specimens were highly damaged by oxide-scale spalling.

For both alloys investigated, crack initiation was always observed only after an incubation period, which is dependent upon the strain range and the maximum temperature of the thermal cycle. All cracks have been initiated at the wedge tip surface and propagated towards the bulk. For all tests, an initial increase in the crack growth rate (CGR) was followed by a pronounced decrease when the crack length was between 2–3 mm [4,9]. The exact beginning of this crack growth retardation depends upon $\Delta\varepsilon_{mec}$ and T_{max}.

The reduction of thermal gradients in the specimen depth and subsequent decrease in thermal strains and stresses at the crack tip is the main reason for crack growth retardation on blade-shaped specimens. For longer cracks, the lower temperature and the lower interaction with oxygen should also be considered as an additional cause. No significant difference in the CGR was observed between SRR99 and CMSX-4 for the same TF test conditions.

Thermal fatigue life

The thermal fatigue life depends not only on mechanical properties, microstructure and environment, but also on the crack initiation and crack propagation criteria. The definition of fatigue life is a matter of controversy even for simple isothermal LCF experiments. The total fatigue life (N_f) is generally divided into crack initiation (N_i) and crack propagation (N_p) periods [13].

For TF blade-shaped specimens the definition of TF-life is still more complicated because a drastic crack growth retardation occurs for long cracks and the tests can not be conducted up to total fracture of specimens. Therefore, an engineering TF-life (N_f) corresponding to the number of thermal cycles to form a 1 mm crack depth was defined. On the other hand, a crack initiation life (N_i) was defined as the number of cycles required to form a crack of 0.1 mm depth. Both high strain and low strain results have shown that N_i is about 50% to 95% of N_f (depending upon alloy and mechanical strain range). Therefore, a major part of the TF-life of both alloys is spent to initiate a detectable crack.

N_i is reported as a function of the maximum cycle temperature T_{max} and of the mechanical strain range, $\Delta\varepsilon_{mec}$, in Figs. 10a and 10b respectively. For a given T_{max}, N_i is about 7 to 20 times shorter for both superalloys under the high strain TF rig in comparison to the low strain TF rig, Fig. 10a. While similar TF crack-initiation lives are found for both alloys under high strain cycles (a factor ≈ 2), the TF crack-initiation life of SRR99 is about six times shorter than that of CMSX-4 under low strain cycles.

For CMSX-4, a unique $\Delta\varepsilon_{mec}-N_i$ curve was observed, Fig. 10b, suggesting that for the test conditions investigated, the thermal fatigue life of CMSX-4 is not much dependent on T_{max}. On the other hand two different $\Delta\varepsilon_{mec}-N_i$ curves for high strain and low strain loading are observed for SRR99. For the same $\Delta\varepsilon_{mec}$ the thermal fatigue life decreases considerably with increasing the maximum cycle temperature, which should be related to the lower oxidation resistance of SRR99 compared with CMSX-4 [14].

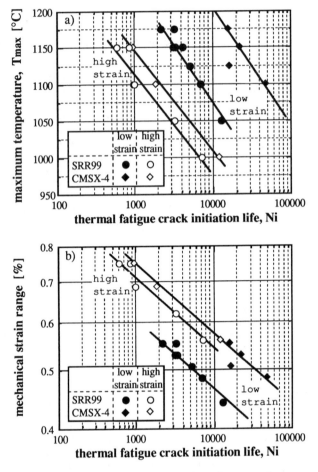

FIG. 10--TF-crack initiation life (N_i) as a function of maximum cycle temperature T_{max} (a), and mechanical strain range $\Delta\varepsilon_{mec}$ (b).

DISCUSSION

TF loading and TF crack initiation are complex processes where intrinsic and extrinsic parameters should be considered. Intrinsic factors include resistance to aggressive environments (oxidation or corrosion), microstructure features (grain size, dendrite arm spacing, size and stability of γ'-phase), crystallographic orientation (in DS or SC alloys), size of residual microporosities, etc.

The main extrinsic parameter is the distribution of the transient thermal gradients in the specimen, which imposes transient thermal strains and stresses to any element of a TF specimen. Increasing T_{max} increases the thermal gradients in the specimen, resulting in wider temperature-mechanical strain loops (higher damage per cycle) at the wedge tip and lower TF lives. For the same reason, the crack initiation life under high strain cycles is shorter as compared to the low strain cycles for a given T_{max}.

The temperature-strain histories in TF induce a complex loading on the TF specimen. An element of volume at the wedge tip is first subjected to a compressive thermal shock in the beginning of heating. Then it is maintained under compressive loading at high temperature up to the maximum temperature of the thermal cycle. In the last step, the element is subjected to a severe tensile thermal shock in the beginning of cooling. The maximum loading in compression and tension appear at intermediate temperatures (about 500°C to 950°C). During compression at high temperatures, a complex interaction of fatigue-creep-oxidation weakens the mechanical properties of the alloys (in particular in the sub-surface zone) which will be later subjected to a tensile loading at lower temperatures, where the oxide is less ductile and is prone to spalling due to the internal strains.

The γ-matrix of nickel-based superalloys is "reinforced" by a high volume fraction of intermetallic γ'-particles. During thermal cycling (as well as during isothermal oxidation) a "composite-like" layer consisting of an oxide-scale and its corresponding γ'-depleted zone is formed at the wedge tip of the polished TF specimens [9]. Cast microporosities open to the surface, which are sites of crack initiation, show the same sequence of oxide-scale and the γ'-depleted zone. Figure 11 shows a sample of CMSX-4 after isothermal oxidation for 1 hour at 1100°C. Although there is no constraint under isothermal exposure, this figure illustrates the complex mechanism involved in crack initiation during TF cycling.

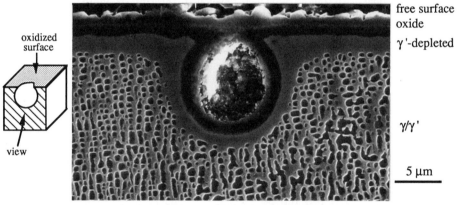

FIG. 11--Polished and etched cross section of a CMSX-4 sample showing the γ'-depleted zone and an oxidized cast microporosity open to the surface (1 hour at 1100°C).

The micro-mechanical behaviour and the mechanical resistance of this "composite-like" layer, which is in fact a "sub-surface process zone", determines the crack initiation life of the alloys. The overall resistance of this composite layer depends on the resistance of the oxide-scale, of the γ'-depleted zone and of the interface between the oxide scale and γ'-depleted zone (adherence resistance). A higher strength γ'-depleted zone should improve the resistance of this "composite", which in turn should enhance the crack initiation life in TF. Microhardness measurements within the γ'-depleted zone have revealed a "softening" of this zone during thermal TF loading [9,15]. This softening is less pronounced in CMSX-4 than in SRR99. The second generation SC alloy CMSX-4 shows a high level of balanced mechanical properties [16] and oxidation resistance through the beneficial effects of adding rhenium, which tends to stabilize the strengthening γ'-phase [17].

Thermal fatigue and thermo-mechanical fatigue: An integrated approach

Thermal and thermo-mechanical fatigue tests are two complementary laboratory methods to assess materials behaviour under high temperature non-isothermal fatigue. However, TF or TMF behaviour and lives of nickel-based superalloys are generally compared solely to the available high temperature isothermal LCF results. An "integrated approach", where a *measured* (instead of calculated) temperature-strain hysteresis loop in TF is used as the basic cycle for TMF investigation, makes possible the comparison between TF and TMF under similar temperature-strain histories.

The measurement of stresses is difficult on real blades (as on the TF specimens under the test conditions reported here). Therefore, these stresses are generally calculated by FEM. These calculations are quite complex and the results depend on several factors, like the discretization model used, the stress-strain constitutive laws employed (elastic, elasto-visco-plastic, etc.) and the thermal as well as mechanical boundary conditions considered. For DS and SC alloys the situation is still more complex as the relevant anisotropic constitutive laws need to be verified. By the proposed integrated approach, however, the stress history on the wedge tip of TF specimens can, to some extent, be *experimentally* evaluated.

As an example, the mechanical strain measured in the inner 20 mm wedge tip of a SRR99 TF specimen under a TF test between 200°–1100°C (Fig. 7) was applied under TMF. A hollow cylindrical specimen was used (details of TMF tests are given in [18]). The measured stresses are reported in Fig. 12. During the test a positive mean stress as well as a negative residual plastic deformation were developed. This corresponds to the shortening of the TF specimen wedge tip that is observed during TF tests [4].

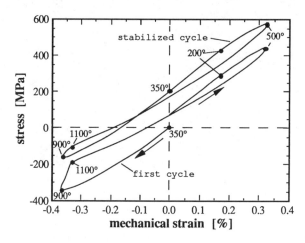

FIG. 12--Stress-mechanical strain loops obtained in TMF of SRR99 for a measured strain-temperature TF-cycle (200°–1100°C, Fig. 7).

To compare crack initiation and total lives of TF and TMF, T_{min} of the TMF cycle was maintained at 600°C due to TMF test limitations (the total period of a TMF cycle between 200°–1100°C without forced cooling is too long). Therefore, the temperature-mechanical strain cycle was simplified, as illustrated in Fig. 13.

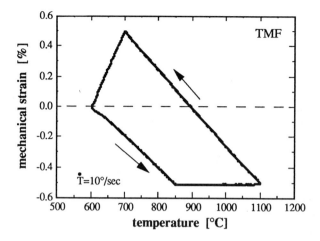

FIG. 13--TMF temperature-mechanical strain cycle.

The replica method was used to determine the crack initiation and propagation in TMF [18]. Like in TF, cracks initiate always on microporosities. The same criteria used in TF for N_i and N_f were defined for TMF. The fatigue lives obtained for SRR99 under TF and TMF are compared in Fig. 14 for $T_{max}=1100°C$. As it can be seen, under the test conditions used, TF loading results in a shorter life for both N_i and N_f criteria.

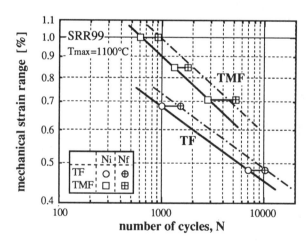

FIG. 14--Crack initiation N_i and total life N_f of SRR99 in TF and TMF tests, maximum cycle temperature 1100°C.

Three differences in the test parameters between TF and TMF should be highlighted:

- First, the strain rates during thermal shocks in TF are much higher than the maximum TMF strain rate.

- Second, the difference in the geometry of the specimens: the TF specimen wedge tip radius (0.25 mm) is much smaller than the external radius of TMF specimens (5.5 mm). Therefore, the wedge tip is more sensitive to the notch effect of a cast microporosity, what means that a higher stress intensity factor is expected in a TF specimen.

- Third, the temperature amplitudes ($\Delta T = T_{max} - T_{min}$) are different, 900°C for the TF-cycle and 500°C for the TMF cycle. Given the "composite" nature of the microstructure near the surface, the mismatch of thermal expansion between the different phases will enhance the interfacial stress experienced. As this effect increases with increasing ΔT, TF tests tend to be more damaging, which is consistent with the TF and TMF fatigue lives observed in the present work.

CONCLUSION

The thermal fatigue behaviour of CMSX-4 (containing rhenium) and SRR99 was investigated on blade-shaped single edge wedge specimens. The strain was measured at the wedge tip of the TF specimens. Temperature-mechanical strain cycles with different mean strain values and strain ratios were obtained. The strain distribution on the edge was presented. Further, TF crack initiation life and total life depend on the strain cycle, which itself is temperature gradient dependent. On the basis of TF strain measurements, a new TMF cycle is introduced. An integrated approach for TF and TMF investigations was proposed. Preliminary investigations show that under the test conditions used, TF cycling is more damaging than TMF. TF crack initiation mechanisms for both superalloys were identified. Finally, the higher TF crack initiation resistance of CMSX-4 is explained by its higher oxidation resistance combined with a higher mechanical resistance of its γ'-depleted zone and γ/γ'-microstructure.

ACKNOWLEDGEMENTS

The authors would like to acknowledge MTU Munich GmbH, Germany, the Swiss National Foundation and the Swiss Priority Program on Materials for supporting this investigation. C. Engler-Pinto thanks also the Secretary for Science and Technology of Brazil (CNPq) for granting his PhD scholarship. Prof. B. Ilschner is acknowledged for many fruitful discussions and support.

REFERENCES

[1]   Coney, M.H., "Thermal Fatigue Cycling in Turbine Blades," High Temperature Technology, Vol. 8 No 2, May 1990, pp 115-120.

[2]   Glenny, E., Thermal and High Strain Fatigue, The Institute of Metals, London, 1967, pp 346-363.

[3]   Rézaï-Aria, F., François, M. and Rémy, L., "Thermal Fatigue of MAR-M509 Superalloy - I. The Influence of Specimen Geometry," Fatigue Fract. Engng. Mater. Struct. 11, No. 5, 1988, pp 277-289.

[4]   Meyer-Olbersleben, F., Dissertation thesis No. 1255, Swiss Federal Institute of
       Technology, Lausanne, 1994.

[5]   Mallet, O., Kaguchi, H., Ilschner, B., Nikbin, K., Rézaï-Aria, F. and Webster,
       G.A., "Influence of Thermal Boundary Conditions on Stress-Strain Distribution
       Generated in Blade-Shaped Samples," accepted to be published in International
       Journal of Fatigue, Materials, Structures, Components.

[6]   Woodford, D.A. and Mowbray, D.F., "Effect of Material Characteristics and Test
       Variables on Thermal Fatigue of Cast Superalloys. A Review," Mater. Sci. and Eng.
       16, 1974, pp 5-43.

[7]   Boone, D.H. and Sullivan, C.P., Fatigue at Elevated Temperature, ASTM STP 520,
       1973, pp 401-415.

[8]   Beck, C.G. and Santhanam, A.T., Thermal Fatigue of Materials and Components,
       ASTM STP 612, 1976, pp 123-140.

[9]   Meyer-Olbersleben, F., Goldschmidt, D. and Rézaï-Aria, F., Proceedings of the
       Seventh International Symposium on Superalloys, 1992, TMS, Seven Springs
       Mountain Resort, Champion, Pennsylvania, 1992, pp 785-794.

[10]   Bernstein, H.L., Grant, T.S., McClung, R.C. and Allen, J.M., Proceedings of
       Thermomechanical Fatigue Behaviour of Materials, ASTM STP 1186, Philadelphia,
       1993, pp 211-238.

[11]   Malpertu, J.L. and Rémy, L., "Influence of Test Parameters on the Thermal-
       Mechanical Fatigue Behaviour of a Superalloy," Metallurgical Transactions A, Vol.
       21A, Feb. 1990, pp 389–399.

[12]   Kraft, S, Zauter, R. and Mughrabi, H., "Aspects of High-Temperature Low Cycle
       Thermomechanical Fatigue of a Single Crystal Nickel-base Superalloy," Fatigue Fract.
       Engng. Mater. Struct. Vol. 16, No. 2, 1993, pp 237-253.
[13]   Maiya, P. S., "Considerations of Crack Initiation and Crack Propagation in Low
       Cycle Fatigue," Scripta Metallurgica, Vol. 9, 1975, pp 1141–1146.

[14]   Göbel, M., Rahmel, A., Schütze, M., "The Isothermal-Oxidation Behavior of Several
       Nickel-Base Single-Crystal Superalloys with and without Coatings," Oxidation of
       Metals, Vol. 39, Nos.3/4, 1993, pp 231-261.

[15]   Meyer-Olbersleben, F., Goldschmidt, D. and Rézaï-Aria, F., Proceedings of the
       International Conference on Corrosion-Deformation Interactions CDI'92, Editions de
       physique, Les Ulis, Fontainebleau, 1993, pp 543-551.

[16]   Thomas, M.C., Helmink, R.C., Frasier, D.J., Whetsone, J.R., Harris, K., Erickson,
       G.L., Sikkenga, S.L. and Eridon, J.M., Proceedings of Materials for Advanced
       Power Engineering 1994, Kluwer Academic Publishers, Part II, Liège, 1994, pp
       1075-1098.

[17]   Giamei, A.F. and Anton, D.L., "Rhenium Additions to a Ni-Base Superalloy : Effects
       on Microstructure," Metallurgical Transactions A, 16A (1985), pp 1997-2005.

[18]   Engler-Pinto Jr., C.C., Härkegård, G., Ilschner, B., Nazmy, M.Y., Noseda, C. and
       Rézaï-Aria, F., Proceedings of Materials for Advanced Power Engineering 1994,
       Kluwer Academic Publishers, Part I, 1994, pp 853-862.

Johan Bressers[1], Jens Timm[1], Stephen J.Williams[2], Andrew Bennett[2] and Ernst E.Affeldt[3]

EFFECTS OF CYCLE TYPE AND COATING ON THE TMF LIVES OF A SINGLE CRYSTAL NICKEL BASED GAS TURBINE BLADE ALLOY

REFERENCE: Bressers, J., Timm, J., Williams, S., Bennett, A., and Affeldt, E., ''Effects of Cycle Type and Coating on the TMF Lives of a Single Crystal Nickel Based Gas Turbine Blade Alloy,'' Thermomechanical Fatigue Behavior of Materials: Second Volume, ASTM STP 1263, Michael J. Verrilli and Michael G. Castelli, Eds., American Society for Testing and Materials, 1996 pp. 56-67.

ABSTRACT: Strain controlled thermo-mechanical fatigue cycles simulating the temperature-strain-time history at critical locations of blades of advanced aero gas turbines are applied to the single crystal nickel based alloy SRR99 in the uncoated and aluminide coated conditions. The TMF cycle selection includes a -135°lag cycle and an in-phase cycle, with various R-ratios, $T_{min}=300°C$, and $T_{max}=1050°C$ and $850°C$, respectively. The cycle-specific stress response is analyzed and discussed in terms of the accumulation of inelastic strain during the TMF tests, and the inelastic strain build-up is correlated with the cyclic hardening / softening behaviour. The number of cycles for initiating microcracks is measured by means of a computer vision system. Various modes of crack initiation and crack growth are observed and correlated with the TMF cycle type, with the strain range imposed, and with the ductile/brittle behaviour of the coating. The differences in TMF lives are discussed in terms of the material and TMF parameters

KEYWORDS: Single crystal nickel based alloy, aluminide coating, thermomechanical fatigue, microcrack initiation, fatigue life.

Single crystal nickel based alloys are widely used as blade materials in aero gas turbine blades because of their excellent resistance to high temperature deformation. Tailoring of their chemical composition towards improving the high temperature strength has resulted in a reduced oxidation resistance. Coatings are applied in order to provide the blades with adequate protection against environmental degradation. Aluminide coatings, formed by diffusing aluminium into the nickel based alloy by means of the pack aluminizing process are commonly

---

[1]Institute for Advanced Materials, Joint Research Centre, P.O.Box 2, 1755 ZG Petten, The Netherlands.

[2]Rolls-Royce plc, P.O.Box 31, Derby DE24 8BJ, United Kingdom.

[3]MTU Motoren und Turbinen Union München, Postfach 50 06 40, 80976 München, Germany

used on gas turbine blades.

The major cause of failure in current single crystal blades of aero gas turbines is thermally induced stresses, which result from thermal strains over the blade thickness caused by temperature gradients during heating and cooling cycles. Isothermal mechanical fatigue data, which are traditionally used for blade design purposes, do not account for the damage and failure processes occuring in blades exposed to thermal fatigue cycles. Moreover, blades provided with aluminide coatings are expected to display a non-isothermal fatigue behaviour because of the ductile-brittle transition of the coating at intermediate temperatures. Actual blade behaviour is more closely simulated by thermo-mechanical fatigue (TMF) tests, which are designed to reproduce the temperature and strain cycles seen by critical volume elements of the blade. As opposed to isothermal fatigue testing, the coated test specimen is repeatedly cycled through the ductile-brittle transition temperature of the coating during TMF cycling, which gives rise to different damage and failure mechanisms.

In a number of studies the nature and the extent of the influence of the presence of the coating on the mechanical performance of nickel based superalloys has been investigated. Either detrimental or beneficial/neutral effects on specific, isothermally-measured mechanical properties of the substrate were reported [1-8]. When subjected to TMF cycling, the presence of a coating can be detrimental [9] or beneficial [10-11] to the TMF life. However, the sparse data reported so far are not comparable in terms of coating types, substrates, temperatures and loading cycles, making an interpretation of the results rather difficult. As part of a study into the TMF behaviour of single crystal nickel based alloys in the uncoated and aluminide coated conditions, the effect of the TMF cycle type and the influence of the presence of the coating on the TMF life, and the associated crack initiation and failure modes have been investigated and are reported here and in a companion paper in this volume [11].

EXPERIMENTAL DETAILS

Materials

Single crystal SRR99 of industrial quality has been selected as the substrate material for all of the TMF testing reported in this paper. Its chemical composition is given in Table 1. Following the solution heat treatment of the 25 mm diameter single crystal blanks, these are machined into TMF test specimens. Half of the test specimens are high activity pack aluminized prior to the coating diffusion treatment of 1 hr at 1100°C in an Ar atmosphere, and a 16 hr ageing heat treatment at 870°C in Ar. The other half of the test specimens remain uncoated, but are given

TABLE 1-Chemical composition of SRR99 (in wt.%)

Al	Co	Cr	Mo	Ta	Ti	W	C	Ni
5.3	4.8	8.25	0.5	2.65	2.05	9.25	.015	bal.

the same heat treatment as the coated samples inclusive of a mock coating treatment, in order to achieve a similar microstructure as in the coated substrate. The crystallographic orientation of the long axis of the test specimens is within 10° of <001>. The microstructure of SRR99 consists of approximately 70 vol.% of γ' phase in a γ matrix, with an average size of 0.6-0.7 μm. In addition W carbides and (Ta,Ti) carbides are observed, along with interdendritic porosity. The main constituent of the coating is βNiAl which contains fine precipitates of W_xC. In the diffusion zone

βNiAl, $\gamma'$and W_xC coexist. The transition to the substrate material is formed by a thin layer of γ'.

Thermo-mechanical Fatigue Test Procedure

The thermo-mechanical fatigue tests are carried out on a computer controlled servohydraulic testing machine. The test specimen is heated by means of direct induction. Control of the testing machine and of the high frequency generator, and data acquisition are performed by means of a dedicated computer system, using a graphical programming language for constructing hierarchically ordered command and data acquisition software modules.

The TMF test specimen has a cylindrical end geometry and a solid gauge length with a rectangular cross section of 12mmx3mm in order to enable observation of the surface by means of an optical microscope during testing. The edges of the gauge length section of the coated (and uncoated) specimens are rounded in order to promote good adherence of the coating. The surface finish in the gauge length of the substrate material is $R_a=0.05$ µm. Coated specimens are tested in the as received condition ($R_a=0.5$ µm) without any further surface modification treatment.The coating cross section is taken into account when calculating stresses on the coated specimens.

The sample temperature is controlled by means of a thermocouple spotwelded to the specimen surface outside the gauge length, in order to inhibit premature crack formation in the gauge section as the result of spot welding damage. A calibration procedure is used to obtain the correct temperature at the centre of the gauge length, in conjunction with a pyrometer for double checking of the temperature. Strains are measured by means of a longitudinal extensometer, with the limbs spanning a gauge length of 10 mm.

All tests are carried out in total strain control between mechanical strain limits. The temperature is varied linearly with time and synchronously in-phase, or with a 135° phase lag, with respect to the mechanical strain. The mechanical strain ε_m is defined as

$$\varepsilon = \varepsilon_m - \varepsilon_{th} \tag{1}$$

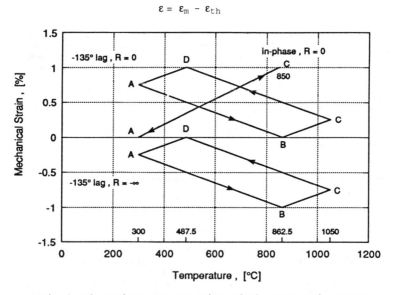

Fig.1-Schematic representation of the TMF cycle types

where ε and ε_{th} are the total and thermal components of the strain, respectively. The two cycle types are shown schematically in Fig.1. Tests with different R-ratios are performed i.e.R=-∞ and R=0 in the case of the -135°lag tests, and R=0 in the case of the in-phase tests.

The minimum cycle temperature is 300°C. The maximum cycle temperature varies from 850°C in the in-phase tests to 1050°C in the -135°lag tests. Generalized plane-strain as well as 3D Finite Element calculations confirm that these cycle types mimic the strain-temperature conditions at the most critical blade locations. For example, the -135°lag, R=-∞ tests simulate a hot spot condition.

In order to approach in-service heating and cooling rates of blades as realistically as possible, a compromise between high rates and an acceptable temperature gradient over the specimens' gauge length had to be made. Specimen heating and cooling rates were set at 25°C/s and 12.5°C/s respectively, resulting in a maximum temperature gradient over the gauge length of ±15°C. The corresponding maximum deviation from zero of the stress measured during cycling under thermal strain control is well within ±10 MPa. Forced cooling during the larger part of the downward branch of the cycle is required in order to enable the cooling rate of 12.5°C/s to be achieved. Cycle periods are 90 s and 66s for the out-of-phase and in-phase tests, resulting in mechanical strain rates of approximately 3×10^{-5} s^{-1} and 1.5×10^{-5} s^{-1} during the heating and cooling parts of the cycle respectively . In view of the modest strain rate sensitivity of the material [12] and of the small changes in strain rate involved, the effect on the mean stress of halving the strain rate between the heating and cooling ramps is estimated to be less than 20 MPa.

Prior to each TMF test the temperature dependence of the E-modulus for the test specimen involved is measured at intervals of 100°C, up to the maximum temperature of the TMF cycle. The thermal expansion of the specimen is recorded as a continuous function of the temperature whilst heating and cooling the specimen at the same rate as in the actual TMF test. All tests are started at 300°C.

Crack Initiation and Crack Growth

One of the reasons for adopting a flat specimen geometry is related to the use of a computer vision system for the non-intrusive, in-situ monitoring of the initiation and growth of microcracks during the TMF tests. The surface of the test specimen is optically imaged at preselected cycle numbers, at a magnification which enables microcracks approximately l=15-30 μm long to be detected under the oxidizing conditions of the test. The images are digitized and stored for post processing. Image acquisition occurs in a fully automated, computer controlled manner. All the operational parameters of the system are user defined.

RESULTS AND DISCUSSION

Stress Response

Fig.2 shows the evolution of the characteristic cycle stresses with cycle number N at points A to D (see Fig.1) for each of the three TMF cycle types, measured on an uncoated specimen at a mechanical strain range $\Delta\varepsilon_m$=1%. Similar changes of the characteristic cycle stresses are observed at smaller mechanical strain ranges, although the changes are less explicit in terms of magnitude. The stress response is not influenced by the presence of the coating.

In the -135°lag, R=-∞ test the material exhibits a saturation stage which sets in after a fraction of the cyclic life of the order of N/N_f=0.1 has elapsed. Prior to saturation a pronounced shift of all the cycle stresses towards more positive values is noted. The stress range assumes a constant value of nearly 1100 MPa from the first cycle

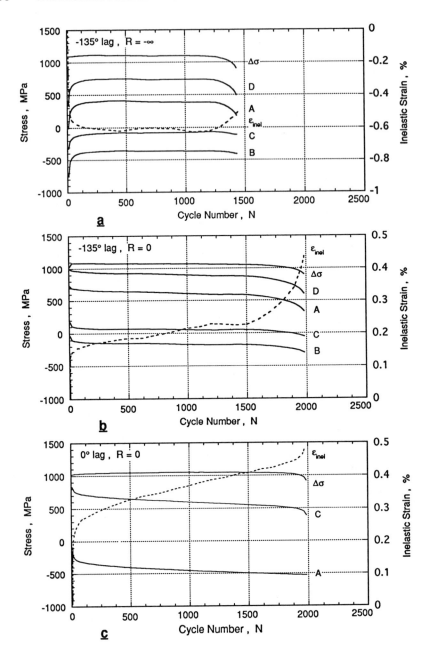

Fig.2- Evolution of the characteristic cycle stresses with number of cycles for the various TMF cycle types;$\Delta\varepsilon_m$=1%; uncoated specimen

onwards.In the -135°lag, R=0 test and in the in-phase test a primary
stage of tensile softening / compressive hardening is observed, which is
more pronounced in the latter. Subsequently the material response changes
to another regime of either gradual or pronounced tensile softening /
compressive hardening in the -135°lag, R=0 and in the in-phase test
conditions respectively, from N/N_f = 0.1 onwards. The corresponding
stress range in the -135°lag, R=0 test condition slightly increases to
nearly 1100 MPa before stabilizing at approximately N/N_f =0.02. In the
in-phase test the stress range increases slighty up to appoximately 1050
MPa. Fig.2 also displays the evolution with the cycle number N of the
momentary value of the inelastic strain at the end point A (see fig.1) of
each cycle. The changes of the inelastic strain within each cycle are not
shown.
      The pronounced cyclic hardening / softening between cycle 1 and
N/N_f =0.1 is the result of the accumulation of inelastic deformation,
caused by plasticity and/or creep. Fig.3, which documents the stresses
and temperatures as a function of time in the first cycle of each TMF
cycle type for a strain level of $\Delta\varepsilon_m$=1%, will be used to illustrate this
point. The TMF cycle stresses are compared to the stress required for
achieving 0.1% plastic strain (0.1% PS) at the corresponding temperature.
The stress values for 0.1% PS are derived from a series of tensile tests
performed at a strain rate of 3.10^{-5} s^{-1} over the range of temperatures
covered by the TMF experiments [12]. In addition the creep rate
corresponding to the momentary TMF cycle stress and temperature is
plotted. The creep rates are determined by linear interpolation, using a
data base of primary creep rates averaged over 1 s hold times in
isothermal fatigue tests on SRR99 [12]. Inelastic deformation occurs when
the cycle stress equals the 0.1% PS level and/or when it reaches a level
where the creep contribution becomes significant. Because SRR99 displays
elastic-ideal plastic behaviour over most of the temperature range
considered, the (arbitrary) choice of the 0.1% PS stress as a flow
criterion does not influence the argument. In the -135°lag,R=-∞ test
(fig.3a), large scale plastic deformation in compression starts at a
temperature of nearly 850°C, just prior to reaching the condition marked
B in fig.1,and ending just beyond T_{max} (point C). The high average creep
rate shown in fig.3 results from the combination of high stress and
temperature during this part of the cycle. However, for the present
argument it is irrelevant whether the inelastic strain accumulation is
caused by plastic deformation or by creep. The pronounced primary
hardening then is the result of re-establishing the prescribed mechanical
strain levels during subsequent cycles. However, with increasing cycle
number the compressive cycle stresses which drive the inelastic
deformation rapidly drop to small absolute values, eventually ceasing to
cause inelastic strain build-up from N/N_f =0.1 onwards.
      Plastic deformation / creep in the high temperature part of the TMF
cycle drives the primary tensile softening / compressive hardening stage
in the in-phase test, see fig.2c. At the end of the primary stage the
maximum stress has dropped below the stress for 0.1% PS, but relatively
high values in excess of 500 MPa persist up to N/N_f=0.9. Consequently,
the steady accumulation of the inelastic strain between N/N_f =0.1 and
N/N_f =0.9 continues, primarily as the result of creep deformation,
causing the pronounced tensile softening / compressive hardening
throughout the TMF life.
      In the -135°lag, R=0 cycle test the inelastic strain accumulation
is primarily due to plastic deformation in tension in the temperature
range 650°C to 350°C, fig.3b. Because the stresses in the high
temperature part of the cycle are of the order of 100 MPa or less
(fig.2b), the contribution of creep to the inelastic strain accumulation
is small. In addition, the limited extent of the tensile softening /
compressive hardening following the primary softening / hardening stage
suggests a limited role of creep recovery processes during the high
temperature part of the cycle.

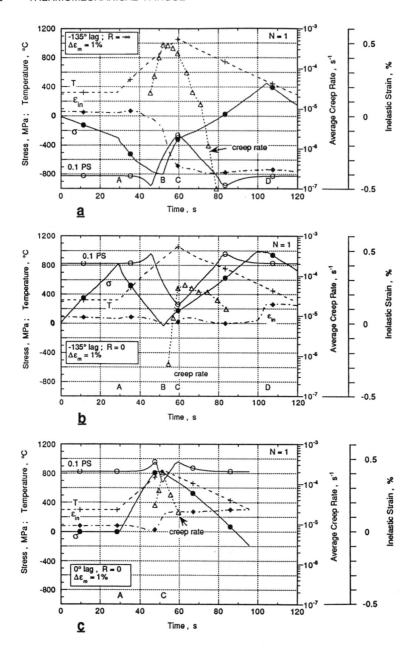

Fig.3-Evolution of temperature, stress and inelastic strain with time during the first cycle of the TMF tests at $\Delta\varepsilon_m$=1%. The stress for 0.1% plastic deformation (0.1% PS) and the average primary creep rate are included for comparison.

TMF lives

    A relatively small fraction of the TMF life is spent in initiating microcracks. Initiation is defined as the cycle number N_i where cracks with a length on the surface of $l=30\mu m$ are measured, a definition which is dictated by the resolution of the computer vision system. The TMF life is defined as the cycle number N at which the maximum cycle stress has decreased to 2/3 of the saturation stress level.The fraction of life required for initiation ranges from 1% to 5% in nearly all the -135°lag tests of coated and uncoated SRR99. Precise N_i data are not available for the in-phase tests because of the sub-surface initiation of cracks. A conservative estimate of N_i/N_f, based on the observation of the first surface breaking of sub-surface initiated cracks, suggests values of the order of <30%.

    Effect of TMF cycle type on the life of uncoated SRR99--Since the larger fraction of the TMF life is spent in crack growth which is a stress controlled process, the lives of the uncoated SRR99 are plotted in fig.4 as a function of the stress range $\Delta\sigma$ at $N_f/2$. The differences in life at and in excess of $\Delta\sigma=1000$ MPa are considered to be within the limits of scatter. At the small strain ranges the in-phase cycle yields a substantially longer life when compared to the lives of the -135°lag cycles. A similar correlation is observed when the results are plotted as a function of the mechanical strain range.

    Apart from the disparities in terms of stress response and deformation between the various cycle types, pronounced differences also exist in terms of crack initiation and crack growth mechanisms. In the -135°lag cycle tests multiple cracks initiate at the wide surfaces of the test specimen, prior to N/N_f =0.05. The process of crack initiation appears frequently to be associated with the preferential oxidation of microstructural features located in stringers along the long axis of the test specimen, resulting in oxide hillocks protruding from the surface, which are cracked. There is some evidence that these preferred sites of crack initiation, which are located in the interdendritic zones, are associated with casting porosity and/or carbides.

    The process of crack initiation is nearly complete at the end of the primary hardening stage in the R=-∞ test, but continues up to N/N_f =0.3-0.4 in the R=0 test. Cracks grow individually until at approximately N/N_f =0.5 crack extension by a process of coalescence starts contributing to the extension of the cracks. Crack growth starts off in stage II, changing to stage I later on. The area fraction of the fracture surface covered by stage I is approximately 0.6 at the high stress ranges for both R-ratios. At long lives stage I/stage II area fractions are 0.4 and 0.1 for R=0 and R=-∞, respectively. The crack tip environments observed in specimen cross sections suggest a growth mechanism whereby oxide films, repeatedly formed in the high temperature part of the TMF cycle, are disrupted by crack extension in the intermediate/low temperature part of the cycle where the stress is at and near its maximum. In the in-phase tests cracks initiate at pores in the bulk of the specimen, growing in stage II in an environment shielded from oxidation during life fractions of N/N_f =>0.5. Subsequent growth is also largely in stage II, but surface piercing of some of the sub-surface growing cracks, and their exposure to the air environment result in an increase of the crack growth rate. The lower maximum cycle temperature, and the initiation and the growth of cracks in the oxygen free environment are believed to lead to the longer lives associated with the in-phase tests relative to the -135°lag tests.

    Effect of the coating on the life of SRR99--The influence of the presence of the aluminide coating on the TMF life vs. stress range relationships of SRR99 is shown in fig.5 for the various cycle types. The presence of the coating results in a relatively small yet consistent reduction in life in the -135°lag, R=-∞ tests. For the -135°lag, R=0 tests a nearly 5-fold reduction is noted at $\Delta\sigma=1000$ MPa, reducing to

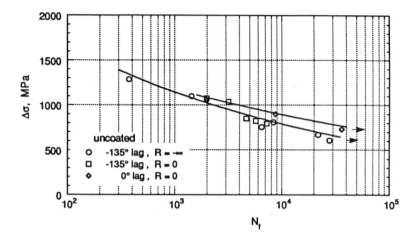

Fig.4-Comparing the TMF lives of uncoated SRR99 as a function of the stress range.

either a small, or to no decrease relative to the uncoated condition at the lower end of the investigated stress range. In the in-phase tests equal lives are measured at $\Delta\sigma=1000$ MPa, whereas the presence of the coating leads to a life reduction at the lower stress range levels.
    In the $-135°$lag tests the brittle fracture strain of the coating at temperatures below 500°C is exceeded when $\Delta\varepsilon_m=>0.8\%$ [11] (corresponding to $\Delta\sigma=>800$ MPa), leading to an entirely different mode of crack initiation relative to the uncoated test specimens in both the R=0 and R=$-\infty$ tests. Very long, shallow cracks spanning the entire width of the test specimen are formed. An example of these cracks is shown in figs.2 and 5 of ref.[11] of this volume. The cyclic stress intensity factor associated with these line initiated cracks is larger than the crack growth driving force associated with the semi-circular cracks growing from the point initiation sites in uncoated specimens [13]. Moreover, the crack extension length of the semi-circular crack required for covering equal fracture surface areas is approximately twice that of the line crack, which is observed to nearly uphold its crack front shape when growing. Finally the initial crack depths a_0 are different. Point initiation of cracks results in a_0 values smaller than the a_0 value of a line initiated crack, because the brittle nature of the coating fracture leads to cracks with $a_0=30$ µm (coating thickness). Coated test specimens consequently display the higher crack growth rates and corresponding shorter lives.
    At stress ranges $\Delta\sigma<800$ MPa cracks initiate in the point mode at the surface on the wide faces of both coated and uncoated specimens, at fractions of life N_i/N_f of 0.2 or less. Examples of point initiated cracks on the side face and on the fracture surface of a coated test specimen are shown in figs. 4 and 6 of ref.[11] of this volume. In both the uncoated and coated specimens, the major fraction of the life is spent in crack growth in the bulk when $\Delta\sigma<800$ MPa. Hence the differences in life between coated and uncoated SRR99 are expected to vanish, which indeed is suggested by the data in figs. 5a and 5b.
    As is the case for the uncoated test series, cracks initiate invariably at casting pores in the bulk of the coated test specimens in

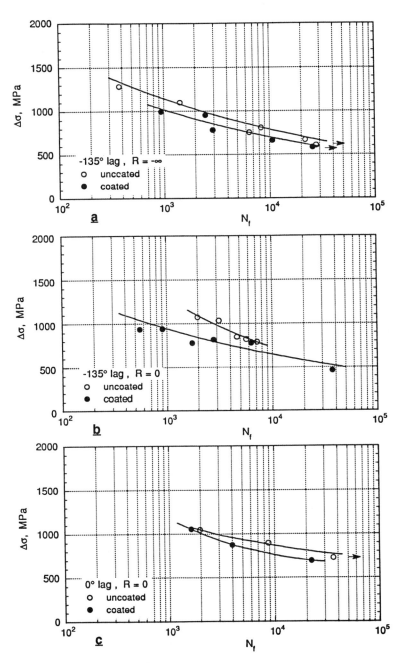

Fig.5-Comparing the TMF lives of aluminide coated versus uncoated SRR99 for various TMF cycle types, plotted as a function of the stress range.

the in-phase test series, and grow in an air free environment until the associated crack fronts pierce the specimen surface. Hence similar lifetimes are expected for TMF tests on coated and uncoated material, as is indeed observed at the highest stress range in fig.5c. The reductions in life at stress ranges inferior to 1000 MPa are therefore not caused by the presence of the coating, whose strain to fracture exceeds 10% at the T_{max} of the cycle [11]. The phenomenon responsible for reducing the lifetimes rather is related to the formation of a through crack at one of the 3 mm edges of the specimen relatively early in the lifetime, resulting from the surface breaking of critically located cracks initiated at sub-surface pores. In the uncoated specimen tested at a stress range of about 900 MPa on the other hand, inspection of the fracture surface and of the computer vision images shows that cracks have been growing in a stable, non-surface breaking mode during a fraction of the lifetime of the order of 55%. In the latter condition lower growth rates are anticipated, based on considerations of the absence of chemical interactions of the crack tips with oxygen, of smaller stress intensity factors and, possibly, of crack shielding effects.

CONCLUSIONS

1.TMF testing of the single crystal nickel base alloy SRR99, using cycle types of -135°lag in combination with R=0 and R=-∞ between minimum and maximum cycle temperatures of 300° C and 1050°C, and of in-phase, R=0 cycles between 300°C and 850°C, leads to a cycle-specific hardening-softening behaviour.

2.Inelastic strain is accumulated during TMF cycling as the result of plastic deformation and/or creep in specific parts of the cycle. The cycle-specific hardening/softening behaviour is controlled by the inelastic strain accumulation process.

3.Pronounced differences exist between the mode of microcrack initiation in the uncoated condition, in particular between the -135°lag and the in-phase cycles. In the -135°lag tests semi-circular microcracks initiate at the surface within life fractions N_i/N_f =0.05, as opposed to the in-phase tests where cracks invariably initiate in the bulk at pores. A major reason for the longer lives associated with the in-phase cycle at the smaller stress ranges is the anticipated lower growth rate of cracks in an oxygen free environment, and the lower maximum temperature of the in-phase TMF cycle relative to the -135°lag cycles .

4.The presence of an aluminide coating reduces the TMF lives of SRR99 at the high end of the applied stress range in -135°lag tests. The reduction relative to uncoated SRR99 is related to the brittle failure of the coating, and to the associated higher crack growth rates and the smaller crack extension lengths required for specimen failure. At the lower end of the applied stress range similar crack initiation and crack growth mechanisms operate in uncoated and aluminide coated material, giving rise to comparable TMF lives. The reduction in TMF lives of coated relative to uncoated specimens at medium and small applied stress ranges in the in-phase tested SRR99 is not caused by a coating related process, but rather is explained on the basis of porosity related fracture mechanics considerations.

Acknowledgments

This work is part of the Brite-Euram project BE3338-89 with financial support from the European Commission, coordinated by Rolls-Royce (UK). Financial support of the IAM-JRC by Rolls-Royce and by MTU is also gratefully acknowledged. The authors thank J.Estevas-Guilmain, R.De Cat, E.Fenske and K.Schuster for their assistance with the testing and fractography investigations.

REFERENCES

[1] Strang, A.and Lang, E., "Effect of coatings on the mechanical
     properties of superalloys", in R.Brunetaud et al. (eds.), High
     Temperature Alloys for Gas Turbines 1982, D.Reidel Publishing
     Co., 1982, pp 469-506.

[2] Strangman, T.E., Fuji, M. and NGuyen-Dinh X., in Gell et al.
     (eds.), Superalloys 84, The Metal Society, AIME, 1984, pp 795-
     804.

[3] Wood, M.I. and Restall, J.E.,"The mechanical properties of coated
     nickel based superalloy single crystals", in W.Betz et al. (eds.),
     High Temperature Alloys for Gas Turbines and other Applications,
     D. Reidel Publishing Co., 1986, pp 1215-1226.

[4] Grünling, H.W., Schneider K. and Singheiser,L., " Mechanical
     properties of coated systems", Mater. Sci. Eng., Vol. 88,   1987,
     pp 177-189.

[5] Veys J.-M. and Mevrel, M., "Influence of protective coatings on
     the mechanical properties of CMSX 2 and Cotac 784", Mater. Sci.
     Eng., Vol.88, 1987, pp 253-260.

[6] Veys, J.-M. and Mevrel,R., "Creep and high cycle fatigue
     properties of coated CMSX 2", in T.Khan and A.Lasalmonie (eds.),
     Advanced Materials and Processing Techniques for Structural
     Applications, ONERA, 1988, pp 168-178.

[7] Au, P., Dainty, R.V. and Patnaik, P.C. in T.S.Sudarshan and
     D.G.Bhat (eds.), Isothermal low cycle fatigue properties of
     diffusion aluminide coated nickel and cobalt based superalloys,
     Surface Modification Technologies III,   TMS, 1990, pp 729-748.

[8] Bain, K.R. "The effect of coatings on the thermo-mechanical
     fatigue life of a single crystal turbine blade material",
     AIAA/SAE/ASME/ASEE 21st Joint Propulsion Conference, Monterey
     (Ca), July 8-10,1985.

[9] Bernard,H. and Remy, L.,"Thermal mechanical fatigue damage of
     an aluminide coated nickel base superalloy", in H.Exner and
     V.Schumacher (eds.), Advanced Materials and processes EUROMAT
     1989,1989, pp 529-534.

[10] Guedou J.-Y.and Honnorat, Y. "Thermo-mechanical fatigue of
     turboengine blade superalloys", in Thermomechanical Fatigue
     Behaviour of Materials, ASTM STP 1186, ed. H.Sehitoglu, San
     Diego, Oct.1991, pp 157-175.

[11] Martinez-Esnaola,J.M., Arana,J.M., Bressers,J., Timm,J.,
     Martin-Meisozo,A., Bennett,A., and Affeldt,E.,"Crack initiation
     in an aluminide coated single crystal during thermo-mechanical
     fatigue", this volume.

[12] Rolls Royce plc, private communication

[13] Stress Intensity Factors Handbook, ed.Y Murakami, Pergamon Press,
     1987.

José M. Martínez-Esnaola,[1] María Arana,[1] Johan Bressers,[2] Jens Timm,[2] Antonio Martín-Meizoso,[1] Andy Bennett,[3] and Ernst E. Affeldt[4]

CRACK INITIATION IN AN ALUMINIDE COATED SINGLE CRYSTAL DURING THERMOMECHANICAL FATIGUE

REFERENCE: Martínez-Esnaola, J. M., Arana, M., Bressers, J., Timm, J., Martín-Meizoso, A., Bennett, A., and Affeldt, E. E., **"Crack Initiation in an Aluminide Coated Single Crystal During Thermomechanical Fatigue,"** Thermomechanical Fatigue Behavior of Materials: Second Volume, ASTM STP 1263, Michael J. Verrilli and Michael G. Castelli, Eds., American Society for Testing and Materials, 1996.

ABSTRACT: Two different modes of coating crack initiation are observed in an aluminide coated single crystal nickel base alloy upon thermomechanical fatigue loading with a cycle where the lag of the strain cycle behind the temperature cycle was 135°. Minimum and maximum cycle temperatures were 300°C and 1050°C, respectively. At applied mechanical strain ranges in excess of 0.8% the coating fails in a brittle manner, initiating a network of nearly equidistant parallel cracks. As opposed to this line initiation mode, a point initiation mode of multiple thumbnail crack origins is found for low strain ranges. Uncoated test specimens fail by point initiation at all strain ranges. The line initiation behaviour can be rationalized using a fracture mechanics approach based on energy release rate concepts to estimate the strain (or stress) to first coating crack and to predict the coating crack density during the test. This model highlights the effects of various parameters, such as coating thickness and temperature, on the failure mode. The influence of the ductile-brittle transition temperature and the stress-temperature coupling on the failure mode of the coating is discussed and the model predictions are compared with the experimental observations.

KEYWORDS: thermomechanical fatigue, aero engines, superalloy, single crystal, coating, crack initiation, surface crack.

[1] Centro de Estudios e Investigaciones Técnicas de Guipúzcoa (CEIT), Paseo Manuel de Lardizábal 15, 20009 San Sebastián, Spain.

[2] Institute for Advanced Materials - Joint Research Centre, P.O. Box 2, 1755 ZG, Petten, The Netherlands.

[3] Rolls-Royce plc, P.O. Box 31, Derby DE24 8BJ, United Kingdom.

[4] Motoren- und Turbinen-Union MTU, Dachauer Straße 665, D-8000 München 50, Germany.

INTRODUCTION

Single crystal nickel base superalloys for use in blades of aero gas turbines are coated in order to provide them with an adequate protection against oxidation and loss of mechanical performance. Coatings can be classified into two groups. The first group is that of diffusion aluminide coatings, obtained by a pack aluminization process in which aluminium diffuses into the superalloy to form a coating with nominal composition NiAl. The metallurgical and mechanical properties of the coatings obtained by this process are highly affected by the substrate on which they are applied. The second group is that of the MCrAlY type coatings, where M is usually Fe, Ni, or Co; in this case, the chemical or metallurgical interaction with the substrate can be neglected.

During in-service loading, critical locations of the blade are subjected to their maximum tensile loads at relatively low temperature which may result in premature failure of the coating. Coating cracking is important in terms of loss of environmental protection against corrosion and oxidation, and because of the effect of the coating on the mechanical properties of the substrate. Since the peak tensile strains at critical locations on turbine blades are likely to occur at relatively low temperatures, it is necessary to study the ductility properties of coatings, especially for aluminide coatings which exhibit low ductility and high ductile-brittle transition temperature (DBTT) [1, 2].

In this paper, the fracture strain under TMF conditions of a diffusion aluminide coating applied on a single crystal nickel base superalloy is studied, and compared to the fracture strains measured during tensile testing of the same substrate-coating combination as a function of temperature. The strain to first coating cracking is predicted using a fracture mechanics approach based on an energy release rate concept, and the predictions are compared to the experimental observations.

MATERIAL AND EXPERIMENTAL CONDITIONS

The substrate material is a single crystal nickel base superalloy, SRR99, produced by Rolls-Royce plc (U.K.). The material is coated by means of a pack cementation process in a powder rich in aluminium during approximately 4 hours at 870°C. A further heat treatment (1100°C in inert atmosphere) is required to complete the diffusion process. Finally, the substrate-coating composite is aged (16 hours at 870°C) for optimizing the mechanical properties by precipitation of γ' particles in the substrate.

High activity aluminide processes produce a microstructure composed of two main phases: β-NiAl and γ'-Ni$_3$(Al, Ti). The percentage of each of these phases depends on the substrate and coating characteristics and on the aluminizing process. In the present material, the process results in coatings about 30 μm thick.

The material is tested under thermomechanical fatigue conditions between 300°C and 1050°C in strain control, with the loading axis within 10° of the <001> orientation. Different strain ranges are applied with a frequency of 0.66 cycles/minute, at strain ratios R = 0 and R = −∞. The lag of the applied strain cycle behind the temperature cycle is 135° with 30 s heating up and 60 s cooling down times. The test specimen has a cylindrical end geometry and a solid gauge length with a rectangular cross-section of 12 mm × 3 mm. The edges of the gauge length section are rounded in order to promote good adherence of the coating. The reason for adopting a flat specimen geometry is related to the use of a computer vision system for the non-intrusive in-situ monitoring of the initiation and growth of microcracks during the tests. The test specimen is heated by means of a high frequency induction coil, and forced cooling is used. Strains are measured by means of a longitudinal extensometer, with the limbs spanning a gauge length of 10 mm. The mechanical strain, ε_m, is defined as $\varepsilon_m = \varepsilon - \varepsilon_{th}$, where ε and ε_{th} are the total and thermal components of the strain, respectively. More details of the experimental procedure can be found in Ref. [3]. As an example, Fig. 1 shows the temperature-strain cycle for a test with 1% mechanical strain range.

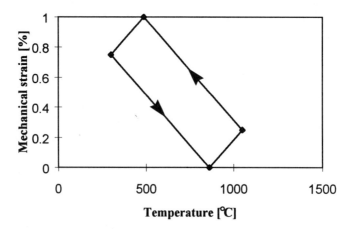

FIG. 1--Temperature-strain cycle for a test with 1% mechanical strain range, $R = 0$.

Images of the sample surface are recorded during the test by means of a computer vision system. The gauge length (of 10 mm × 9 mm) is divided into 30 sectors of 2.57 mm × 1.78 mm (6 rows by 5 columns), which results in a resolution of 5 μm per pixel. Images are digitized and stored in a computer for review and measurement after testing. The computer vision system allows the in-situ detection of surface crack initiation during the test, as well as the measurement of surface crack growth rates.

EXPERIMENTAL OBSERVATION OF CRACK INITIATION

In the aluminide coated SRR99 two types of crack initiation are observed, namely line initiation and point initiation [3]. Typical examples are shown in Figs. 2 and 3. Figure 2 is the post-mortem surface of a specimen tested at 1.4% mechanical strain range and $R = 0$, and shows a system of parallel cracks perpendicular to the loading direction. The cracks tend to be more or less equidistant, and macroscopically straight. This fracture mode is typical of coated specimens subjected to strain ranges above 0.8 %. Examination of computer vision records confirms that crack initiation takes place in the early stages of the test for $R = 0$ and later in life for $R = -\infty$. Details of the characteristics of some tests showing line initiation cracking are summarized in Table 1. This initiation mode has been associated with brittle fracture of the coating [2, 4]. Figure 3 shows the post-mortem surface of a specimen tested at 0.65% mechanical strain range and $R = -\infty$. Apparently, it is difficult to notice any surface cracks due to the high degree of oxidation. Observations by means of an optical microscope of a higher resolution reveal a distribution of several small cracks which follow an irregular path, as shown in Fig. 4. These cracks initiate later in time (between 200 and 2000 cycles, depending on the mechanical strain range) and their growth during the test can be observed by means of the computer vision system. This initiation mode takes place in specimens tested at mechanical strain ranges below 0.8% and is associated with ductile fracture of the coating. The initiation sites are related to structural defects within the coating and at the coating-substrate interface [2, 4, 5].

In the uncoated SRR99 cracks invariably initiate in the point initiation mode [3].

TABLE 1--Characteristics of tests with line initiation cracking.

test	Test Conditions $\Delta\varepsilon_m$ [%]	R	Statistics (*) crack spacing [mm]	standard deviation	Crack Initiation Conditions cycle number	peak stress [MPa]
S9	1	0	0.29	0.06	1	850
S9-11	1	0	0.588	0.16	1	900
6AKLM/4	1	0	0.411	0.1	1	900
S9-14	0.8	0	1.708	0.07	1	830
6AKLN/4	1	$-\infty$	0.3	0.04	20	630
6AKLN/5	0.8	$-\infty$	1.2	0.33	600	600

(*) cracks measured over the whole gauge length.

INFLUENCE OF CRACK INITIATION MODE ON TMF LIVES

Brittle fracture, as described above, reduces TMF lives relative to uncoated specimens by a factor of up to 5 [3]. The difference is not as pronounced in situations where ductile fracture occurs. This behaviour can be explained using simple calculations in terms of fracture mechanics.

The line initiation mode produces crack fronts approximately parallel to the specimen surface, as shown in Fig. 5, whereas the point initiation mode produces semi-elliptical surface cracks, see Fig. 6. The stress intensity factor range, ΔK, can be written as

$$\Delta K = F\Delta\sigma\sqrt{\pi a} \tag{1}$$

where $\Delta\sigma$ is the stress range, a is crack depth and F is a factor which depends on the crack shape: $F = 1.12$ for an edge crack (line initiation mode) and $F = 0.66$ for the deepest point on the front of a semicircular crack (point initiation mode).

FIG. 2--Line initiation cracking in a test with $\Delta\varepsilon_m = 1.4\%$, $R = 0$.

FIG. 3--Point initiation cracking in a test with $\Delta\varepsilon_m = 0.65\%$, $R = -\infty$.

FIG. 4--Point initiation cracking in a test with $\Delta\varepsilon_m = 0.65\%$, $R = -\infty$ (detail).

FIG. 5--Crack fronts parallel to the specimen surface in a test with line initiation mode.

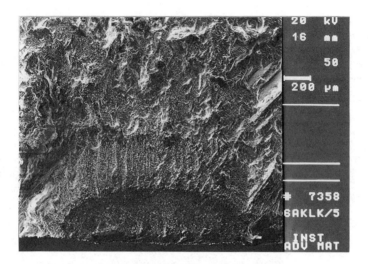

FIG. 6--Typical semi-elliptical surface crack in a test with point initiation mode.

If a simple Paris' equation is assumed for crack growth, the lives spent in propagating such cracks between the same initial and final depths are related through the expression

$$\frac{N_{f\,\text{line}}}{N_{f\,\text{point}}} = \left(\frac{F_{\text{point}}}{F_{\text{line}}}\right)^n \qquad (2)$$

where n is the exponent in the Paris' law. A typical value of the Paris' exponent between 2 and 3 would produce a ratio of cycles to failure between 0.34 and 0.2, respectively. This simple approach indicates that the line initiation mode by itself could lead to lives between 3 and 5 times shorter than those of uncoated specimens where only point initiation occurs. Experimental results are in agreement with these calculations [3]. In addition, line initiation occurs earlier in life than point initiation and smaller final depths are required for failure of the specimen (defined by a 33% load drop). It should be pointed out that only in the short time laboratory tests do the coated specimens have a shorter life. In realistic engine environments of longer times per cycle, the coating is necessary to obtain the required lives, i.e., the coated material will then exhibit longer lives than would uncoated material.

STRAIN TO FRACTURE OF THE ALUMINIDE COATING

Mechanical properties of an aluminide coating are highly influenced by the characteristics of the substrate on which it is applied. It is then very difficult to find appropriate literature references of strain to fracture corresponding to a specific coating-substrate system. Figure 7 shows sets of data of strain to fracture reported in the literature

[6-9] corresponding to high activity aluminide coatings applied on various nickel base superalloys. The differences among the different groups of data are not negligible.

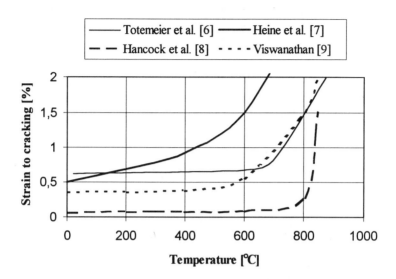

FIG. 7--Fracture strains of high activity aluminide coatings applied on various nickel base superalloys.

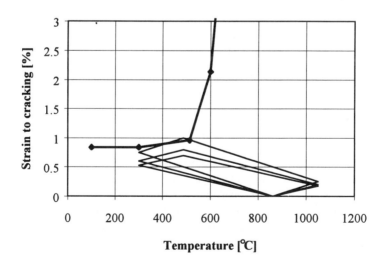

FIG. 8--Fracture strain of an aluminide coating on SRR99 (tensile test results) and strain-temperature loops in the coating at mechanical strain ranges $\Delta\varepsilon_m = 0.7, 0.8$ and 1%.

Tensile tests with the same specimen geometry as that used for TMF experiments were carried out in order to determine the strain to fracture under monotonic loading and to compare the results with those obtained under TMF conditions. Tests were carried out at 300, 490, 600 and 700°C. The coating fracture strain resulting from the tests (calculated as the remote applied strain) is shown in Fig. 8. The strain to fracture at low temperature is about 0.8%. TMF tests would be expected to produce line initiation cracking if the peak strain is above this critical strain to fracture. This is confirmed by the experiments [3].

Figure 8 also sketches the mechanical strain-temperature loops in the coating during three different TMF tests with applied strains of 0.7%, 0.8% and 1%. As expected, the cycle with the highest tensile strain always produces a brittle fracture of the coating, the cycle with the lowest tensile strain produces a ductile fracture of the coating, and the intermediate cycle can produce both fracture modes.

INFLUENCE OF THE COATING THICKNESS ON THE STRAIN TO FRACTURE

A dependence of the strain to fracture on the coating thickness has been reported by some authors [6, 8]. The equilibrium temperature at which the system coating-substrate is free of residual stresses also influences the strain to coating fracture.

To quantify these effects the fracture mechanics analysis of multiple cracking in coatings developed by Nairn and Kim [10] has been applied. This model relates the in-situ fracture toughness, independent of non-material parameters (e.g., thickness), to the applied strain to first crack and to subsequent cracking. The model solves the problem of the brittle fracture of the coating under the following assumptions: (a) elastic strain and stress fields, (b) the tensile stresses in each layer are independent of the thickness direction, and (c) the propagation of each crack is instantaneous from one side of the specimen to the other.

By using the equilibrium equations, the conditions of continuity of transverse and shear stresses across the layer boundaries, and the principle of the minimum complementary energy, it is possible to obtain an expression for the energy release rate as a function of the crack density. The energy release rate, G, for the first crack is given by [10]

$$G = \left[E_c \varepsilon_0 + C_1(\alpha_s - \alpha_c)\Delta T\right]^2 C_2 t_c \tag{3}$$

where ε_0 is the applied strain, α_s and α_c are the linear coefficients of thermal expansion for the substrate and the coating, respectively, and ΔT is the temperature increment measured from the reference state (i.e., the equilibrium temperature at which there are no residual thermal stresses). C_1 and C_2 are constants depending on the Young's modulus, Poisson's ratio and thickness of the substrate (E_s, v_s and t_s) and of the coating (E_c, v_c and t_c).

For the subsequent cracks:

$$G = \left[E_c \varepsilon_0 + C_1(\alpha_s - \alpha_c)\Delta T\right]^2 C_2 t_c f(D) \tag{4}$$

where $f(D)$ is a function of the crack density, D. The expressions of the constants C_1 and C_2 and the function $f(D)$ involved in equations (3) and (4) are rather complicated and can be found in [10].

The material properties used in the calculations have been obtained by fitting to experimental measurements, except for the Young's modulus of the coating, E_c, which has been calculated using data reported by Hancock et al. [8] for β-NiAl diffusion coatings obtained in a high activity aluminizing pack at 900°C on a commercial purity nickel substrate. These are as follows:

$$\alpha_s = 7.62 \times 10^{-5} + 8.40 \times 10^{-9}\, T$$
$$\alpha_c = 1.02 \times 10^{-5} + 3.73 \times 10^{-9}\, T$$
$$E_s = 1.34 \times 10^{11} - 4.03 \times 10^7\, T + 2.72 \times 10^4\, T^2 - 31.00 T^3 \qquad (5)$$
$$E_c = 3.53 \times 10^{11} - 3.79 \times 10^8\, T + 1.34 \times 10^5\, T^2$$
$$v_s = v_c = 0.3$$

where T is the temperature in °C, α_s and α_c are in °C^{-1}, and E_s and E_c are in Pa. The expressions for α_s and α_c represent mean values between 20°C and the actual temperature, T.

RESULTS

The first analysis is focused on estimating the coating fracture toughness as a function of temperature. Using a strain to first crack of 0.8% at 487.5°C (the theoretical temperature at which the peak strain is reached in the present tests) in equation (3), a fracture toughness of 6 MPa√m is obtained. As mentioned before, there is a lack of appropriate data for this type of coating in the literature. As a first approximation, Noebe et al. [11] give the fracture toughness of an intermetallic component of Ni and Al. This is fitted between 4 and 20 MPa√m, increasing with the γ' percentage. The fracture toughness obtained in the present analysis, 6 MPa√m, is between these limits and constitutes a reasonable value for NiAl [11].

The ductile to brittle transition temperature (DBTT) was set between 550°C and 650°C, see Fig. 8. The high temperature fracture toughness is not relevant in our analysis and was fixed at 50 MPa√m [11]. Figure 9 shows the fracture toughness calculated using equation (3) and the above DBTT.

Once the fracture toughness is estimated the strain to fracture for different coating thicknesses can be calculated. Figure 10 shows the strain to fracture obtained by using equation (3) for different coating thicknesses. The results show a dependence on the coating thickness that has been also reported in the literature [8].

The influence of the residual thermal stresses has been considered by assuming the diffusion temperature (1100°C) as that at which the system is free of residual stresses. It should be pointed out that this is only a first approximation to the problem since there are no references on the order of magnitude of the residual stresses produced by high activity aluminizing processes. The model predicts differences as large as 100% in the fracture strain when different equilibrium temperatures are assumed.

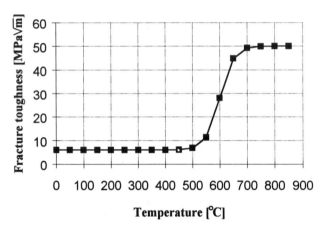

FIG. 9--Estimation of the coating fracture toughness.

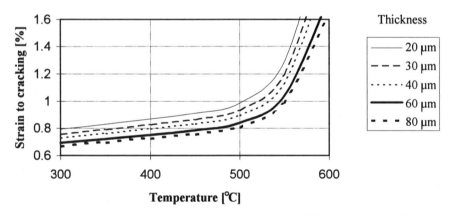

FIG. 10--Strain to fracture as a function of temperature for different coating thicknesses.

The model could be used to predict crack density as a function of the applied strain using equation (4) for a specific temperature and coating thickness. However, at strains above 0.8%, the substrate reaches the yield stress and consequently the analysis, which assumes elastic conditions, loses prediction capability for high strain ranges. Figure 11 shows the crack density predicted by the model versus the applied strain range, for a coating thickness of 30 μm. Crack densities measured in different TMF tests are also shown for comparison. A considerable scatter is found for nominally identical tests and even within each test (see Table 1). The slight disorientation of the specimens from the theoretical <001> direction and the inhomogeneous plastic strains developed in each test, according to Schmid's law, can affect the crack density.

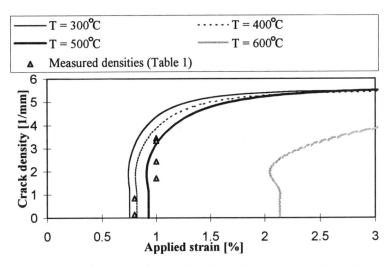

FIG. 11--Prediction of crack density as a function of the applied strain for a 30 μm thick coating at different temperatures, and comparison with the experimental observations.

CONCLUSIONS

Crack initiation in an aluminide coated single crystal during thermomechanical fatigue has been investigated. TMF lives in coated specimens are highly influenced by the crack initiation mode in the coating (point or line); reductions in life of a factor of up to 5 could be expected when compared to the uncoated material.

Results of tensile tests carried out to estimate the strain to fracture of the coating-substrate system of this investigation are in agreement with the experimental observations of TMF tests.

A fracture mechanics approach has been used to study the influence of different parameters on the coating fracture strain. In particular a high dependence on the coating thickness has been found, in agreement with some experimental observation of other authors. The model also accounts for the effect of the residual thermal stresses generated at the interface. The specific selection of the equilibrium temperature strongly influences the prediction of strain to fracture which can vary by as much as 100% with different assumed equilibrium temperatures.

ACKNOWLEDGMENTS

This work is part of the Brite-Euram project BE3338-89 with financial support from the European Commission, coordinated by Rolls-Royce plc (UK). Financial support of CEIT and of IAM-JRC by Rolls-Royce plc and MTU is also gratefully acknowledged. One of the authors (M.A) wants to thank the Departamento de Educación, Universidades e Investigación of the Basque Government for the grant received.

REFERENCES

[1]   Saunders, S.R.J. and Nicholls, J.R., "Coatings and surface treatments for high temperature oxidation resistance", Materials Science and Technology, Vol. 5, August 1989, pp. 780-798.

[2]   Bernard, H. and Rémy, L., "Ductile-brittle transition of an aluminide coating on IN100 superalloy", Proceedings of the Conference on High Temperature Materials for Power Engineering 1990, Liège, Belgium, Part II, pp. 1185-1194.

[3]   Bressers, J., Timm, J., Williams, S., Bennett, A. and Affeldt, E.E., "Effects of cycle type and coating on the TMF lives of a single crystal nickel based gas turbine blade alloy", in this volume.

[4]   Zhang Detang, "Effect of ductile-brittle transition temperature of Al-Si coating on fatigue properties of Ni-base superalloys", Acta Metallurgica Sinica Series A, Vol. 3, 1990, pp. 435-438.

[5]   Wood, M.I., "Mechanical interaction between coatings and superalloys under conditions of fatigue", Surface and Coatings Technology, Vol. 39/40, 1989, pp. 29-42.

[6]   Totemeier, T.C., Gale, W.F. and King, J.E., "Fracture behaviour of an aluminide coating on a single crystal nickel base superalloy", Materials Science and Engineering, Vol. A169, 1993, pp. 19-26.

[7]   Heine, J.E., Warren, J.R. and Cowles, B.A., "Thermal mechanical fatigue of coated blade materials", Final Report, WRDC-TR-87-4102, United Technologies Corporation, Pratt & Witney, West Palm Beach, Florida, 1989, pp. 13-20.

[8]   Hancock, P., Chien, H.H., Nicholls, J.R. and Stephenson, D.J., "In situ measurements of the mechanical properties of aluminide coatings", Surface and Coatings Technology, Vol. 43/44, 1990, pp. 359-370.

[9]   Viswanathan, R., "Damage Mechanisms and Life Assessment of High Temperature Components", ASM International, Metals Park, Ohio, 1989, pp. 443-448.

[10] Nairn, J.A. and Kim, S.R., "A fracture mechanics analysis of multiple cracking in coatings", Engineering Fracture Mechanics, Vol. 42, No. 1, 1992, pp. 195-208.

[11] Noebe, R.D, Bowman, R.R. and Nathal, M.V., "Physical and mechanical properties of the B2 compound NiAl", International Materials Reviews, Vol. 38, No. 4, 1993, pp. 193-232.

Johan Bressers[1], José M. Martínez-Esnaola[2], Antonio Martín-Meizoso[2], Jens Timm[1], and María Arana-Antelo[2]

COATING EFFECTS ON CRACK GROWTH IN A SINGLE CRYSTAL NICKEL BASED ALLOY DURING THERMO-MECHANICAL FATIGUE

REFERENCE: Bressers, J., Martínez-Esnaola, J. M., Martín-Meizoso, A., Timm, J., and Arana-Antelo, M., ''Coating Effects on Crack Growth in a Single Crystal Nickel Based Alloy During Thermo-Mechanical Fatigue,'' Thermomechanical Fatigue Behavior of Materials: Second Volume, ASTM STP 1263, Michael J. Verrilli and Michael G. Castelli, Eds., American Society for Testing and Materials, 1996.

ABSTRACT: Single crystal specimens of the nickel based alloy SRR99 with [001] orientation are subjected to TMF cycling between 300°C and 1050°C, using a -135°lag between temperature and mechanical strain, and various R ratios. The in-situ observation of the test specimen by means of a computer vision system during TMF testing enables the initiation of microcracks to be detected when cracks have reached lengths of approximately 30 µm at the surface. High densities of point initiated micro-cracks are generated during the early stages of the TMF life. Crack increment measurements reveal marked differences between the growth rates of individual micro-cracks. The presence of a nickel-aluminide coating consistently reduces the TMF life, due largely to higher growth rates of the main crack in the substrate. In order to estimate the influence of initial crack distribution and density on life, a two dimensional crack shielding model is incorporated in a computer programme which is used to simulate the growth of interacting, parallel surface cracks. The effect of various parameters, including the density and the initial distribution of the microcracks, on the life spent to grow a crack to a pre-defined length, are discussed and the model predictions are compared with the experimental observations.

KEYWORDS: Thermo-mechanical fatigue, aero engines, micro-cracking, superalloy, single crystal, coating, crack shielding, multiple cracking, life assessment, surface crack.

[1]Institute for Advanced Materials, Joint Research Centre, P.O.Box 2, 1755 ZG, Petten, The Netherlands.

[2]Centro de Estudios e Investigaciones Técnicas de Guipuzcoa (CEIT), Paseo Manuel Lardizábal 15, 20009 San Sebastián, Spain.

INTRODUCTION

Single crystal nickel base superalloys for use in blades of aero gas
turbines are coated in order to provide them with an adequate
protection against oxidation. An evaluation of the effect of the presence
of the coating on the lifetime of the blade is of interest in order to
optimize the design of the substrate/coating system in terms of oxidation
resistance-lifetime considerations. Thermo-mechanical fatigue (TMF)
testing is well suited for establishing the effect of a coating on
lifetime in in-service conditions, because the test mimics the thermal
fatigue loading situation of blades and the corresponding failure
mechanisms in critical locations.

In a companion paper in this volume [1] the influence of the
presence of a high activity aluminide coating on the TMF life of single
crystal SRR99 is reported. The presence of the coating reduces the TMF
life in -135°lag, R=0 and R=-∞ tests over the entire stress range
investigated, although at small stress ranges the differences in life
between coated and uncoated SRR99 are relatively small, yet consistent.
At stress ranges in excess of approximately 800 MPa the reduction in life
relative to the uncoated SRR99 is related to the brittle failure of the
coating (see also [2]), which gives rise to the initiation of long
shallow cracks (line initiation), as opposed to the point initiated semi-
circular cracks in uncoated material. The higher crack growth rates
associated with line cracks, and the smaller crack extension length
required for specimen failure lead to the shorter lives of the aluminide
coated specimens [1,2]. For stress ranges $\Delta\sigma$<800 MPa, the mechanism of
multiple point initiation of cracks is observed both in uncoated and in
aluminide coated SRR99. In these tests, semi-circular cracks initiate and
propagate through the coating and the coating affected zone very early in
the life, spending by far the largest fraction of life in propagating
through the substrate material. Hence differences in the TMF lives of
uncoated and coated SRR99 are expected to be very small in this stress
range regime, and in particular no differences are anticipated to exist
between the numbers of cycles required to propagate cracks through the
SRR99 material itself. Yet consistent differences are observed to exist
experimentally. The question then arises whether these differences in
crack propagation can be explained by indirect effects related to the
presence of the coating. In this paper potential causes for the observed
behaviour are analyzed and discussed. Crack shielding in particular will
be emphasised, because experimentally observed differences in crack
densities between coated and uncoated SRR99 suggest that crack shielding
may operate to a different extent in bare and coated specimens.

EXPERIMENTAL RESULTS

For details about the SRR99 material and the aluminide coating the reader
is referred to ref [1] in this volume, which also highlights the TMF
testing procedure and the technique used for measuring the number of
cycles for crack initiation and the crack extension. The test conditions
and the test results relevant to the discussion are listed in Table 1.
All the results pertain to a TMF cycle with a -135°lag between strain and
temperature, and minimum and maximum cycle temperatures of 300°C and
1050°C, respectively. Heating and cooling rates are 25°C/s and 12.5°C/s.

All the specimens referred to in Table 1 initiate multiple cracks in the point mode which extend, to a first approximation, along semi-circular crack fronts. Crack initiation is defined as the cycle number N_i where a crack with a surface length l_o =30 μm (depth a_0=15 μm) is first observed. The values of N_i, the corresponding life fractions taken up by crack initiation and the total lives are listed in Table 1. In this study cracks initiate within 1-15% of the lifetime, leaving more than 85% of the life to be consumed in growing a crack to failure of the test specimen. The total life of the coated test specimens is consistently smaller than the life of the uncoated samples although the difference is small for the R=0, $\Delta\varepsilon_m$=0.7% test. The reduction in TMF life caused by the presence of the nickel-aluminide coating is corroborated by other test results in [1] as well as by literature data [3]. The limited number of test results listed in Table 1 has been selected from the larger data base in [1] because the maximum stress in the cycle and the stress range during the entire TMF test in each of the uncoated/coated specimen pairs are nearly equal, as illustrated by the corresponding values at half life in Table 1. Therefore the driving force for crack growth resulting from the externally applied stress is equal within each pair.

The crack initiation and growth history of specimen LL4 is shown in Fig.1, illustrating the multiplicity of cracks and the discontinuous nature of the crack initiation process. Analysis shows that pronounced differences exist between the growth rates of individual micro-cracks. It seems logical to assume that multiple cracking will induce a huge competition for crack propagation and that crack shielding will play an important role in determining individual crack growth rates. Differences in crack densities and/or initial crack configuration between specimens

TABLE 1--<u>Test results</u>

Spec.	cond.	R	$\Delta\varepsilon_m$(%)	σ_{max}(MPa) ($N_f/2$)	$\Delta\sigma$(MPa) ($N_f/2$)	N_i	N_f	N_i/N_f	$N_f{}^c/N_f{}^u$
LL3	unc.	0	0.8	706	822	857	5702	0.15	0.49
LR1	coat.	0	0.8	703	817	235	2809	0.08	
LL5	unc.	0	0.7	666	791	250	7206	0.04	0.87
SP1	coat.	0	0.7	656	778	660	6325	0.10	
LL4	unc.	-∞	0.7	420	667	270	21820	0.01	0.49
LM5	coat.	-∞	0.7	436	661	585	10625	0.05	

therefore are expected to influence the TMF life, inclusive of that fraction of the total life which is consumed by crack growth through the part of the specimen substrate material not directly influenced by the presence of the coating. At this point a distinction must indeed be made between the coating and coating affected zone of the substrate on the one hand, and the substrate material on the other hand. Obviously the number of cycles for crack initiation and the crack growth rate in the coating and in the coating affected zone of the substrate will differ from the corresponding values in the equivalent thickness layer of an uncoated specimen. The thickness of the coating and of the coating affected zone is 65-70 μm. It consists of a main coating layer, a coating diffusion zone, a thin layer of γ' or interconnected γ/γ' and a zone of anomalously rafted γ' in the γ matrix of the substrate [4]. By adopting an initial crack depth a_o =100 μm (equivalent to l_o =200 μm) and by calculating the number of cycles required to grow the main crack to the depth corresponding to specimen failure, the direct influence of the presence of the coating on the TMF life is singled out whilst any indirect influence due to coating induced differences in crack density and related effects still persist. Table 2 lists the number of cycles required to reach a depth of the main crack a_o =100 μm (N_i100) and to drive the main crack from that initial depth to specimen failure (N_f-N_i100). All the data refer to the main crack i.e. the crack which leads to failure of the test specimen. An exception to this data is specimen SP1. In this particular case the main crack initiated and grew at the backside of the test specimen, remaining invisible for optical imaging during the largest part of the life.

Where full data are available, Table 2 shows that the ratio of the lives consumed for crack growth in the actual substrate material of the coated and uncoated specimens is approximately 1/2. This conclusion remains intact when selecting an initial crack length a_o =150 μm instead

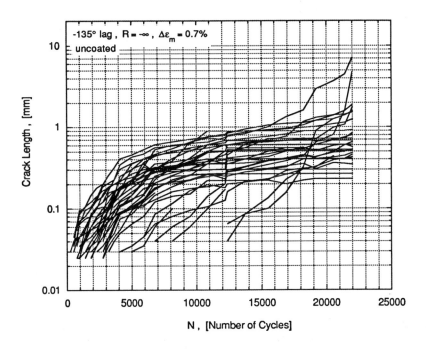

FIG.1--Initiation and extension as a function of cycle number of all the cracks observed in the gauge length (front side) of specimen LL4.

TABLE 2--Crack growth data

Spec.	cond.	R	$\Delta\varepsilon_m$(%)	N_i100	N_f-N_i100	(N_f-N_i100)/N_f [(N_f-N_i150)/N_f]	crack dens.(cm^{-2})
LL3	unc.	0	0.8	1500	4202	0.74 [0.60]	33
LR1	coat.	0	0.8	969	1840	0.65 [0.63]	32
LL5	unc.	0	0.7	1501	5704	0.79 [0.75]	194
SP1	coat.	0	0.7	3000	3325	0.53*[0.45]*	13
LL4	unc.	-∞	0.7	3397	18423	0.84 [0.81]	38
LM5	coat.	-∞	0.7	1081	9543	0.90 [0.86]	8

* based on longest crack on front face of test specimen.

of a_o =100 μm, see Table 2. Moreover, observation of crack fronts on the
fracture surfaces of coated specimens suggests no accelerating or
decelerating effects of the presence of the coating zone and associated
layers on the crack front shape, which also justifies the choice of a_o
=100 μm. Also note that all the tests in Tables 1 and 2 correspond to TMF
cycles with the same temperature-mechanical strain phasing and that the
differences in the resulting stresses between coated and uncoated
specimens are not significant. In addition, it is reasonable to assume
that the direct influence of the coating on the lives spent in
propagating a crack from an initial depth a_o =100-150 μm to failure can
be neglected since no differences in chemical composition and
microstructure are measured between uncoated and coated specimens beyond
depths of 65-70 μm.

The statistical analysis of these data indicates that the reduction
in life in the coated specimens relative to the uncoated specimens cannot
be attributed to the inherent scatter of the substrate. Table 3 shows two
different comparisons between the number of cycles consumed in crack
propagation in the substrate of coated and uncoated material. If the
scatter of the substrate were responsible for the experimental
differences in life between coated and uncoated specimens, the expected
mean value of

$$(N_f-N_i100)\text{uncoated} \; / \; (N_\cdot-Ni_i100)\text{coated}$$

should be 1. However, the analysis shows that this hypothesis is not
acceptable from a statistical point of view for a 95% confidence interval
since the value 1 is outside this interval.

Other indirect effects associated with the presence of the coating
still remain as potential causes to explain the observed behaviour. These
include the residual tensile stress field resulting from differences in
thermal expansion coefficient between coating and substrate materials in
coated specimens (and related crack opening effects), and differences in
the initial crack configuration and/or in the crack densities between
coated and uncoated samples, see Table 2. The latter aspects will be

TABLE 3--<u>Statistical analysis of data in Table 2</u>

Tests	$\frac{(N_f-N_i100)\text{uncoated}}{(N_f-N_i100)\text{coated}}$	$\frac{\log(N_f-N_i100)\text{uncoated}}{\log(N_f-N_i100)\text{coated}}$
LL3 / LR1 LL5 / SP1 LL4 / LM5	2.284 1.715 1.931	1.110 1.067 1.072
Mean, $\overline{X}$	1.977	1.083
95% C.I. for the mean*	1.977±0.713	1.083±0.058

* $\overline{X} \pm t_{0.95,n-1}\sigma_{n-1}/\sqrt{n}$, where $t_{0.95,n-1}$ is the Student distribution with n-1 degrees of freedom, σ_{n-1} is the standard deviation, and n=3.

investigated in this paper. A computer model which simulates the interactive effect of multiple crack growth has been developed in order to (i) investigate the differences existing between the growth rates of individual cracks, see Fig.1, and (ii) estimate the influence of crack distribution and density on the life spent to propagate the main crack to a pre-defined length.

TWO-DIMENSIONAL MODEL OF CRACK SHIELDING

As a first approximation, a 2-D crack growth simulation is presented. It is expected that a 2-D model will reflect accurately the behaviour of line-initiated cracks and will constitute a first approximation for point-initiated micro-cracks. Line-initiated parallel surface cracks are frequently observed in coated TMF testpieces when tested at high strain ranges (an example is shown in ref. [2], Fig. 2).
In the model a fixed number of initial micro-cracks can be positioned periodically or at random in a given length (specimen gauge length). Initial microcrack size distributions are either uniform, normal or log-normal (with a given mean and standard deviation). Two types of boundary conditions can be simulated: periodic boundary conditions to model a small portion of the surface (with a high density of micro-cracks) within a larger body, and free boundary conditions (i.e., there

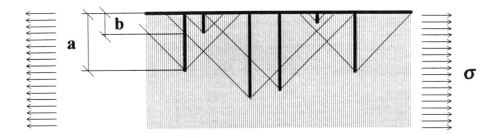

FIG.2--Crack shielding according to a scheme of shadows at 45°.

is no crack outside the selected area), which are useful when modelling
the whole specimen gauge length, assuming that no large crack appears at
the specimen shoulders. Although the latter is more accurate, the
calculations become very time consuming if the micro-crack density is
high. To illustrate the idea of crack shielding, consider the schematic
depicted in Fig. 2 where each crack is assumed to relax the stresses in a
region defined by its shadows at 45°. Then the stress intensity factor,
K_I, can be calculated using the Green's function [5] of a 2-D edge crack
in a semi-infinite isotropic solid as:

$$K_I = \sigma\sqrt{\pi a}\int_b^a \frac{g(x/a)}{\sqrt{a^2 - x^2}}dx \tag{1}$$

where σ is the remote applied stress, a and b determine the region of the
crack subjected to loading, see Fig.2, and $g(x/a)$ is the Green's function
for this geometry given in [6] in closed-form. The anisotropy of SRR99 is
thought not to be relevant in estimating crack shielding effects.
Although this 45° shadow scheme will be further refined, some implicit
assumptions of this first approximation are retained in the actual
solution. In particular, the total shielding for a given crack is
determined by only one of the rest of the cracks. According to this
model, only the crack producing the largest shielding is taken into
account and no additional shielding due to other cracks is considered.
Obviously this is not altogether correct, since the other cracks
contribute to reduce the applied stress intensity factor. It will be
shown that this assumption leads to a very convenient simplification of
the problem.

Crack growth is assumed to obey a Paris' law of the form

$$\frac{da}{dN} = C(\Delta K_I)^m \tag{2}$$

An adaptive step Euler's method [7] is used to integrate Eq. (2).
The integration step, ΔN, is selected in terms of the crack growth
relative to the actual crack size. A value of 10% extension of the crack
depth per step was selected as the default value for the calculations,
yielding a reasonable computing time and an over-estimation in life
prediction of typically less than 5%.

This simple model of crack shielding is modified by using a more
accurate computation of crack interaction and of the resulting stress
intensity factors. Isida [8] obtained the applied stress intensity
factors for two edge cracks in a semi-infinite plane under uniform
tension. These solutions are used to refine the computation of the
applied stress intensity factors, maintaining the approximation that each
crack is only affected by shielding from one other crack. The programme
considers all the possible crack couples for a given crack and retains
the smallest stress intensity factor resulting from the computations. The
hypothesis for crack interaction (by couples) represents a lower bound
for shielding, and hence it is a conservative assumption for the applied
stress intensity factor.

RESULTS AND DISCUSSION

The standard conditions chosen in the calculations are as follows:

-100 microcracks on a length of 9 mm (the gauge length of the actual TMF
  specimen)
-cracks arranged at random on the specimen gauge length,
-crack sizes distributed according to a log-normal distribution function
  with mean size of 1 μm, and standard deviation of 5 x 10^{-2} μm (the
  deepest microcrack in a sample of 100 cracks/9 mm is about 1.34 μm)
-fatigue effective stress of 782 MPa,
-fatigue crack growth according to a Paris' equation, with exponent,
  m = 2, and constant, C = 1 x 10^{-9} (where da/dN is in m/cycle and ΔK_I is
  in MPa√m),
-free boundary condition (no cracks considered outside the gauge length),
-limit crack depth of 1 mm (a loss of a 33% of the specimen cross-
  section).

As stated above, although the geometry of the problem arising from
multiple point initiation cracking corresponds to a 3-D configuration, 2-D
crack growth is simulated in the model. In order to use parameters
representative of tests resulting in a 2-D situation (line initiation),
the selected fatigue effective stress of 782 MPa corresponds to the
stress range of an actual test producing line initiated cracks. Note that
this value is not very different from those of Table 1. The default
values for the parameters of Paris' equation are chosen to match the
number of cycles to failure of that particular test. The influence of the
exponent m will be discussed below. Other conditions are explicitly
stated where used. Series of 20 runs are made, using 20 different seeds
for the pseudo-random number generation to cope with the intrinsic
scatter usually associated with experimental results (seeds are taken
from tables of random numbers [9]). The effect of a number of parameters
is discussed below. Figures 3(a)-(d) represent crack depths versus number
of cycles for different initial crack densities. The initial microcracks
are dispersed at random in 9 mm. The initial micro-crack sizes are
normally distributed with 1 μm mean and 0.2 μm standard deviation. In
Figs. 3(a)-(d), the dotted line represents the behaviour of the non
shielded reference crack.In the case where 1000 cracks are modelled, only
the last 100 propagating cracks are plotted for clarity. The results of
Figs.3(a)-(d) indicate that multiple crack interaction reasonably
explains the differences observed in TMF experiments among the growth
rates of individual cracks, see Fig.1. Crack shielding appears to be
responsible for the deceleration of crack growth in a situation of
multiple cracking compared to that of a single crack subjected to the
same remote loading. Tables 4.a and 4.b summarize the obtained results
for a gauge length of 9 mm. The extreme and mean values of the crack
depth and the extreme and mean values of the life listed on the same row
do not necessarily correspond. A life improvement of about 10% relative
to the single crack situation is obtained upon introducing as many as one
thousand initial micro-cracks in the gauge length, for a Paris' exponent
of 2. The effect is similar for a Paris' exponent of 3. The experimental
scatter in TMF experiments is usually much larger than 10%. The limited
effect of the crack density on TMF life is confirmed by the
experimentally measured crack density data listed in Table 2. TMF life
reductions are indeed observed in uncoated-coated specimen pairs

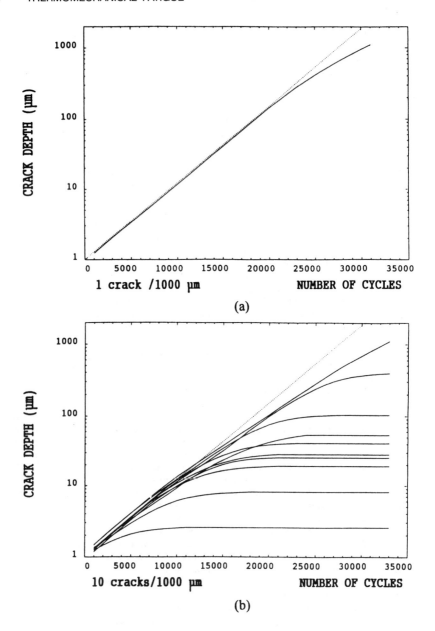

(a)

(b)

FIG.3--Crack depth versus number of cycles for different crack densities, Δσ=782 MPa. (a) 1 crack/9 mm, (b) 10 cracks/9 mm, (c) 100 cracks/9 mm, (d) 1000 cracks/9 mm.

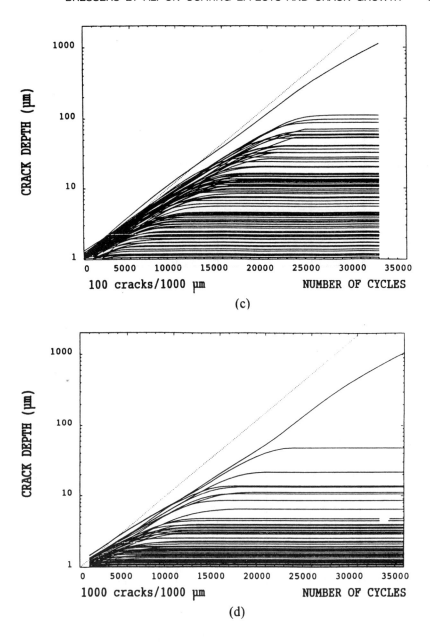

FIG.3--Continued

TABLE 4.a--The effect of microcrack density on fatigue life,
Paris' exponent m=2

Summary of results from	1 cracks	/ 9 / mm	10 cracks	/ 9 / mm	100 cracks	/ 9 / mm	1000 cracks	/ 9 / mm
20 different runs	deepest initial crack, μm	Life, cycles	deepest initial crack, μm	Life, cycles	deepest initial crack, μm	Life, cycles	deepest initial crack, μm	Life, cycles
Minimum	0.84	2940	1.04	2992	1.23	3036	1.35	3078
Maximum	1.14	3070	1.27	3200	1.52	3380	1.65	3429
Mean	1.018	2989	1.154	3092	1.336	3208	1.445	3285
Unshielded crack	...	2989	...	2989	...	2989	...	2989

irrespective of the ratio of the associated crack densities, suggesting
that other factors are responsible for the life reductions. In order to
estimate the effect of microcrack arrangement, two microcrack
dispositions are considered: regularly spaced microcracks and microcracks
distributed at random within the gauge length. Table 5 summarizes the
results of the computer model. The periodic microcrack disposition
provides an insignificant life improvement over the random arrangement.
    The influence of the initial microcrack size distribution is
evaluated using uniform, normal and log-normal distributions with a

TABLE 4.b--The effect of microcrack density on fatigue life, Paris'
exponent m=3

Summary of results from	1 crack	/ 9 / mm	10 cracks	/ 9 / mm	100 cracks	/ 9 / mm	1000 cracks	/ 9 / mm
20 different runs	deepest initial crack, μm	Life, cycles	deepest initial crack, μm	Life, cycles	deepest initial crack, μm	Life, cycles	deepest initial crack, μm	Life, cycles
Minimum	0.84	4024	1.04	3809	1.23	3517	1.35	3484
Maximum	1.14	4699	1.27	4362	1.52	4287	1.65	4191
Mean	1.018	4269	1.154	4062	1.336	3899	1.445	3919
Unshielded crack	...	4269	...	4269	...	4269	...	4269

constant mean size of 1 μm. The results are summarized in Table 6. Note
that a normal or log-normal distribution with a standard deviation equal
to 0 is the case of a uniform distribution. A small life reduction is
obtained upon increasing the standard deviation, mainly because of the
deeper initial crack size.

TABLE 5--The effect of microcrack arrangement, 100 cracks/9mm

Summary of results from	PERIODIC		At RANDOM	
20 different runs	deepest initial crack, μm	Life, cycles	deepest initial crack, μm	Life, cycles
Minimum	1.24	3051	1.23	3036
Maximum	1.52	3346	1.52	3380
Mean	1.326	**3251**	1.336	**3208**
Unshielded crack	...	2989	...	2989

TABLE 6--The effect of initial crack size distribution, 100 cracks/mm

Summary of results from 20 different runs	UNIFORM 1μm cracks	NORMAL Mean = 1 μm St.dev.=0.2 μm		LOG - Mean = St.dev. =	NORMAL 1 μm 5x10^{-2} μm
	Life, cycles	deepest initial crack, μm	Life, cycles	deepest initial crack, μm	Life, cycles
Minimum	3170	1.36	2984	1.23	3036
Maximum	3369	1.73	3 328	1.52	3380
Mean	**3263**	1.502	**3154**	1.336	**3208**
Unshielded 1 μm crack	2989	...	2989	...	2989

The computer simulation of crack growth in a 9 mm long testpiece,
which corresponds to the gauge length of the actual TMF test specimens,
suggests that neither crack density, nor spatial crack disposition or
initial crack size distribution lead to shielding factor effects
sufficiently large to explain the experimentally observed life reductions
of coated versus uncoated samples.

Other indirect effects related to the presence of the coating, such as the occurrence of tensile residual stress fields resulting from the difference in thermal expansion coefficient of SRR99 and NiAl coating, are currently being investigated as possible reasons for crack growth rate acceleration in the deep zones of the substrate.

CONCLUSIONS

-135°lag TMF cycling of uncoated and aluminide coated single crystal specimens of SRR99 at stress ranges smaller than 800 MPa in R=0 and R=-∞ conditions between 300°C and 1050°C creates a multitude of point initiated surface cracks in the early stages of the lifetime. The total lifetime of the coated specimens is consistently smaller than the life of the uncoated specimens, notwithstanding similar levels of the stress range and of the maximum cycle stress which control the crack growth dominated life fraction. Analysis of the microcrack growth history shows that the reductions in lifetime caused by the presence of the coating can not be accounted for by differences related to the number of cycles for initiating and growing of the main crack through the coating and coating-affected zone of the substrate, and through its equivalent layer, in terms of thickness, of the uncoated specimen. The differences in TMF lifetime are indeed dominated by the difference in cycle numbers required to grow cracks with a depth of 100-150 µm to the specimen failure condition through the part of the SRR99 not directly affected by the coating. Therefore indirect coating related effects are held responsible for the observed coating induced reduction of the TMF life. A possible cause is the crack density related interaction between multiple cracks, a hypothesis which was tested by means of the computer simulated growth of 2-D cracks. The main results of the simulation are
        -the observed random behaviour of multiple cracks in TMF tested specimens is reasonably well explained in terms of crack shielding.
        -with specimens having gauge lengths as actually used in the TMF tests, the crack leading to fracture departs very little from the behaviour of a reference unshielded crack, all the other cracks playing a minor role on the component life. Factors such as crack disposition, initial crack size distribution and crack density bear only minor effects on TMF life. The predicted minor effect of crack density on TMF life is corroborated by experiment.
        Therefore another indirect effect related to the presence of the coating is held responsible for the acceleration of crack growth in the substrate of coated material relative to the uncoated material. One could speculate that the difference in thermal expansion coefficient between SRR99 and the NiAl coating, and the corresponding residual tensile stress field in the substrate (and related crack opening effects), is a potential mechanism by which the observed behaviour could be explained. Further investigations to confirm this hypothesis are ongoing.

ACKNOWLEDGMENTS

This work is part of the Brite-EuRam project BE3338-89 with financial support from the European Commission, coordinated by Rolls-Royce plc (U.K.). Financial support of IAM-JRC and of CEIT by Rolls-Royce plc and MTU is also gratefully acknowledged. One of the authors (M.A.) wants to thank the Departamento de Educación, Universidades e Investigación of the

Basque Government for the grant received.

REFERENCES

[1]   Bressers J.,Timm J.,Williams S.J.,Bennett A. and Affeldt E.,
        "Effects of cycle type and coating on the TMF lives of single
        crystal based gas turbine blade alloy",in this volume.

[2]   Martínez-Esnaola, J. M., Arana-Antelo, M., Bressers, J., Timm, J.,
        Martin-Meisozo, A., Bennett, A., and Affeldt, E., "Crack
        Initiation in an Aluminide Coated Single Crystal During
        Thermomechanical Fatigue", in this volume.

[3]   Heine J.E., Warren J.R. and Cowles B.A., "Thermal mechanical
        Fatigue of Coated Blade Materials", Final report, WRDC-TR-87-
        4102, United Technologies Corporation, Pratt & Whitney, West Palm
        Beach, Florida, 1989.

[4]   Bressers J., Arrell D.J., Ostolaza K. and Vallés J.L., "Effect of a
        coating on the precipitate rafting in a single crystal superalloy",
        submitted to Seventh International Conference on Mechanical
        Behaviour of Materials, The Hague, May 28-June 2, 1995.

[5]   Murakami, Y. et al. (Edts.), Stress Intensity Factors Handbook,
        Vol. 1, Pergamon Press, 1987.

[6]   Hartranft, R.J. and Sih, G.C., Methods of Analysis and Solutions of
        Crack Problems, Ed. G.C. Sih, Noordhoff International Publishers,
        Holland, 1973.

[7]   Press, W.H., Flannery, B.P., Teukolsky, S.A. and Vetterling, W.T.,
        Numerical Recipes. The Art of Scientific Computing, Cambridge
        University Press, Cambridge, 1986.

[8]   Isida, M., "Tension of a half plane containing array cracks,
        branched cracks and cracks emanating from sharp notches",
        Transactions of the JSME, 45, 1979, pp. 303-317.

[9]   Spiegel, M.R., "Manual de fórmulas y tablas matemáticas", Serie
        de compendios Schaum, McGraw-Hill, 1970, Madrid.

Sam Y. Zamrik, [1] Daniel C. Davis, [1] and Lee C. Firth [1]

ISOTHERMAL AND THERMOMECHANICAL FATIGUE OF TYPE 316
STAINLESS STEEL

REFERENCE: Zamrik, S. Y., Davis, D. C., and Firth, L. C., **"Isothermal and
Thermomechanical Fatigue of Type 316 Stainless Steel,"** Thermomechanical
Fatigue Behavior of Materials: Second Volume, ASTM STP 1263, M. J. Verrilli and
M. G. Castelli, Eds., American Society for Testing and Materials, 1996.

ABSTRACT: Discussed is the thermomechanical fatigue (TMF) life of Type 316
stainless steel under in-phase and out-of-phase cycling and a comparison with the
isothermal fatigue case. The TMF tests were conducted in air over a temperature range of
399°C (750°F) to 621°C (1150° F) at several mechanical strain ranges: $\Delta\varepsilon_m = 0.4\%$
through 1.0%. The isothermal fatigue data were from total strain range ($\Delta\varepsilon_m = 0.35\%$,
0.5%, 0.68% and 1.0%) tests conducted at 621°C (1150° F). The results show that in-
phase TMF is the most damaging cycle for Type 316 stainless steel at the high strain
range levels ($\Delta\varepsilon_m > 0.7\%$), followed by isothermal fatigue cycling at 621°C (1150° F) and
then out-of-phase TMF cycling. The out-of-phase TMF life curve was observed to cross
over the isothermal fatigue life curve at $\Delta\varepsilon_m \cong 0.6\%$, and the trend of the data indicate that
the out-of-phase TMF life curve will cross over the in-phase TMF life curve at $\Delta\varepsilon_m <$
0.4%, and become the most damaging cycle type at the lower strain range levels.

KEYWORDS: thermomechanical fatigue, Type 316 stainless steel, creep, oxidation.

The widespread use of stainless steels in the fabrication of components such as in
nuclear reactors and chemical refineries is indicative of the their attractive properties for
elevated temperature applications. Under these operating conditions component life may
not be limited by strength as the primary design parameter, but life may be limited by
various damage mechanisms such as creep, fatigue or oxidation, which can act
independently or in combination. Creep damage, for example, is a time dependent
process which depends on the loading history and temperature; whereas, fatigue damage
is generated by cyclic loadings and the process is primarily time independent. Corrosion,
which results from surface oxidation predominantly at high temperatures, enhances the
fatigue process. When these three damage mechanisms act in combination, a creep-
fatigue-oxidation interaction exists.

Fatigue life and damage assessment for materials have generally been based on
isothermal test results. These fatigue tests are usually conducted at the highest design or

---

[1] Professor, Associate Professor and Graduate Student, respectively, Department of Engineering Science
and Mechanics, The Pennsylvania State University, University Park, PA 16802.

operating temperature to activate those damage mechanisms that could occur during service life. It is expected these test conditions would yield the most conservative fatigue life predictions. However, it has been observed with a number of engineering materials that the most conservative fatigue life may not be under isothermal test conditions but under thermomechanical fatigue (TMF) where combined cycles of both temperature and mechanical strain are applied. Hence, under TMF the deformation process and life are dependent on both the temperature range and the stress/strain range. There are two basic TMF cycle types that are most often used to assess life: In-Phase (IP) TMF and Out-of-Phase (OP) TMF. An IP TMF cycle is where the temperature and applied mechanical strain of the cycle reach their maximum and minimum values at the same time. The OP TMF cycle is where the maximum applied mechanical strain and lowest temperature of the cycle reach their values at the same time and the minimum strain and highest temperature values of the cycle occurs at the same time. Normally, constant strain rate triangular waveform type cycles are used.

There have been numerous research studies on the thermomechanical fatigue life of engineering materials, both experimental and analytical, over the past 20 years. Most of these TMF studies have focused on the Nickel-base superalloys, for example [1-10], because of their wide use in gas turbine and jet engine applications, and the austenitic stainless steels, Types 304 and 316 [11-14], because of their usage in nuclear power plant designs. Some studies have attempted to develop TMF life prediction methods which are based on established isothermal fatigue life prediction approaches such as strainrange partitioning, Ostergren model, damage rate equations and modified linear damage rules [12-19]. A review of these and many more studies of the fatigue life of engineering materials would show that in nearly half the cases TMF life was less than the comparable isothermal fatigue life [20].

However, for the austenitic stainless steels, Type 304 stainless steel has received the most attention in TMF studies, with few exceptions [19, 21]. The current study will present some results of an ongoing investigation of the thermomechanical fatigue life assessment of Type 316 stainless steel, and compare these results with isothermal fatigue test results [22]. In the current study the mechanical strain cycling is uniaxial; however, ultimately this research will be extended to assessing the TMF life of Type 316 stainless steel under biaxial strain cycling where the authors have conducted extensive research on the isothermal fatigue and creep-fatigue life of this material under combined axial-torsional strain cycling [23-27].

MATERIAL AND TEST PROCEDURE

Material and Specimens

Type 316 stainless steel material used in this study was provided by the Oak Ridge National Laboratory in the form of 2.54 cm (1 in.) diameter rods which were hot rolled at 1180°C (2156°F). From this material specimens of two different dimensions were machined as shown in Figs. 1a and 1b. Subsequent treatments of both types of specimens were identical. The gage section was polished through the use of low stress machining techniques to minimize the residual stresses produced by machining. The

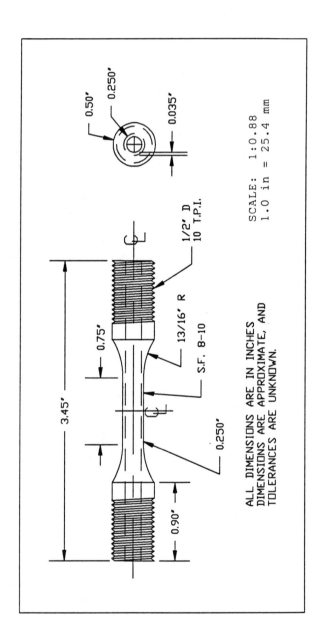

0.50'
0.250'
0.035'

1/2' D
10 T.P.I.

13/16' R

S.F. 8-10

0.75'

3.45'

0.250'

0.90'

SCALE: 1:0.88
1.0 in = 25.4 mm

ALL DIMENSIONS ARE IN INCHES
DIMENSIONS ARE APPROXIMATE, AND
TOLERANCES ARE UNKNOWN.

FIG. 1a -- TMF Specimen Used with Forced Cooling

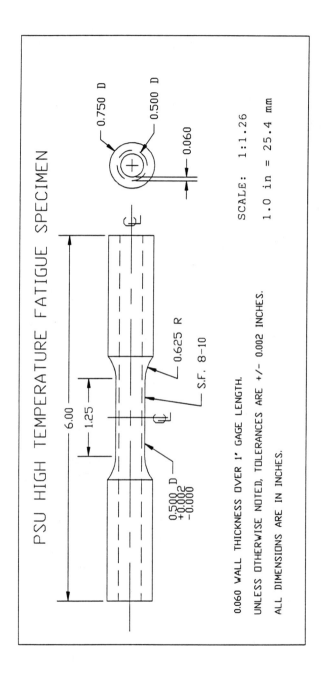

FIG. 1b -- TMF Specimen Used with Natural Convection Cooling

specimens were then solution annealed at 1065°C (1950°F) in Argon for 30 minutes and air cooled. To remove the resulting oxide and completely eliminate any residual surface stresses the specimens were then electropolished. The specimen shown in Fig. 1a was used in TMF testing in which heating was by induction and cooling by forced air (F/A) through the bored center hole, and the specimen shown in Fig. 1b was used in TMF testing in which heating was by induction and cooling by natural convection (N/C). Two specimen designs were needed in order to accommodate the different experiment setups used for the two types of specimen cooling methods.

Experimental Facility and Test Procedure

A schematic of the thermomechanical experimental test facility is shown in Fig. 2. The facility consists of a servo-hydraulic fatigue machine capable of conducting uniaxial and combined axial-torsional mechanical strain/stress cycling. Only the axial loading mode was used in this study. A digital system controller controlled both the mechanical and thermal closed loops via software commands installed on a local laboratory computer. Specimen heating was done using a 2.5 kW high frequency induction generator. Along the specimen gage section three independently adjustable coils were used to maintain an acceptable dynamic temperature gradient during testing, while five thermocouples along the gage section monitored the temperature distribution. One thermocouple just outside the upper end of the gage section was used as feedback to the temperature controller. Axial strain was measured using an air-cooled extensonmeter with wedge probes placed over the specimen gage section.

There were three steps in the procedure for initiating and conducting IP and OF TMF tests using the facility described above. These procedures are illustrated in the schematics shown in Figs. 3. The first step was the initial specimen heat-up under zero load and stabilization at the mean temperature of the temperature cycle range. The second step involved the start up of the thermal cycle where previously determined command signal increments were sent to the induction heater, through the temperature controller, to heat the specimen to the prescribed maximum and minimum amplitudes and at the prescribed frequency (f) of the temperature cycle waveform. The temperature feedback signals were output to a strip chart plotter where they could be monitored with respect to time. The temperature distribution along the specimen gage section was maintained to within ± 5 - 10°C. In the final step the strain increments of the mechanical cycle, beginning at the zero mean strain level, were added to the temperature cycle to initiate either the IP or OP TMF testing. During testing stress range versus cycles, mechanical and total strain range versus load hysteresis, and specimen gage section temperature distribution data were stored in a computer database and plotted on strip chart and x-y plotters as appropriate.

RESULTS AND DISCUSSION

In Table 1 are the isothermal continuous cycling fatigue data at 621°C (1150°F) for Type 316 stainless steel obtained from some previous studies [22, 27, 28]. In Table 2

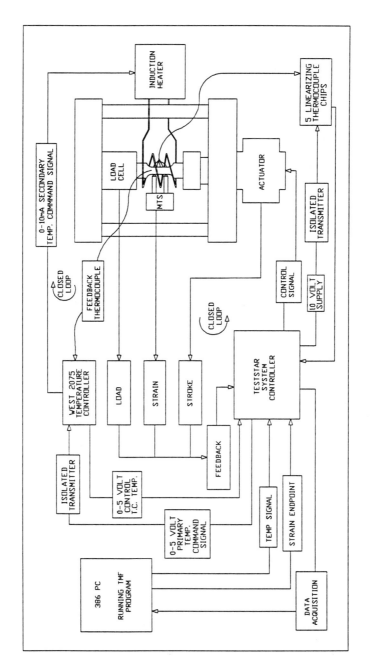

FIG. 2 -- Schematics of TMF Testing Facility

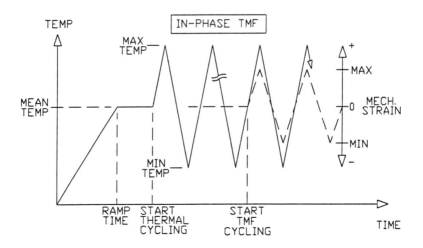

FIG. 3a -- In-Phase TMF Testing Start-Up and Testing Procedure

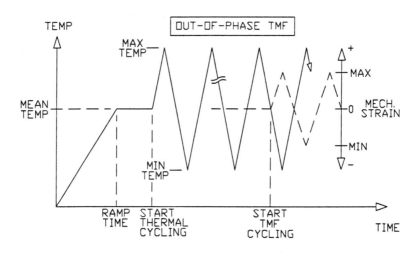

FIG. 3b -- Out-Of-Phase TMF Testing Start-UP and Testing Procedure

Table 1 -- <u>Isothermal Fatigue Life Data - 621°C (1150°F) in air.</u>

Total Strain Range $\Delta\varepsilon_m$, %	Cycles to Failure N_f	Cycle Frequency f, cpm
1.00	868	10
0.68	3,200	10
0.50	4,000	10
0.35	42,780	10

Table 2 -- <u>In-Phase TMF Life Data - 399°C (750 °C) to 621°C (1150°F) in air.</u>

Mechanical Strain Range $\Delta\varepsilon_m$, %	Cycles to Failure N_f	Cycle Frequency f, cpm
1.00	800	1.5
0.80	1,025	0.33
0.60	2,549	0.33
0.50	4,607	1.5

Table 3 -- <u>Out-Of-Phase TMF Life Data - 399°C (750°C) to 621°C (1150°F) in air.</u>

Mechanical Strain Range $\Delta\varepsilon_m$, %	Cycles to Failure N_f	Cycle Frequency f, cpm
1.00	1,507	1.5
0.82	1,996	1.5
0.80	2,295	0.33
0.60	5,712	0.33
0.50	6,115	1.0
0.50	6,392	0.33
0.40	10,388	1.0

and 3 are the life data from IP and OP TMF tests of Type 316 stainless steel for a temperature range of 399°C (750°F) to 621°C (1150°F), and at the applied mechanical strain ranges noted. The TMF tests which used forced air cooling (F/A) had a cycle frequency of 1.0 or 1.5 cpm, while the TMF tests conducted using natural convection (N/C) cooling had a cycle frequency of 0.33 cpm. However, the differences in the test frequencies showed no significance on the TMF life of this material. From published data [29] on Type 304 stainless steel at 650°C (1200°F), no effect on fatigue life was observed over several decades of mechanical strain rates. Assuming a similarity between these two austenitic stainless steels in this respect, could explain why no difference in the TMF life of Type 316 stainless steel was observed at the different frequencies used in these tests.

<u>Analysis and Characterization</u>

Total mechanical strain range ($\Delta\varepsilon_m$) versus cycles to failure (N_f) curves for the isothermal [28] and TMF data are plotted in Fig. 4. Failure is defined as a 10% change in load from the mid-life load amplitude. Each of the strain range versus cycles to failure curves in Fig. 4 could follow the classical Manson-Coffin equation by taking into account the plastic and elastic strain range components. The fatigue life curves show that the

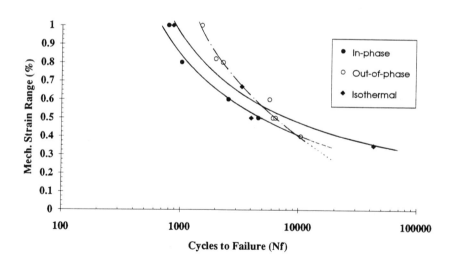

FIG. 4 -- Mechanical Strain ($\Delta\varepsilon_m$) vs. Cycles to Failure (N_f) Life Curves

most damaging cycle type at the high strain ranges ($\Delta\varepsilon_m > 0.7\%$) was IP TMF, followed by isothermal fatigue (IF) and then OP TMF. However, at $\Delta\varepsilon_m \cong 0.6\%$ the OP TMF life curve crosses over the IF life curve and tends to show a probable crossover of the IP TMF life curve at $\Delta\varepsilon_m < 0.4\%$, and hence, would become the most damaging of these cycle types for Type 316 stainless steel at lower strain ranges. In a previous TMF study on Type 304 stainless steel, Majumdar [12] found that IP TMF cycling was most damaging in comparison to IF and OP TMF cycling at all strain ranges tested. In TMF testing of superalloys the OP TMF life curves have been observed to crossover the IP TMF life curve and become the most damaging TMF cycle type [7]. The effects of oxidation on the specimen surfaces was greater in OP TMF because the oxide layer, formed at the high temperatures, is brittle at the low temperature and crack under the tensile load. These cracks then cause stress concentrations at the oxide-base material interface and hence became fatigue crack initiators [16]. Other TMF studies by Jaske [30] and Neu and Sehitoglu [16] on carbon steels have observed that OP TMF cycling can have a more detrimental effect on life than IP TMF cycling.

Typical stabilized stress versus mechanical strain and stress versus total thermal-mechanical strain hysteresis loops are shown respectively in Figs. 5 for each an IP and OP TMF test of Type 316 stainless steel conducted over a temperature range of 399°C (750°F) to 621°C (1150° F). The applied mechanical strain range ($\Delta\varepsilon_m$) for both TMF tests was 0.8%. The features of these two hysteresis loops are similar to those discussed in many previous TMF studies referenced. One obvious feature is that the IP TMF thermal-mechanical hysteresis loop is considerably wider than the OP TMF loop. This occurs because thermal strains are respectively added to and subtracted from the mechanical strains to obtain the total thermal-mechanical strain. For this IP TMF test the thermal-mechanical strain range ($\Delta\varepsilon_{TM}$) was measured as 1.2% at cycle # 477 and 0.36% for the OF TMF test at cycle # 1961. These hysteresis loops show none of the distortions at the maximum and minimum stress (load) amplitude points which have been observed to show up after long term TMF cycling because of a loss of the original phasing relation between the temperature cycle and mechanical strain cycle [31]. Another feature of these IP and OP TMF hysteresis loops, not found in comparable isothermal hysteresis loops of this material [28], is the asymmetry of the tensile and compressive stress (load) areas of the loops. This asymmetry can be attributed to the differences in the material's strength and ductility at the two temperature extremes of the cycle. The mean stress effect is further illustrated in the stress versus cycles plots of these two tests shown in Figs. 6. The mean stress is consistently negative for IP TMF cycling and positive for OP TMF cycling. The stress-cycles plots show that at the higher mechanical strain ranges, the material hardens and then slightly softens before final failure. At the lower mechanical strain ranges the material continues to harden or maintain hardness until failure.

A number of methods have been developed to characterize fatigue data for life prediction and assessment. The hysteretic energy method is adopted in this study to characterize the TMF behavior of Type 316 stainless steel has been used previously as a basis for life prediction models [32-34]. From the stabilized mid-life hysteresis loop, the energy per volume cycle or density can be calculated and fit to the energy life relation by:

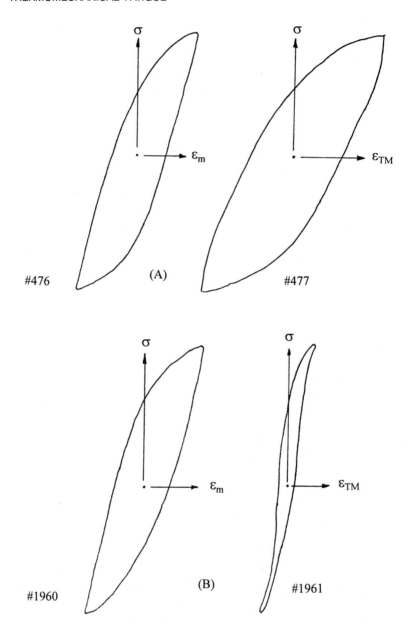

FIG. 5 -- (A) IP TMF Hysteresis Loops, $\Delta\varepsilon_m$ = 0.80% (Cycles 476 & 477, respectively); (B) OP TMF Hysteresis Loops, $\Delta\varepsilon_m$ = 0.80% (Cycles 1960 & 1961, respectively).

0.80% In-phase (0.33 cpm)

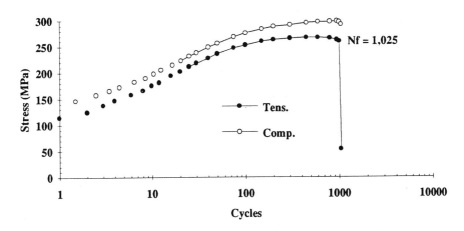

FIG. 6a -- Stress vs. Cycles Plots, In-Phase Test at $\Delta\varepsilon_m = 0.80\%$

0.80% Out-of-phase (0.33 cpm)

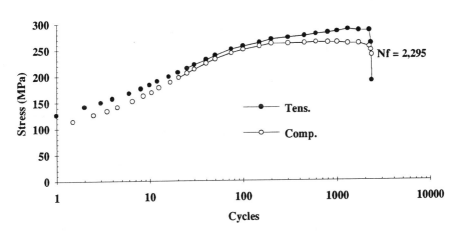

FIG. 6b -- Stress vs. Cycles Plots, Out-Of-Phase Test at $\Delta\varepsilon_m = 0.80\%$

$$\frac{\text{Energy}}{\text{Volume Cycle}} = \Delta U = \frac{W}{AL} = \int \sigma \; d\varepsilon = A_o (N_f)^a \qquad (1)$$

where:

ΔU = Energy per volume cycle, Joules/m^3
W = Work or area of the hysteresis loop, Joules
A = Specimen cross-sectional area, m^2
L = Specimen gage length, m
σ = Stress, MPa
$d\varepsilon$ = Strain derivative
A_o = Constant
N_f = Cycles to failure
a = Constant

The hysteretic energy approach is illustrated in Fig. 7 where some interesting aspects of the characterizations of the IP and OP TMF test data can be observed. The first is that the IP TMF hysteretic energy life line is lower than the OP TMF line which suggest more damage. The second aspect is that the lines are not parallel suggesting different types of damage mechanisms which can be revealed by the microstructural evaluations. The two hysteretic energy life lines also show a trend toward crossover at low values.

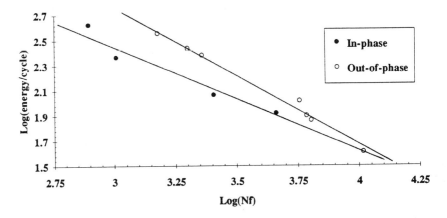

Characterization of TMF data using hysteretic energy

FIG. 7 -- Hysteretic Energy Life Curves
(Energy in units: in-lbs = 0.113 Joules)

Microstructural Evaluation and Analysis

The fracture surfaces of all TMF tested specimens were examined using a scanning electron microscope (SEM) and some of these fractographs are given in Figs 8 through 11. Some of the specimens tested in TMF were cooled by forced air and their fractographs can be compared to specimens cooled by natural convection. Figs. 8 and 9 of IP TMF tests indicate severe intergranular cracking along with ductile crack propagation between crystallographic facets. This intergranular fracture can be attributed to creep type damage since the maximum tensile strain occurs at the maximum temperature. Although creep is much more dominant in hold-time tests [22], the temperature range at which these TMF tests were conducted was high enough to precipitate $M_{23}C_6$ carbides, which tend to weaken grain boundaries. The higher temperature during the tension portion of the cycle also provides energy for cross-slip and climb, which makes cavitation at the grain boundaries easier [35]. Another reason can be attributed to environmental attack at the boundaries causing microscopic stress concentrations to develop. The fracture mode of all specimens tested in OP TMF showed strictly transgranular cracking with striations and regions of secondary cracking as shown in Figs. 10 and 11. In comparison, the fracture surfaces of the 621°C (1150°F) isothermal fatigue tests of Type 316 stainless steel conducted at total strain ranges: $\Delta\varepsilon_m$ =0.5% and 1.0%, were both strictly transgranular. Only when creep was introduced by imposing a 30 minute hold-in-tensile strain on the $\Delta\varepsilon_m$ =0.5% isothermal continuous cycle test did the fracture mode become intergranular [22].

Frequency and Cooling Effects on TMF Tests

The effects of TMF testing at different cyclic frequencies of 1.0 and 1.5 cpm, when forced cooling was used, and 0.33 cpm when natural convection cooling was used, appeared insignificant when the total fatigue life was compared (Fig. 4). However, specimens tested using forced cooling, whether IP or OP TMF, showed multiple crack initiation sites along the inner surfaces of the tubular specimen as shown in Fig. 12, but the effect seemed more pronounced during OP TMF cycling. This would lead one to believe that the forced air down the center of the specimen produced a thermal gradient through the thickness of the specimen's wall. With the inner surface cooler, a tensile biaxial stress state will arise since the material adjacent to the inner surface is warmer and constrains the inner surface from contraction. The magnitude of the biaxial stress state depends on the severity of the radial thermal gradient producing thermal cracking. Hence, the multiple initiation sites can be attributed to thermal cracking of the inner surface. Specimens tested using natural convection cooling did not show extensive cracking or an extensive number of initiation sites on the inner surface of the specimen. However, future TMF testing of Type 316 stainless steel in this research will only use a natural convection cooling method.

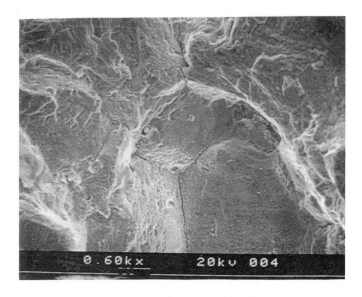

FIG. 8 -- IP TMF, N/C Cooling, $\Delta\varepsilon_m = 0.60\%$, f = 0.33 cpm, $N_f = 2,549$.

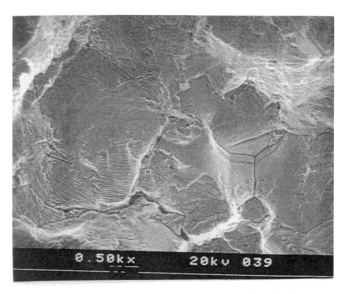

FIG. 9 -- IP TMF, F/A Cooling, $\Delta\varepsilon_m = 1.0\%$, f = 1.5 cpm, $N_f = 787$.

FIG. 10 -- OP TMF, N/C Cooling, $\Delta\varepsilon_m$ = 0.80%, f = 0.33 cpm, N_f = 2,295.

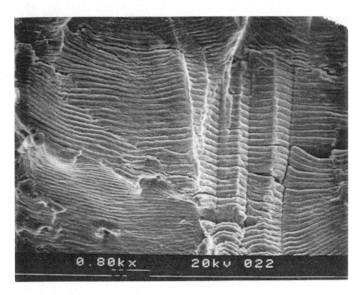

FIG. 11 -- OP TMF, F/A Cooling, $\Delta\varepsilon_m$ = 0.40%, f = 1.0 cpm, N_f = 10,388.

SUMMARY

Thermomechanical fatigue life of Type 316 stainless steel depends on the type of cycle and the applied mechanical strain and temperature ranges. For Type 316 stainless steel, the most damaging cycle type at the high mechanical strain range level ($\Delta\varepsilon_m$ >0.7%) is In-Phase TMF where the fracture mode is predominantly intergranular. The Out-of-Phase TMF cycle is less damaging and the fracture mode is transgranular with low cycle fatigue features. At high strain ranges the isothermal fatigue (IF) tests at 621°F (1150°F) show a reduction in life in comparison to the OP TMF tests. However, at $\Delta\varepsilon_m \cong 0.6\%$ the OP TMF life curve crosses over the IF life curve to potentially become a more damaging cycle type than isothermal fatigue, but not as damaging as IP TMF cycling. It is suspected that at $\Delta\varepsilon_m< 0.4\%$ the OP TMF fatigue life curve would cross over the IP TMF curve and would become the most detrimental of the cycle types studied. This cross over of the life curves is believed due to the modulus behavior at the temperature extremes and surface oxidation damage occurring over the long exposure period. Crack growth is still believed to be the dominant portion of OP TMF life as compared to crack initiation, so the multiple crack initiators observed on the specimen inner surfaces would not significantly influence life.

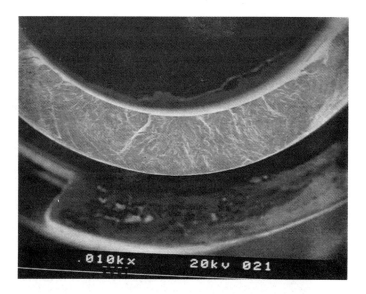

FIG. 12 -- Multiple Initiation Sites Along Inner Surface, OP TMF, F/A Cooling, $\Delta\varepsilon_m = 0.40\%$, f = 1.0 cpm, $N_f = 10,388$.

ACKNOWLEDGMENTS

The research support from the Pressure Vessel research Council (PVRC) under Grant # 94-09 and from the National Science Foundation under Grant # MSS-9215694 is greatly appreciated, as well as the Oak Ridge National Laboratory for donating the material. Also, the authors acknowledge the technical advice on testing procedures by M. G. Castelli and J. R. Ellis, NASA-LERC. This type of research, initiated by PVRC and NSF, is emerging as a significant area for fatigue study in advanced coating materials where life prediction methodology is a major factor in the development of the coating system.

REFERENCES

1. Malpertu, J. L. and Remy, L., "Thermomechanical Fatigue Behavior of a Superalloy," Low Cycle Fatigue, ASTM STP 942, H. D. Solomon, G. R. Halford, L. R. Kaisand, and B. N. Leis, Eds., American Society for Testing and Materials, Philadelphia, 1988, pp. 657-671.

2. Cook, T. S., Kim, K. S., and McKnight, R. L., "Thermal Mechanical Fatigue of Cast Rene 80," Low Cycle Fatigue, ASTM STP 942, H. D. Solomon, G. R. Halford, L. R. Kaisand, and B. N. Leis, Eds., American Society for Testing and Materials, Philadelphia, 1988, pp. 692-708.

3. Marchant, N., L'Esperance, G., and Pelloux, R. M., "Thermal-Mechanical Cyclic Stress-Strain Responses of Cast B-1900+Hf," Low Cycle Fatigue, ASTM 942, H. D. Solomon, G. R. Halford, L. R. Kaisand, and B. N. Leis, Eds., American Society for Testing and Materials, Philadelphia, 1988, pp. 638-656.

4. Rau, C. A., Jr., Gemma, A. E., and Leverant, G. R., "Thermal-Mechanical Fatigue Crack Propagation in Nickel- and Cobalt-Base Superalloys Under Various Strain-Temperature Cycles," Fatigue at Elevated Temperatures, ASTM STP 520, American Society for Testing and Materials, Philadelphia, 1973, pp. 166-178.

5. Gemma, A. E., Langer, B. S., and Leverant, G. R., "Thermomechanical Fatigue Crack Propagation in an Anisotropic (Directionally Solidified) Nickel-Base Superalloy," Thermal Fatigue of Materials and Components, ASTM STP 612, D. A. Spera and D. F. Mowbray, Eds., American Society for Testing and Materials, Philadelphia, 1976, pp. 199-213.

6. Hopkins, S. W., "Low-Cycle Thermal Mechanical Fatigue Testing," Thermal Fatigue of Materials and Components, ASTM STP 612, D. A. Spera and D. F. Mowbray, Eds., American Society for Testing and Materials, Philadelphia, 1976, pp. 157-169.

7. Sehitoglu, H. and Boismier, D. A., "Thermo-Mechanical Fatigue of Mar-M247: Part 1 - Experimental & Part 2 - Life Prediction," Journal of Engineering Materials and Technology, January 1990, Vol. 2, pp. 68-89.

8. Castelli, M. G., Miner, R. V., and Robinson, D. N., "Thermomechanical Deformation Behavior of a Dynamic Strain Aging Alloy, Hastelloy X," Thermomechanical Fatigue Behavior of Materials, ASTM STP 1186, H. Sehitoglu, Ed., American Society for Testing and Materials, Philadelphia, 1993, pp. 106-125.

9. Guedou, J.-Y. and Honnorat, Y., "Thermomechanical Fatigue of Turbo-Engine Blade Superalloy," Thermomechanical Fatigue Behavior of Materials, ASTM STP 1186, H. Sehitoglu, Ed., American Society for Testing and Materials, Philadelphia, 1993, pp.157-175.

10. Remy, L., Bernard, H., Malpertu, J L. and Rezai-Aria, F., "Fatigue Life Prediction Under Thermal-Mechanical Loading in a Nickel-Base Superalloy," Thermomechanical Fatigue Behavior of Materials, ASTM STP 1186, H. Sehitoglu, Ed., American Society for Testing and Materials, Philadelphia, 1993 pp. 3-16.

11. Zauter, R., Petry, F., Christ, H.-J., and Mughabi, H., "Thermomechanical Fatigue of the Austenitic Stainless Steel AISI 304L," Thermomechanical Fatigue Behavior of Materials, ASTM 1186, H. Sehitoglu, Ed. American Society for Testing and Materials, Philadelphia, 1993, pp. 70-90.

12. Majumdar, S., "Thermomechanical Fatigue of Type 304 Stainless Steel," Thermal Stress, Material Deformation and Thermo-Mechanical Fatigue, ASME PVP-Vol. 123, H. Sehitoglu and S. Y. Zamrik, Eds. American Society for Mechanical Engineers, New York, 1987, pp. 31-36.

13. Halford G. R. and Manson, S. S., "Life Predictions of Thermal-Mechanical Fatigue Using Strainrange Partitioning," Thermal Fatigue of Materials and Components, ASTM STP 612, D. A Spera and D. F. Mowbray, Eds., American Society for Testing and Materials, 1976, pp. 239-250.

14. Kuwabara, K., Nitta, A., and Kitamura, T., "Thermal-Mechanical Fatigue Life Prediction in High-Temperature Component Materials for Power Plants," Advances in Life Prediction Methods, D. A. Woodford and J. R. Whitehead, Eds. American Society of Mechanical Engineers, New York, 1983, p. 131.

15. Berstein, H. L., "A Evaluation of Four Creep-Fatigue Models for a Nickel-Base Superalloy," Low Cycle Fatigue and Life Prediction, ASTM STP 770, C. Amzallag, B. N. Leis, and P. Rabbe, Eds., American Society for Testing and Materials, Philadelphia, 1982, pp. 105-134.

16. Neu, R. W. and Sehitoglu, H., "Thermomechanical Fatigue, Oxidation and Creep: Part II. Life Prediction," Metallurgical Transactions A, Vol. 20A, September 1989, pp. 1769-1782.

17. Ramaswamy, V. G., Stroufer, D. C., VanStone, R. H., and Laflen, J. H., "Modeling Thermomechanical Cyclic Response with a Unified State Variable Constitutive Equation," Thermal Stress, Material Deformation and Thermo-Mechanical Fatigue, ASME PVP-Vol. 123, H. Sehitoglu and S. Y. Zamrik, Eds. American Society for Mechanical Engineers, New York, 1987, pp. 57-64.

18. Halford, G. R. and Saltsman, J. F., "Calculation of Thermomechanical Fatigue Life Based on Isothermal Behavior," Thermal Stress, Material Deformation and Thermo-Mechanical Fatigue, ASME PVP-Vol. 123, H. Sehitoglu and S. Y. Zamrik, Eds. American Society for Mechanical Engineers, New York, 1987, pp. 9-21.

19. McGaw, M. A., "Cumulative Damage Concepts in Thermomechanical Fatigue," Thermomechanical Fatigue Behavior of Materials, ASTM STP 1186, H. Sehitoglu, Ed., American Society for Testing and Materials, Philadelphia, 1993, pp. 144-156.

20. Halford, G. R., Low-Cycle Thermal Fatigue, NASA Technical Memorandum 87225, February 1986.

21. Shi, H. J., Robin, C., Pluvinage, G., "Thermal-Mechanical Fatigue Lifetime Prediction of an Austenitic Stainless Steel," Advances in Fatigue Lifetime Predictive Techniques: Second Volume, ASTM STP 1211, M. R. Mitchell and R. W. Landgraf, Eds., American Society for Testing and Materials, Philadelphia, 1993, pp. 105-116.

22. Zamrik, S. Y. and Davis, D. C., "A Ductility Exhaustion Approach for Axial Fatigue-Creep Damage Assessment Using Type 316 Stainless Steel," Journal of Pressure Vessel Technology, Transactions of ASME, Vol. 113, May 1991, pp. 180-186.

23. Zamrik, S. Y., Davis, D. C., and Kulowitch, P. C., "Failure Modes in a Type 316 Stainless Steel Under Biaxial Strain Cycling," Advances in Lifetime Predictive Techniques, ASTM STP 1122, M. R. Mitchell and R. W. Landgraf, Eds., American Society for Testing and Materials, Philadelphia, 1991, pp. 299-318.

24. Zamrik, S. Y. and Davis, D. C., "Cyclic Stress Relaxation in Multiaxial Creep-Fatigue and Damage Assessment of a Type 316 Stainless Steel," Proceedings, 7th International Conference of Pressure Vessel Technology, Vol. 1, 1992, pp. 613-625.

25. Zamrik, S. Y., Mirdamadi, M., and Davis, D. C., "A Proposed Model for Biaxial Fatigue Using the Triaxiality Factor Concept," Advances in Multiaxial Fatigue, ASTM STP 1191, D. L. McDowell and R. Ellis, Eds., American Society for Testing and Materials, Philadelphia, 1993, 85-105.

26. Zamrik, S. Y., Mirdamadi, M., and Davis, D. C., "Ductility Exhaustion Criterion for Biaxial Fatigue-Creep Interaction in Type 316 SS at 1150°F (621°C), Material Durability/Life Prediction Modeling - Materials for the 21st Century, ASME PVP-Vol. 290, S. Y. Zamrik and G. R. Halford, Eds., American Society of Mechanical Engineers, New York, 1994, pp. 107-134.

27. Mirdamadi, M., "Biaxial Fatigue Life Prediction Theory at Elevated Temperatures Under Continuous and Hold-Time Strain Cycling," Ph. D. Thesis, The Pennsylvania State University, June 1990.

28. Zahiri, F., "Microstructural Responses of Nickel-Base Superalloy Waspaloy and Type 316 Stainless Steel to Axial and Torsional Strain Cycling and Creep-Fatigue Interaction," M.S. Thesis, The Pennsylvania State University, 1987.

29. Conway, J. B., Stenz, R. H., and Berling, J. T., "Fatigue, Tensile and Relaxation Behavior of Stainless Steels," Tech. Information. Center, Office of Information Services, USAEC, 1975..

30. Jaske, C. E., "Thermal-Mechanical, Low-Cycle Fatigue of AISI 1010 Steel," Thermal Fatigue of Materials and Components, ASTM STP 612, D. A. Spera and D. F. Mowbray, Eds., American Society for Testing and Materials, Philadelphia, 1976, pp. 170-198.

31. Castelli, M. G. and Ellis, J. R., "Improved Techniques for Thermomechanical Testing in Support of Deformation Modeling," Thermomechanical Fatigue Behavior of Materials, ASTM STP 1186, H. Sehitoglu, Ed., American Society for Testing and Materials, Philadelphia, 193, pp. 195-211.

32. Ostergren, W. J., "A Damage Function and Associated Failure Equations for Predicting Hold Time and Frequency Effects in Elevated Temperature Low Cycle Fatigue," Journal of Testing and Evaluation, Vol. 4, No. 5, 1976, pp. 327-339.

33. Sandox, B. I., Fundamentals of Cyclic Stress and Strain, The University of Wisconsin Press, 1972.

34. Ellyin, F. and Kujawski, D., "Plastic Strain Energy in Fatigue Failure," Journal of Pressure Vessel Technology, Transactions of ASME, Vol. 106, 1984, pp. 342-347.
35. Suresh, S, in Fatigue of Materials, Davis, E. A. and Ward, I. M., Eds. Cambridge University Press, NY, 1991, pp. 385-390.

Masafumi Yamauchi,[1] Tomomi Ohtani,[2] and Yukio Takahashi [3]

THERMAL FATIGUE BEHAVIOR OF A SUS304 PIPE UNDER
LONGITUDINAL CYCLIC MOVEMENT OF AXIAL TEMPERATURE
DISTRIBUTION

REFERENCE: Yamauchi, M., Ohtani, T., and Takahashi, Y., "**Thermal Fatigue
Behavior of a SUS304 Pipe Under Longitudinal Cyclic Movement of Axial
Temperature Distribution**," Thermomechanical Fatigue Behavior of Materials:
Second Volume, ASTM STP 1263, Michael J. Verrilli and Michael G. Castelli, Eds.,
American Society for Testing and Materials, 1996.

ABSTRACT: In a structural thermal fatigue test which imposed an oscillating axial
temperature distribution on a SUS 304 pipe specimen, different crack initiation lives were
observed between the inner and the outer surfaces, although the values of the von-Mises
equivalent strain range calculated by FEM inelastic analysis were almost the same for both
surfaces. The outer surface condition was an in-phase thermal cycle and an almost uniaxial
cyclic stress (low hydrostatic stress). The inner surface condition was an out-of-phase
thermal cycle and an almost equibiaxial cyclic stress (high hydrostatic stress). A uniaxial
thermal fatigue test was performed under the simulated conditions of the outer and inner
surfaces of the pipe specimen. The in-phase uniaxial thermal fatigue test result was in good
agreement with the test result of the pipe specimen for the outer surface. The out-of-phase
uniaxial thermal fatigue test which simulated the inner surface condition, showed a longer
life than the in-phase uniaxial test, and thus contradicted the result of the structural model
test. However, the structural model test life for the inner surface agreed well with the
uniaxial experimental measurement when the strain range of the inner surface was corrected
by a triaxiality factor.

KEYWORDS: Thermal fatigue, Main vessel, FBR, Type 304 stainless steel, Crack
initiation, Biaxial stress state, Hydrostatic stress, Triaxiality factor

[1]Research Manager, Nagasaki R&D Center, Mitsubishi Heavy Industries, Ltd.,
Nagasaki 851-03, JAPAN
[2]Engineer, Kobe Shipyard and Machinery Works, Mitsubishi Heavy Industries, Ltd.,
Kobe 652, JAPAN
[3]Senior Research Engineer, Central Research Institute of Electric Power Industry,
Komae-shi, Tokyo 201, JAPAN

INTRODUCTION

In Japan, a demonstration Fast Breeder Reactor (FBR) which follows the prototype FBR Monju has been under development and many R&D activities focusing on the reliability improvement of the FBR components have been carried out. Creep-fatigue is the most prominent and design limiting potential failure mode in high temperature components such as a reactor main vessel, pipings and so on. Extensive work has been carried out to develop an inelastic design procedure as well as the rule based on an elastic analysis used for Monju. The work includes several types of structural model test which simulate the critical components in the FBR. These structural model test data have been used to verify an evaluation method and to confirm a safety margin in the design procedures.

The main vessel in the FBR has a liquid sodium level. The wall of the vessel below the sodium level is heated by hot liquid sodium, but the upper part which is exposed to low-temperature nitrogen cover gas is cooler and has a sharp temperature gradient in the longitudinal direction. This temperature distribution in the wall has an inflection near the surface of the liquid sodium where the stress also reaches a maximum value. The vessel wall is subjected to cyclic thermal stress due to a cyclic movement of the temperature distribution during start-up and shut-down of the system, and this may cause ratchetting and creep-fatigue damage. Therefore much research has focused on the development and the verification of the evaluation procedure for the deformation and crack initiation behavior in the vessel wall exposed to the cyclic movement of longitudinal temperature distribution [1-6].

In this paper, the results of a thermal fatigue test which imposed an oscillating axial temperature distribution on a SUS 304 pipe specimen are described. This test simulated the service conditions of the main vessel in the vicinity of a liquid-sodium free surface in the FBR described above.

DEVELOPMENT OF STRUCTURAL MODEL TEST APPARATUS

Figure 1 shows a schematic of the test apparatus developed to conduct a structural model test which simulates the service condition of the vessel near the liquid sodium level in a FBR. The apparatus is mainly composed of two systems, a temperature control system and a specimen moving system.

A schematic of the temperature control system is shown in **Fig. 2**. The outside of the specimen is heated by an induction heating coil and the inside of the specimen is cooled by water. The longitudinal temperature distribution with a sharp gradient can be produced in a specimen by the combination of induction heating and water cooling. The maximum temperature gradient obtained by this system was about 20° C /mm for a pipe specimen with 5mm thickness. Peak temperature is measured and controlled without contact by an infrared pyrometer.

STRUCTURAL MODEL TEST

Material and Specimen

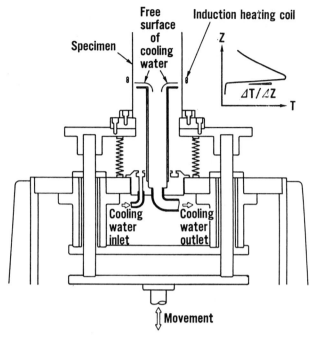

FIG. 1--Structural model test apparatus simulating the condition of the main vessel near liquid sodium level in a FBR

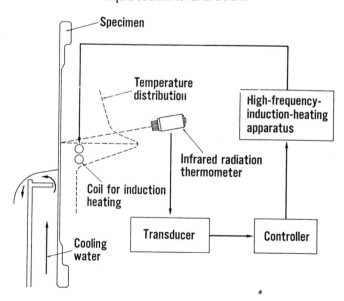

FIG. 2--Schematic of the temperature control system

The material is a commercial SUS304 stainless steel pipe with 165 mm outer diameter and 18.2 mm wallthickness. Cylindrical specimens with 155 mm outer diameter, 5 mm wall thickness and 300 mm length were machined from this material. In this study two specimens were tested. One specimen (specimen No.1) was made of base metal without a weld joint. The other (specimen No.2) had two longitudinal TIG weldments to examine the strength of weld joints. However, the result for the base metal only is described in this paper.

Testing Method and Condition

Tests were conducted in the apparatus described above. The longitudinal temperature distribution in the specimen was measured by about 60 chromel-alumel thermocouples on the outer surface. Crack initiation on the surface was examined by a dye penetration test.

Two specimens were tested under the same conditions. The maximum (peak) temperature was 600 ° C. Specimens were oscilated with an amplitude of ± 30mm and a constant rate of 12mm/min.

Test Results

A measurement of the alteration of the temperature distribution in the specimen wall at cycle N=20 is shown in **Fig. 3**. Both the peak temperature of 600 ° C and the temperature gradient were kept constant throughout the test.

In specimen No.1, crack initiation was detected by a dye penetration test after 3000 cycles on the inner surface and after 4500 cycles on the outer surface and the test was continued until 5500 cycles. In specimen No.2, crack initiation was after 1500 cycles at the inner surface and after 2500 cycles at the outer surface and the test was continued until 4000 cycles. In these tests the crack length detected by a dye penetration test was 0.7 to 2mm on the surface. The crack initiation life for the inner surface was shorter than that for the outer surface for both specimens.

The appearance of the cracks on the outer and on the innner surfaces of the specimen No.1 after the end of the test is shown in **Fig.4**. In the outside surface, the crack direction was circumferential. On the inner surface, the crack planes were randomly distributed, which are typical of a component subjected to equibiaxial cyclic thermal stresses.

SEM fractographs of cracks on the outer and on the inner surfaces are shown in **Fig. 5**. Transgranular cracks were observed on both surfaces which indicated that the influence of aqueous corrosion on the crack initiation life is negligible.

ANALYSIS AND DISCUSSION

Inelastic Stress Analysis

An inelastic (elastic-plastic) FEM stress analysis was conducted for the two structural model tests by using the temperature distribution data shown in Fig. 3. In the analysis, the classical kinematic hardening with a bilinear approximation of the cyclic stress-strain curve was used. The cyclic stress-strain curve was based on the test data from the uniaxial low-

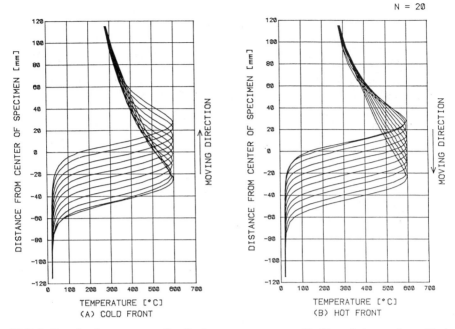

FIG. 3--Result of temperature distribution measurement at N=20 cycle in specimen No.1.

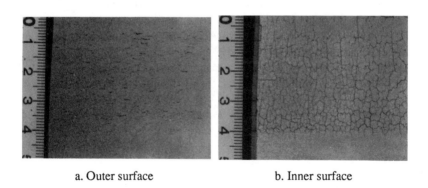

a. Outer surface   b. Inner surface

FIG. 4--Appearance of surface cracks on the outer (a) and the inner (b) surfaces
of specimen No.1

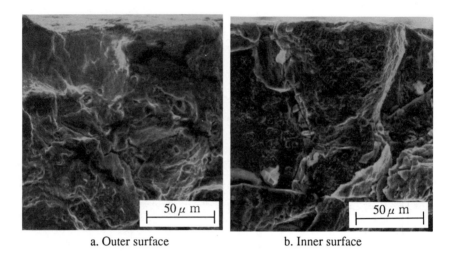

a. Outer surface   b. Inner surface

FIG. 5--SEM fractographs for the outside (a) and the inside (b) surface cracks

cycle fatigue tests at room temperature (RT) and 600° C using the same SUS304 base material. The uniaxial low-cycle fatigue tests were conducted in a strain controll mode at a constant strain rate of 0.1%/s by using smooth bar specimens with 8mm outer diameter and 15mm gage length. The stress-strain characteristics at temperatures between RT and 600° C were estimated using linear interpolation. The weldments in the specimen No.2 were not taken into account in the analysis.

The analysis was performed for a temperature cycle (see Fig.6) of heating up to test temperature (steps 1-50), two thermal cycles (steps 50-290) and cooling down to RT (steps 290-340). The variation of the temperature, the Mises equivalent strain and the triaxiality factor (TF) at the most severe loading point on the specimen as obtained from the analysis, is shown in **Fig. 6** and **Fig. 7** for the outer and the inner surfaces, respectively. In these figures, the sign of the stress with the absolute maximum was given to the Mises equivalent strain and the TF was calculated by the following equation[7].

$$TF = \frac{\sigma_1 + \sigma_2 + \sigma_3}{\frac{1}{\sqrt{2}}[(\sigma_1 - \sigma_2)^2 + (\sigma_2 - \sigma_3)^2 + (\sigma_3 - \sigma_1)^2]^{1/2}} \tag{1}$$

Hysteresis loops of stress and strain at the same location as shown in Figs. 6 and 7 are shown in **Fig. 8**.

The results of the inelastic analysis are summarized in **Table 1**. The thermal fatigue condition of the outer surface is in-phase because temperature and strain reaches their maximum at the same time step. On the other hand, the thermal fatigue condition of inner surface is out-of-phase because strain reaches its maximum when temperature reaches its minimum. The difference in the surface crack appearance between the inner and the outer surfaces described above, could be explained by the different stress states and/or the

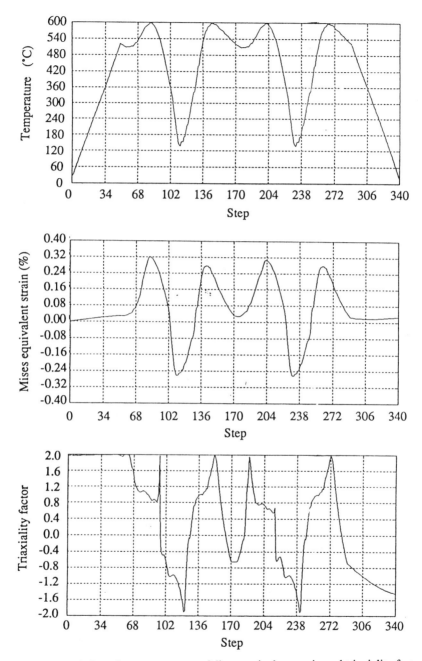

FIG. 6--The variation of temperature, von Mises equivalent strain and triaxiality factor
for the outer surface

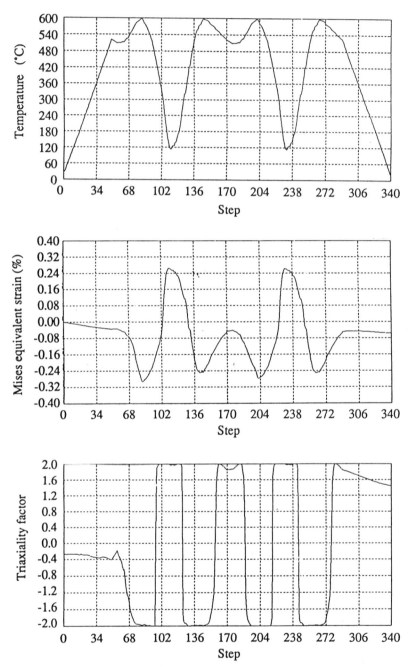

FIG. 7--The variation of temperature, von Mises equivalent strain and triaxiality factorfor the inner surface

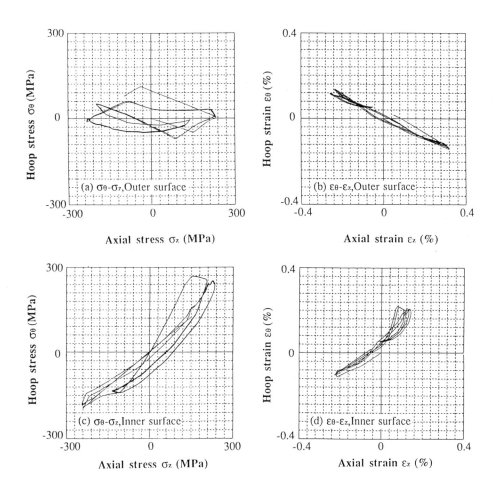

FIG. 8--Hysteresis loops of stress and strain at the outer and the inner surfaces
obtained by inelastic analysis

TABLE 1 --Summary of results of inelastic analysis

	von Mises Equivalent Strain Range (%)	Thermal Fatigue Condition	Stress State	TF (at Peak Strain)
Outer Surface	0.57% (specimen No.1) 0.58% (specimen No.2)	In-phase	Uniaxial	1
Inner Surface	0.54% (specimen No.1) 0.55% (specimen No.2)	Out-of-phase	Biaxial	2

different thermal fatigue phase.

The relation between the Mises equivalent strain range shown in Table 1 and the crack initiation life in the structural model test is plotted in Fig. 9, as well as the uniaxial low cycle fatigue data at a constant strain rate of 0.1%/s and 0.002%/s at 600° C. The data at a constant strain rate of 0.1%/s are the same as those used to obtain a cyclic stress strain curve described above. The strain rate of 0.002%/s is comparable to the strain rate in the structural model test and the tests were carried out to investigate an effect of a strain rate on the fatigue life. The result of the structural model test for the ouer surface was in a good agreement to the uniaxial test results. However, the life for the inner surface was much shorter than in the uniaxial test results, and therefore, it was considered that the difference in crack initiation life was caused by the different thermal fatigue phase and stress state condition as shown in Table 1.

<u>Effect of the Difference of Thermal Fatigue Condition</u>

In order to examine the influence of the difference in the thermal fatigue condition, in-phase and out-of-phase uniaxial thermal fatigue tests were carried out with a strain range of 0.7%, a constant strain rate of 0.002%/s and temperature cycling between 300° C and 600 ° C with a heating/cooling rate of about 51° C/min, which simulated the conditions of the structural model test. The shape of the specimen was the same as those used in a low cycle fatigue test described above.

The result (Nf=1930 cycles) for the in-phase uniaxial test which simulated the outer surface condition, agreed well with the structural model test result. However, the out-of-phase test which simulated the condition of the inner surface, showed a longer life (Nf=3850 cycles) than the in-phase test, and thus contradicted the results of the structural model test. (see Fig.10)

<u>Effect of Stress State</u>

The difference in the crack initiation life between the outer and the inner surfaces was considered to be attributed to the effect of stress state. A number of investigators have applied the triaxiality factor shown in Equation 1 to evaluate the low cycle fatigue life under a multiaxial stress state [7-9]. In order to evaluate the crack initiation life at the inner surface subjected to biaxial stress, the multiaxial factor (MF) procedure proposed by Manson and Halford [7] was used as follows:

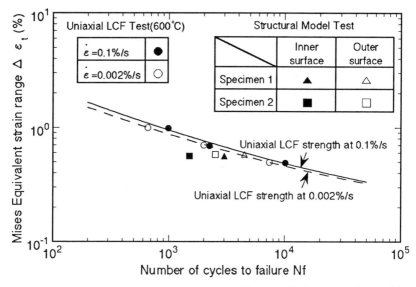

FIG. 9--Comparison of the result of structural model test with the result of uniaxial low-cycle fatigue test by using von Mises equivalent strain range

$$\Delta \varepsilon_{tm} = MF \Delta \varepsilon_p + \Delta \varepsilon_e \qquad (2)$$
$$\text{where } MF = TF \quad \text{for } TF \geq 1 \qquad (3)$$

where $\Delta \varepsilon_{tm}$ is a modified von Mises total strain range. $\Delta \varepsilon_p$ and $\Delta \varepsilon_e$ are the plastic and the elastic components of the von Mises equivalent strain range, respectively, estimated from the results of the FEM inelastic analysis.

The relation between $\Delta \varepsilon_{tm}$ and the crack initiation life at the inner surface is shown in **Fig. 10**, together with the result for the outer surface and the result of uniaxial thermal fatigue test. The result for the inner surface agreed well with the result of the uniaxial out-of-phase thermal fatigue test.

SUMMARY

In a thermal fatigue test of a pipe specimen experiencing a longitudinal cyclic movement of the axial temperature distribution, circumferential cracks and cracks with randomly distributed planes initiated at the outer and the inner surfaces, respectively. The outer and the inner surfaces were essentially uniaxial and biaxial stress states, respectively, and the difference in the crack directions was attributed to the stress state. The outer surface is subjected to an in-phase thermal cycling. The crack initiation life at the outer surface evaluated by a modified von Mises equivalent strain range was in good agreement with the results of the uniaxial in-phase thermal fatigue test and the low-cycle fatigue test at a constant

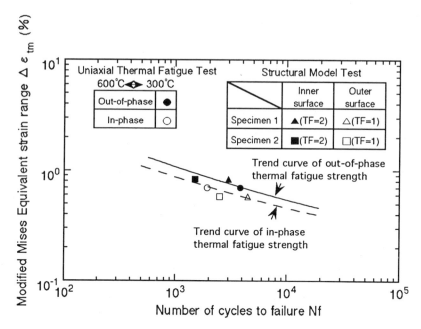

FIG. 10--Comparison of the result of structural model test with the result of uniaxial thermal fatigue test by using the von Mises equivalent strain range modified by a triaxiality factor

temperature, when the life of the structural model was defined by the initiation of the crack of 0.7 to 2 mm length on the surface, which was detected by a dye penetration test. The inner surface was subjected to an out-of-phase thermal cycling. The crack initiation life at the inner surface was shorter than the outer surface, although the von Mises equivalent strain ranges were comparable. The difference was attributed to the difference in stress state. The crack initiation life at the inner surface was evaluated by a modified von Mises equivalent strain range with a triaxiality factor, and the result agreed well with the result of a uniaxial out-of-phase thermal fatigue tests.

ACKNOWLEDGEMENT

This research program was conducted as a part of the project of the Ministry of International Trade (MITI) and Industry, titled "Verification Tests of Fast Breeder Reactor Technology," which has been conducted since 1987 under the sponsorship of MITI in Japan.

REFERENCES

[1]   Igari, T.,Yamauchi, M., Kamishima, Y., and Wada, H., "Experimental Study on

Ratchetting of Cylinder Subjected to Axially Moving Temperature Distribution," Transactions of the Tenth International Conference on Structural Mechanics in Reactor Technology, Vol. E, 1989, pp. 19-24.

[2] Wada, H., Igari, T., and Kitade, S., "Prediction of Plastic Ratchetting of Cylinder Subjected to Axially Moving Temperature Distribution," Transactions of the Tenth International Conference on Structural Mechanics in Reactor Technology, Vol. L, 1989, pp. 201-206.

[3] Igari, T., Kitade, S., Yamauchi, M., Wada, H., and Tashimo, M., "Thermal Ratchetting Behavior of 2-1/4Cr-1Mo Steel," Transactions of the Eleventh International Conference on Structural Mechanics in Reactor Technology, Vol. E, 1991, pp. 227-232.

[4] Kitade, S., Igari, T., Ueta, M., Dousaki, K., Wada, H., Ogawa, K., and Jinbo, M., "Thermal Ratchetting of Cylinder Subjected Short Travel of Axial Temperature Distribution," Creep, Fatigue Evaluation and Leak-Before-Break Assessment PVP, Vol. 266, The American Society of Mechanical Engineers, New York, 1993, pp. 81-88.

[5] Yamauchi, M., Koto, H., Kaguchi, H., and Takahashi, Y., "Evaluation of Creep-Fatigue Design Method by Structural Failure Tests under Thermal Loads," Nuclear Engineering and Desin 153, 1995, pp. 265-273.

[6] Takahashi, Y., "Application of a Two-Surface Plasticity Model for Thermal Ratceting and Failure Life Estimation in Structural Model Tests," Nuclear Engineering and Desin 153, 1995, pp. 245-256.

[7] Manson, S. S. and Halford, R. R., "Multiaxial Low-Cycle Fatigue of Type 304 Stainless Steel," Journal of Engineering Materials and Technology, 1977, pp. 283-285.

[8] Marloff, R. H., Johnson, R. L., and Willson, W. K., "Biaxial Low-Cycle Fatigue of Cr-Mo-V Steel at 1000° C by Use of Triaxiality Factors," Multiaxial Fatigue, ASTM STP 853, K. J. Miller and M. W. Brown, Eds., American Society for Testing and Materials, Philadelphia, 1985, pp. 637-650.

[9] Zamrik, S. Y., Mirdamadi, M., and Davis, D. C., "A Proposed Model for Biaxial Fatigue Analysis Using the Triaxiality Factor Concept," Advances in Multiaxial Fatigue, ASTM STP 1191, D. L. McDowell and R. Ellis, Eds., American Society for Testing and Materials, Philadelphia, 1993, pp. 85-106.

Susan E. Cunningham[1] and Daniel P. DeLuca[2]

ASSESSING CRACK GROWTH BEHAVIOR UNDER CONTINUOUS TEMPERATURE GRADIENTS

REFERENCE: Cunningham, S. E. and DeLuca, D. P., **"Assessing Crack Growth Behavior Under Continuous Temperature Gradients,"** Thermomechanical Fatigue Behavior Of Materials: Second Volume, ASTM STP 1263, M. J. Verrilli and M. G. Castelli, Eds., American Society for Testing and Materials, 1996.

ABSTRACT: A process for automated temperature gradient crack growth testing has been developed and shown to be effective in characterizing crack growth properties which vary as a function of temperature. This method, which controls temperature while correcting for measurable temperature dependent material properties, has been used with DC electric potential drop and compliance techniques for a variety of specimen configurations. Its development has led to varied approaches for temperature dependent crack growth models used in life predictions and has enabled the investigation of temperature dependent fracture modes. Benefits to the aircraft engine industry include better understanding of temperature dependent material behavior, novel temperature-crack growth based design strategies, and reductions in material characterization costs.

KEYWORDS: crack-growth testing, single crystal, crack propagation, fracture, fracture mechanics, temperature gradient, superalloys, fracture modes

INTRODUCTION

Single crystal nickel superalloys are commonly used as turbine airfoil materials in military and commercial gas turbine engines. The unique high temperature capabilities of these alloys provide increased blade performance and durability in the extreme operating environment of the advanced turbine. However, the damage behavior of single crystal materials poses unique problems for life evaluation, particularly in the areas of fracture and fatigue crack propagation.

The fracture characteristics of two single crystal superalloys have been characterized throughout the temperature regime of ambient conditions to 982°C. In the

[1] Mechanics and Materials Specialist, F119 Fan and Compressor Components Center, United Technologies Pratt & Whitney, Government Engines and Space Propulsion, West Palm Beach, FL 33410-9600.

[2] Senior Engineer, Mechanics of Materials Laboratory, United Technologies Pratt & Whitney, Government Engines and Space Propulsion, West Palm Beach, FL 33410-9600.

course of this characterization, these alloys have exhibited various fracture modes, depending on the alloy, operating temperature, stress state and frequency [1-4]. These primary fracture modes, listed as phase decohesion, microscopic octahedral, transprecipitate non-crystallographic (TPNC), macroscopic octahedral, and their variations, are described in detail in [1]. In general, however, the fracture mode is governed by the mobility of dislocations at the crack tip, by the antiphase boundary energy of the precipitate and matrix phases, and by the cross-slip energy. How each of these contribute to describing and predicting what the operative fracture mode is, can be found in [2].

Developing a basic understanding of the fracture behavior of these alloys is a critical aspect of the design process. Accurate life predictions for single crystal alloys require predicting what mode of fracture operates at a particular condition. This is due to the fact that cracks propagate at different rates and along different paths. Take, for example, the case of macroscopic octahedral fracture mode. This fracture mode is one in which cracks propagate along the <111> slip planes, regardless of the Mode I loading orientation. In this mode of fracture, accurate predictions of total life depend on knowing the path of propagation, the mode mixity of the crack, and the crack growth rate along the slip planes.

Temperature plays a key role in the operative fracture mode since thermal energy is tied closely to dislocation mobility. Thus, the conventional practice of characterizing crack growth rate as a function of temperature is confounded by the need to understand the fracture mode as a function of temperature. If the latter characterization can be established, then the ability to predict propagation life for a crack developing in a component during service, is more probable.

Perhaps more important is the role these fracture modes play in analyzing in-service failures. Often in failure analysis, temperatures, steady stresses, and vibratory conditions are not well known and are only assumed or estimated. In these alloys, the characteristics of the fracture features are distinctly different in well defined ranges of operating conditions. By careful interrogation of a post-failure fracture surface, one can distinguish the range of operating conditions by correlating the fracture features to known calibrated ranges of conditions which give those features.

In this paper, we will discuss methods recently developed to assess the role of temperature in determining the operative fracture mode in single crystal materials. This method, though developed specifically for this purpose, can also be used to develop temperature-based design strategies in order to take advantage of crack growth properties which are beneficial at certain conditions and to avoid conditions which are detrimental to behavior. Finally, when this method is used with conventional materials to develop temperature-dependent crack growth models, characterization costs can be reduced due to the fact that fewer tests will be required to characterize the temperature dependence.

AUTOMATED CRACK GROWTH TEST METHODS

Life prediction of aircraft engine components relies to a great extent on empirically determined curves which relate the rate of growth of a crack to the stress and temperature conditions of the cracked part under fatigue loading. Crack growth testing is used extensively to develop characterization curves relating the crack propagation rate per fatigue cycle ($\frac{da}{dN}$) to the stress intensity (K), or stress intensity range (ΔK), at the crack tip:

$$K = \sigma \sqrt{\pi a} \cdot F\left(\frac{a}{W}\right) \tag{1}$$

where σ is the far-field stress applied to the specimen, a is the crack length, and $F\left(\frac{a}{W}\right)$ is a geometry factor which depends, among other things, on the ratio of the crack length to width. The stress intensity range is the difference between the maximum and minimum stress intensity during a fatigue cycle:

$$\Delta K = K(\sigma_{max}, a) - K(\sigma_{min}, a) \tag{2}$$

Conventionally, crack growth testing is performed under constant load and temperature. As the crack grows under constant load, however, the stress intensity increases according to equation (1) and characterization data are obtained over a somewhat arbitrary range of stress intensity values since the stress intensity itself is not controlled during the test.

The development of automated closed loop methods for tracking crack growth data and controlling the test conditions has led to many opportunities for assessing crack growth behavior in control methods other than a standard load-control. For example, the K-gradient method first proposed in [5], and further discussed (with application to the direct current-electric potential drop (DC-EPD) method for tracking crack extension) in [6], enables stress-intensity control (either decreasing, increasing, or constant) according to:

$$\Delta K = \Delta K_o e^{C(a-a_o)} \tag{3}$$

where C is the rate of stress intensity range variation, a is the current crack length, a_o is the initial crack length, and $\Delta K_o = K(\sigma_{max}, a_o) - K(\sigma_{min}, a_o)$. At all values of C, the load is controlled via a feedback loop to give a stress intensity factor defined by the above equation. When $C=0$, the stress intensity remains constant for the duration of the test. The DC-EPD and the compliance methods of tracking crack extension have been used effectively with this algorithm. However, for each of these methods to be effective, the temperature must be constant throughout the test. Otherwise, errors in the crack length and stress intensity will result.

The reason for these errors is related to the methods used for crack length monitoring. Both direct current-electric potential drop and compliance methods are extremely sensitive to temperature variations within the specimen, since metal resistivity (for the DC-EPD method) and elastic modulus (for the compliance method) vary significantly with temperature. For example, in the DC-EPD method, the electric potential drop across the crack is monitored and analytically related to crack length. When the resistivity changes, so does the relative potential drop, causing errors in the crack length computation. Even though crack length values can be corrected post-test by linearizing a set of optical crack length measurements with the potential drop measurements, the stress intensity or load shedding control is not accurately maintained since stress intensity is also functionally tied to the crack length.

EXPERIMENTAL METHOD

To experimentally determine the crack growth rate and fracture mode dependence on temperature, we have developed an approach which avoids these inaccuracies by automatically correcting for errors in crack length and stress intensity induced by the temperature fluctuations. The approach starts with the isothermal calibrations for crack length as a function of either the potential drop across the crack or the compliance change. For example, the DC-EPD method under isothermal conditions analytically relates the crack length to the ratio of the potential drop across the crack (PD) to the potential drop at the starting crack length (PD_o). This relationship includes the geometry of the crack, the initial crack length, and the location of the potential drop measurement probes:

$$a = f\left(\frac{PD}{PD_o}, a_o, y\right) \tag{4}$$

An example is the case of a single edge notch crack growth specimen where the crack length is related to the ratio of $\left(\dfrac{PD}{PD_o}\right)$ according to

$$a = \frac{2W}{\pi} \cos^{-1}\left\{ \frac{\cosh\dfrac{\pi\,y}{2W}}{\cosh\left[\dfrac{PD}{PD_o}\cosh^{-1}\left(\cosh\dfrac{\pi\,y}{2W} \Big/ \cos\dfrac{\pi\,a_o}{2W}\right)\right]} \right\} \tag{5}$$

where a_o is the initial crack length, PD is the measured potential drop across a length $2y$, PD_o is the initial potential drop measured at the crack length a_o, and W represents the width of the specimen [5].

To extend this to the case for varying temperature, we automatically correct for thermally-induced resistivity changes in-situ in the following way. Resistivity change from an initial temperature is determined by measuring the instantaneous change in potential drop at a temperature, $PD_o(T)$, and normalizing this value by the measured potential drop at a known temperature, $PD_o(T_o)$.[3] Values of $\dfrac{PD_o(T)}{PD_o(T_o)}$ vs. $\dfrac{T}{T_o}$ are then plotted to determine the best functional form and fit. Once this relationship is established, $PD_o(T)$ is substituted for PD_o and included in the crack growth monitoring and control algorithm:

$$PD_o(T) = PD_o(T_o) \cdot f\left(\frac{T}{T_o}\right) \qquad (6)$$

A beauty of this approach is that the resistivity depends only on the material and temperature and not on the specimen geometry, crack length, or test load. *Thus, once the temperature dependence is characterized for a particular material, the relationship between* $\dfrac{PD_o(T)}{PD_o(T_o)}$ *and* $\dfrac{T}{T_o}$ *can be used for all subsequent tests of the same material within the characterized temperature range, regardless of the specimen configuration.*

When the corrections for non-isothermal test conditions have been made to either the compliance or potential drop methods, any desired correlation relating temperature to the crack length can be used as a means to control and change the temperature during the test. The process relies on a capable data acquisition and control system, a temperature feedback and control processor, a load feedback and control processor, and a computer interface linking the load and temperature controller to the data acquisition system. As the crack extends, a signal with the revised temperature is sent to the temperature controller which modifies the furnace temperature. A linear crack length-temperature relationship was selected and used for the initial tests described later. Using the form currently encoded as an example, the user selects the intercept (T_1) and slope $\left(\dfrac{dT}{da}\right)$ of the line

$$T = T_1 + \left(\frac{dT}{da}\right)a \qquad (7)$$

to correspond with the desired starting and ending points of the test.

TEMPERATURE-INDUCED FRACTURE MODE CHANGES IN SINGLE CRYSTAL NICKEL ALLOYS

The first step in using the Automated Temperature Gradient Test method with the DC-EPD method is empirically determining the variation of effective resistivity with

---

[3] For the compliance case of crack extension monitoring, the analog to the resistivity change is the change in elastic modulus with temperature. When used with the compliance technique, $PD_o(T_o)$ and $PD_o(T)$ are exchanged with the corresponding initial and instantaneous compliance expressions, $S_o(T_o)$ and $S_o(T)$.

temperature for these materials. Measured variations in PD_o with temperature are shown in Figure 1 for several nickel superalloys, over the temperature range of 50°C to 982°C. The data in Figure 1 represent a total of ten pinned-end single edge notch (SEN) specimens (gage section dimensions: 25.4 × 12.7 × 1.27 mm) and the results show excellent reproducibility of the ratio $\dfrac{PD_o(T)}{PD_o(T_o)}$. Careful control of the test setup and very small increments in temperature minimized any errors due to thermally-induced voltage fluctuations. Excellent reproducibility of the data confirmed the accuracy of the experimental setup. The data fit well to the second order polynomial:[4]

$$\frac{PD_o(T)}{PD_o(T_o)} = 0.86196 + 0.22431\left(\frac{T}{T_o}\right) - 0.086639\left(\frac{T}{T_o}\right)^2 \tag{8}$$

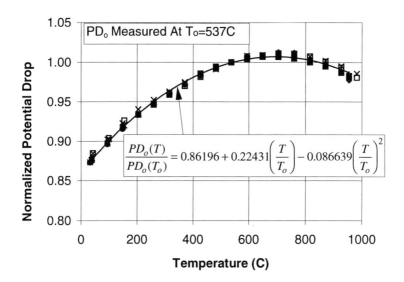

FIG. 1 -- Normalized potential drop variation in three nickel superalloys from room temperature to 981°C. Data from ten specimens are shown.

It is noted that the parabolic behavior of the temperature-potential drop curve reflects the change in resistivity of the *system*. The system response is controlled

---

[4] It is noted that the known temperature normalizing the potential drop, T_o, does not necessarily need to be the *initial* temperature. In fact, depending on the shape of the *PD* vs. *T* curve, it might be prudent to select a temperature in the middle or at the end of the temperature range used in the test.

primarily by the alloy which also exhibits a parabolic resistivity response to temperature. The decrease in normalized potential drop above approximately 700 °C may be caused by other contributors (e.g. oxidation of the lead wires and/or the lead wire attachment, etc.) and reinforces the need to perform the PD-temperature calibrations over the entire temperature range of interest and for the specific experimental setup used.

A summary of conditions for the two variable temperature tests completed in this study is shown in Table 1. Both SEN specimens were tested at a stress ratio $R=0.5$, at 10 Hz in laboratory air. The stress intensity was held constant during each test at values which were carefully selected to give crack growth in regions of primary interest and to avoid any stress intensity driven changes in fracture mode.

TABLE 1--Temperature gradient test matrix.

Specimen	Material	Initial Temperature (^oC)	Final Temperature (^oC)	K_{max} ($MPa\sqrt{m}$)
A	PWA 1480	30	300	7.0
B	PWA 1484	590	65	15.4

The goals of the temperature gradient tests were different, depending on the material and the temperature range of interest. PWA 1480 exhibits a change in fracture mode at room temperature from phase decohesion (low stress intensity) to microscopic octahedral (higher stress intensity). The fracture mode change is associated with a crack growth rate that is constant with stress intensity within the transition region. The goal of the PWA 1480 test was to demonstrate a dependence of the growth rate and fracture mode on the temperature. This was accomplished by initiating the test at a crack growth rate in the phase decohesion mode and incrementally increasing the temperature.

The results of this test are presented in Figure 2 as crack growth rate as a function of both the crack length and the temperature (note the temperature variation is linear with the crack length). Also plotted in this figure is the resultant stress intensity at the crack tip plotted as a function of the crack length. These data are determined post-test by correcting the EPD crack length values with optical crack length measurements performed intermittently during the test. Throughout the entire test, ΔK varied by only 3% from the nominal value of $3.5 MPa\sqrt{m}$.

Consistent with the test objectives, the initial fracture mode observed post-test was phase decohesion. At approximately 100°C, the low energy fracture mode abruptly changed to microscopic octahedral fracture. *Thus, the same mode transition that was produced by holding temperature constant and decreasing the stress intensity was produced by holding the stress intensity constant and increasing the temperature.* This transition to octahedral shearing of precipitates was accompanied by a seven-fold drop in crack growth rate.

Between 100°C and 150°C, the condition was twice briefly reversed as decohesion again became the operative mode of crack growth. However each time octahedral fracture resumed, the crack growth rate abruptly dropped. At 280°C, crack propagation spontaneously arrested. The fracture mode at this point was recognized as an ancillary decohesion-like mechanism which is typically found in transition regions. The arrest of the crack at 280°C confirms the existence of a limiting crack growth condition in this alloy, as described by the parameter ΔK_{lim} [2]. Figure 3 illustrates the different fracture modes observed at critical temperatures in this PWA 1480 temperature-gradient test.

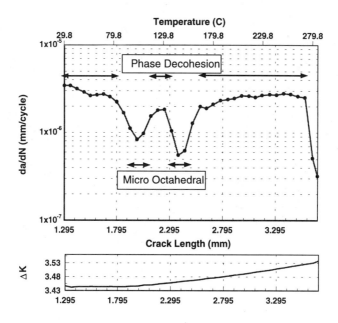

FIG. 2 -- The effect of temperature on crack growth rate in PWA 1480 at low stress intensity from room temperature to 275°C. A fracture mode transition corresponds with the marked change in crack growth rate at approximately 100°C. Crack arrest occurred at approximately 275°C.

A second temperature gradient test was performed on PWA 1484 starting at 590°C. The goal of this test was to identify any temperature-induced fracture mode transitions at intermediate temperatures. This was accomplished by decreasing the

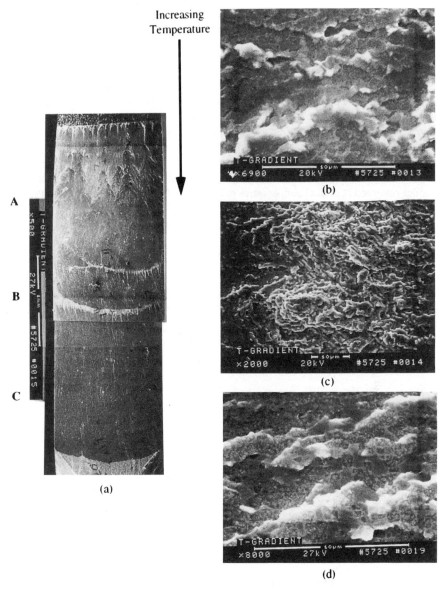

FIG. 3 -- Fracture surface features of PWA 1480 specimen. (a) Macroscopic view show
load shedding during the precrack stage produced the transition from
microscopic octahedral fracture to decohesion at point A. Abrupt transitions
indicated at point B produced a factor of seven decrease in the crack growth
rate. Crack arrest occurred at point C. (b) Detailed microscopy of the phase
decohesion. (c) Transition to microscopic octahedral fracture at 100°C. (d) T
fracture mode at crack arrest is similar to ancillary decohesion.

temperature at a relatively low stress intensity value until the test naturally terminated by crack arrest or until room temperature was reached. Throughout the temperature range of 590°C to 190°C, the stress intensity varied by less than 4% of the nominal ΔK.

Results of the PWA 1484 test are shown in Figure 4. In this case, the crack growth rate was nearly constant with temperature from 490°C to 390°C. This observation reinforces previously obtained results which showed very little difference in crack growth behavior in this alloy between 427°C and 595°C. At approximately 290°C, a fracture mode transition from monoplanar TPNC to microscopic octahedral occurred. This transition was associated with a rapid decrease in crack growth rate. The test terminated in global octahedral fracture at approximately 100°C. Figure 5 illustrates the fracture modes observed on this test specimen.

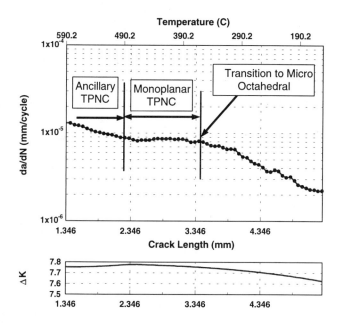

FIG. 4 -- The effect of temperature on crack growth rate in PWA 1484 at low stress intensity from 590°C to 160°C. Crack growth rate is nearly constant with temperature from 490°C to 390°C. At approximately 290°C, a fracture mode transition from TPNC to microscopic octahedral corresponds with a continual decrease in crack growth rate to 160°C. At 160°C, the fracture mode transitioned to macroscopic octahedral.

Decreasing
Temperature

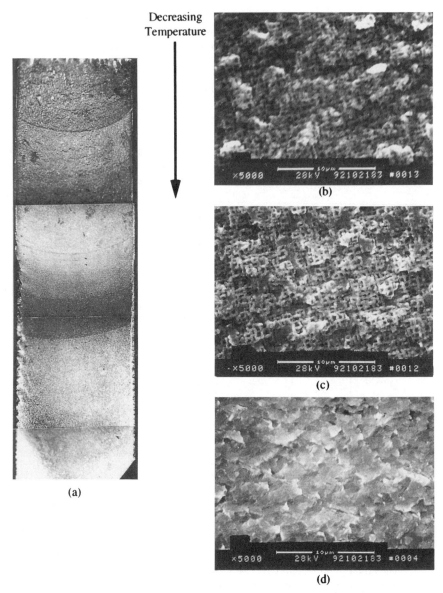

(a)

(b)

(c)

(d)

FIG. 5 -- The fracture surface features of PWA 1484 specimen. (a) Monoplanar to ancillary fracture transition (63×). (b) Heavily oxidized ancillary fracture at 475°C. (c) Ancillary TPNC fracture is observed at 375°C in the constant crack growth region. (d) Microscopic octahedral cracking is exhibited below 145°C.

IMPACT ON LIFE PREDICTION AND CHARACTERIZATION

A new approach to crack growth testing, called the temperature-gradient test, has been presented. This method was used successfully to determine how temperature affects the fracture mode and transitions between fracture modes in two different single crystal nickel alloys. When nonlinear behavior (temperature dependent fracture mode transitions are nonlinear in nature when they affect the crack growth rate) is encountered, the temperature-gradient test is the preferred method of identifying transition temperatures. This method enables explicit identification of the transition temperature with only one test, saving on characterization costs and improving our understanding of material behavior.

Prior to the adoption of the temperature-gradient crack growth test, numerous tests were performed in attempts to identify the fracture mode transition temperatures in single crystal nickel alloys. For example, repeat tests using constant temperature conventional methods succeeded only in bracketing the transition temperature. A single test performed in the manner described in this manuscript provided the exact temperature of fracture mode transition.

The usefulness of this new method goes beyond identifying thermal properties of single crystal fracture modes, however. It can be used as a replacement for methods currently proposed to account for voltage fluctuations induced by temperature variations. Currently, two approaches are recommended to correct for these voltage fluctuations:

1) Instrumenting a "reference" specimen without a crack and placing it in close proximity to the specimen used for testing so that any thermal fluctuations experienced by the cracked specimen will be experienced (and consequently recorded) by the uncracked specimen.

2) Placing "reference" potential drop leads far enough away from the crack (in a single cracked specimen) such that a growing crack has no effect on the potential drop ("far-field" potential drops are recorded).

Both methods use the reference potential drop readings to normalize the potential drop obtained across the crack. The problem with the first case is that often furnaces used in testing are only large enough to accomodate one specimen, not two. The second approach suffers from the fact that the distance required to obtain constant far-field potential drop measurements is so large that it extends well beyond the gage section of the specimen and beyond the stable and well calibrated hot-zone in the furnace.

This temperature-gradient crack growth test method also has practical and potential cost reduction implications for characterization programs for crack growth life prediction. A typical fracture characterization would require four or more temperatures per stress ratio and still may not yield information about nonlinear temperature response. As a result, life prediction estimates could be non-conservatively inaccurate. This is illustrated in Figure 6. Two examples of a linear temperature dependence are shown as representative models of the temperature-gradient test data illustrated previously in Figure

4. Curve 1 could be assumed if two standard constant temperature tests had been performed at 150°C and 300°C, while curve 2 represents linear temperature behavior if two constant temperature tests had been performed at 150°C and 600°C.

The crack growth dependence of each curve was integrated over the duration of the test, and compared with the actual number of cycles to failure. The result is shown in Table 2. Clearly, using the representation of Curve 2 in design calculations would result in life estimates which are non-conservative. However, if temperature values for isothermal crack growth characterization are not carefully selected, this representation could be considered viable. The temperature gradient test method enables prudent selection of temperatures for isothermal tests and permits the development of nonlinear crack growth rate-temperature relationships.

TABLE 2--<u>Model predictions may be inaccurate if linear behavior is assumed.</u>

Prediction Basis	Predicted Cycles to Failure	Predicted/Actual
Curve 1	552,978	1.23
Curve 2	824,028	0.82

The temperature gradient test can reduce crack growth characterization costs. Rather than arbitrarily selecting temperatures at which to perform material characterizations, this one test can describe the temperature response of growth rate at a specific stress intensity range. This could be especially useful for temperature regions where the crack growth rate changes very little. If identified early in the development of the characterization requirements, only one test at a temperature within this range need be run, thus eliminating the need for tests which would yield duplicate results.

This new approach to characterization could be especially useful when characterizing the threshold stress intensity as a function of the temperature. With ΔK set at a value near ΔK_{th}, the near-threshold growth rate as a function of temperature can be obtained over a broad temperature range. Further, if a second temperature gradient test is conducted at a greater stress intensity range, then the growth rate dependence on temperature variation would be bounded. By conducting a third test at an intermediate stress intensity, the dependence could more completely described.

The temperature-gradient test also has implications when considering the effects of environment on fracture mode. When tested in high pressure hydrogen, PWA 1480 exhibits the phase decohesion fracture mode at stress intensity values much higher than observed during air testing [6]. The effect is a temperature dependent increase (up to ten-fold) in growth rate. Isothermal fatigue crack growth tests in 34.5 MPa hydrogen at elevated temperatures are expensive so close bracketing of the transition temperature by performing constant temperature tests is not practical. Plans to adopt the temperature-gradient technique to in-situ high pressure hydrogen testing are in place. This testing is an integral part of the materials characterization effort supporting the NASA SSME alternate turbopump design program.

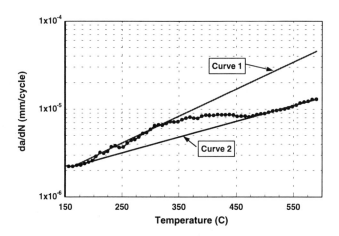

FIG. 6 -- The effect of temperature on crack growth rate in PWA 1484 at low stress intensity from 590°C to 160°C. In addition to the data shown in FIG. 4, the plot illustrates two possible linear interpolative temperature vs. crack growth rate models.

SUMMARY

A new technique for crack growth testing has been described and shown to be effective in characterizing crack growth properties which vary as a function of the temperature. The technique is used with the DC-EPD method of crack length monitoring to vary temperature with crack length according to a user-supplied function and can be used with any specimen geometry appropriate for EPD methods. This technique permits identification of unique characteristics which are dependent on temperature. The method relies on empirically determining the variation of potential drop as a function of temperature and subsequently incorporating this relationship in the PD-crack length calibration. The technique can also be applied to the compliance method for crack length monitoring by using a similar means to vary the initial measured compliance of the specimen.

The temperature gradient test method was used to assess the role of temperature in fracture behavior of single crystal nickel superalloys. Fracture mode transitions previously observed to be caused by changes in stress intensity at isothermal conditions were induced under constant stress intensity conditions by varying the temperature. Understanding this phenomenon will help to develop better fracture models and design capability for turbine engine components using this class of alloys.

The temperature gradient crack growth test method has implications beyond addressing fracture mode dependence on temperature in single crystal alloys. The method can be used to determine temperature regions of accelerated or decelerated propagation rates in conventional materials. It can be used to screen regions where the crack growth may remain relatively constant, so as minimize the number of specimens required for a full temperature dependence characterization. Further, it enables a more accurate representation of the temperature dependence in crack growth modeling, thereby enabling better life prediction estimates.

ACKNOWLEDGMENTS

This work was performed under US Air Force Contract F33657-88-C-0129. Dr. Ted Nicholas of USAF Wright Laboratory - Materials Directorate is acknowledged for his continuing support of tgrowth test, numerous tests were performed in attempts to id contributed valuable insight into fracture mode transition behavior. Mark Poormon and Donald Ogden of Pratt & Whitney Fracture Mechanics Laboratory are gratefully acknowledged for their assistance in developing the temperature gradient testing capability. SEM fractography was ably provided by Fawn Cobia, Mark Spisiak, Gary Panse and Bob Rea.

REFERENCES

[1]   Telesman, J. and Ghosn, L. J., "The Unusual Near-Threshold FCG Behavior of a Single Crystal Superalloy and the Resolved Shear Stress as the Crack Driving Force," Engineering Fracture Mechanics, Vol. 34, 1989, pp 1183-1196.

[2]   Cunningham, S. E., DeLuca, D. P. and Haake, F. K., Crack Growth and Life Prediction in Single Crystal Nickel Superalloys, Volume I, WL-TR-94-4089, August 1994.

[3]   Cunningham, S. E., DeLuca, D. P., Hindle III, E. H., Sheldon, J. W. and Haake, F. K., Crack Growth and Life Prediction in Single Crystal Nickel Superalloys, Volume II, WL-TR-94-4089, August 1994.

[4] Telesman, J. and Ghosn, L. J., <u>Crack Growth and Life Prediction in Single Crystal Nickel Superalloys, Volume III</u>, WL-TR-94-4090, August 1994.

[5] Saxena A., Hudak, S. J., Donald, J. K., and Schmidt, D. W., "Computer-Controlled Decreasing Stress intensity Technique for Low Rate Fatigue Crack Growth Testing," <u>Journal of Testing and Evaluation</u>, JTEVA, Vol. 6, No. 3, May 1978, pp 167-174.

[6] Donald, J. K. and Ruschau, J., "Direct Current Potential Difference Fatigue Crack Growth Measurement Techniques," <u>Fatigue Crack Measurement: Techniques Applications</u>, K. J. Marsh, R. A. Smith, and R. O. Ritchie, eds., Engineering Materials Advisory Service, 1991, pp 11-37.

[7] Johnson, H. H., "Calibrating the Electric Potential Method for Studying Slow Crack Growth," <u>Materials Research and Standards</u>, September 1965, pp 442-445.

[8] DeLuca, D. P., and Cowles, B. A., "Fatigue and Fracture of Single Crystal Nickel in High Pressure Hydrogen," <u>Hydrogen Effects on Material Behavior</u>, N. R. Moody and A. W. Thompson, eds., Warrendale, PA, 1989, pp 603-613.

Steven M. Arnold,[1] Atef F. Saleeb,[2] and Michael G. Castelli[3]

A FULLY ASSOCIATIVE, NONISOTHERMAL, NONLINEAR KINEMATIC, UNIFIED VISCOPLASTIC MODEL FOR TITANIUM ALLOYS

REFERENCE: Arnold, S. M., Saleeb, A. F., and Castelli, M. G., "**A Fully Associative, Nonisothermal, Nonlinear Kinematic, Unified Viscoplastic Model for Titanium Alloys,**" Thermomechanical Fatigue Behavior of Materials: Second Volume, ASTM STP 1263, Michael J. Verrilli and Michael G. Castelli, Eds., American Society for Testing and Materials, 1996.

ABSTRACT: Specific forms for both the Gibb's and complementary dissipation potentials are chosen such that a complete (i.e., fully associative) potential based multiaxial, nonisothermal unified viscoplastic model is obtained. This model possess one tensorial internal state variable (that is, associated with dislocation substructure) and an evolutionary law that has nonlinear kinematic hardening and both thermal and strain induced recovery mechanisms. A unique aspect of the present model is the inclusion of nonlinear hardening through the use of a compliance operator, derived from the Gibb's potential, in the evolution law for the back stress. This nonlinear tensorial operator is significant in that it allows both the flow and evolutionary laws to be fully associative (and therefore easily integrated), greatly influences the multiaxial response under non-proportional loading paths, and in the case of nonisothermal histories, introduces an instantaneous thermal softening mechanism proportional to the rate of change in temperature. In addition to this nonlinear compliance operator, a new consistent, potential preserving, internal strain unloading criterion has been introduced to prevent abnormalities in the predicted stress-strain curves, which are present with nonlinear hardening formulations, during unloading and reversed loading of the external variables. The specific model proposed is characterized for a representative titanium alloy commonly used as the matrix material in SiC fiber reinforced composites, i.e., *TIMETAL* 21S. Verification of the proposed model is shown using "specialized" non-standard isothermal and thermomechanical deformation tests.

KEYWORDS: viscoplasticity, nonlinear hardening, *TIMETAL* 21S, nonisothermal, deformation, multiaxial, correlations, predictions

[1]Research Engineer, Structural Fatigue Branch, National Aeronautics and Space Administration, Lewis Research Center, Cleveland OH, 44135

[2]Professor, Department of Civil Engineering, University of Akron, Akron, OH 44325

[3]Research Engineer, Fatigue & Thermal Group, NYMA, Inc., NASA Lewis Research Center Group, Brook Park, OH 44142

146

NOMENCLATURE

Invariants

Ω	complementary dissipation potential
Φ	Gibb's complementary potential
F	Bingham-Prager threshold function
G	normalized second invariant function
J_2	second invariant of effective deviatoric stress tensor
I_2	second invariant of internal deviatoric stress tensor
H_1	invariant material function

Stresses

σ_{ij}	Cauchy stress tensor
S_{ij}	deviatoric stress tensor
Σ_{ij}	effective deviatoric stress tensor
α_γ	internal state variables (stress-like)
α_{ij}	internal (or back) stress tensor
a_{ij}	deviatoric internal stress tensor
κ, κ_o	drag stress and reference drag stress
$Y(T)$	yield stress

Strains

$e_{ij}, \epsilon_{ij}^I, \epsilon_{ij}^R, \epsilon_{ij}^T$	total, inelastic, reversable, and thermal strain tensors
$\mathcal{A}_\xi$	conjugate internal state variables (displacement-like)
$\mathcal{A}_{ij}$	internal strain tensor

Material Parameters

$\theta_\xi(T), \theta_{rs}(T)$	dynamic thermal recovery operator
$M_{ij}(T)$	thermal expansion tensor
$C_{ijkl}(T), E_{ijkl}(T),$	elastic compliance and stiffness tensor, respectively
$B_0(T), B_1, R_\alpha$	hardening and thermal recovery material parameters
$\mu(T)$	material parameter associated with the viscosity of the material
$\beta(T)$	denotes the extent of strain induced recovery
n, p, q	material exponents
$g(G), f(F), z(T)$	material functions
$\bar{E}(\sigma_{ij}, T), \bar{H}(\alpha_\gamma, T)$	material functions
$\eta, \eta_{\text{tan}}, \hat{E}$	mean coefficient of thermal expansion(CTE), instanteous CTE, Young's modulus, respectively

Miscellaneous

T, T_0	current and reference temperature, respectively
$Q_{ijkl}, (T), L_{ijkl} (T)$	internal compliance and stiffness operators, respectively
δ_{ij}	Kronecker delta function
$\langle \rangle$	Macauley bracket
Hv[ ]	Heaviside unit function
$(\cdot)$	time derivative (or rate) notation

INTRODUCTION

A number of Titanium Matrix Composite (TMC) systems are currently being researched and evaluated for high temperature air frame and propulsion system applications. As a result, numerous computational methodologies for predicting both deformation and life for this class of materials are under development. An integral part of these methodologies is an accurate and computationally efficient constitutive model for the metallic matrix constituent. Furthermore, because of the proposed elevated operation temperatures for which these systems are designed, the required constitutive models must account for both time-dependent and time-independent deformations. To accomplish this we will employ a recently developed complete potential based framework [1] utilizing internal state variables which was put forth for the derivation of reversible and irreversible constitutive equations. This framework, and consequently the resulting constitutive model, is termed complete because the existence of the total (integrated) form of the Gibb's complementary free energy and complementary dissipation potentials are assumed *a priori*. The specific forms selected here for both the Gibb's and complementary dissipation potentials result in a fully associative, multiaxial, nonisothermal, unified viscoplastic model with nonlinear kinematic hardening. Thus this model constitutes one of many models in the **GVIPS** (Generalized VIscoplasticity with Potential Structure) class of inelastic constitutive equations which can be constructed using the generalized framework of Arnold and Saleeb [1].

The particular unified **GVIPS** model of interest in this study possesses one tensorial internal state variable (i.e., the back or internal stress) that is associated with dislocation substructure and an evolutionary law that has nonlinear kinematic hardening and both thermal and strain induced recovery mechanisms. A unique aspect of the present model is the inclusion of nonlinear hardening through the use of a compliance operator (derived from the Gibb's potential) in the evolution law for the back stress. This nonlinear tensorial operator is significant in that it allows both the flow and evolutionary laws to be fully associative (and therefore easily integrated) [2], greatly influences the multiaxial response under non-proportional loading paths [1],[3],[4], and in the case of nonisothermal histories, introduces an instantaneous thermal softening mechanism proportional to the rate of change in temperature. In addition to this nonlinear compliance operator, the new [5] consistent, potential preserving, internal unloading criterion has been utilized to prevent abnormalities in the predicted stress-strain curves, which are present with nonlinear hardening formulations, during unloading and reversed loading.

The primary objective of the present study is to specify material functions and characterize the associated material parameters for the nonisothermal extension of the previously proposed kinematic, isothermal **GVIPS** model [5] for TIMETAL 21S[1], an advanced titanium-based matrix commonly used in TMCs. Although both long and short term behavior is important, capturing the short term (or transient) behavior and rate sensitivity of the material is of primary importance given that the applications of interest are primarily those involving material processing and structural issues in propulsion systems.

The paper begins by briefly summarizing the extension of the complete potential structure to the nonisothermal domain, followed by a multiaxial statement of the newly proposed nonisothermal **GVIPS** model. A discussion regarding the characterization of the proposed model is then followed by numerous results illustrating the predictive capabilities of the model.

COMPLETE POTENTIAL STRUCTURE: NONISOTHERMAL

Here, the basic thermodynamic framework put forth by Arnold and Saleeb [1] is extended to the nonisothermal domain. Expressions for the Gibb's thermodynamic and the complementary dissipation potential functions are assumed in terms of a number of state and internal variables characterizing the changing internal structure of the material. For conciseness, the discussion is limited to a case involving i) small deformations (in which the initial state is assumed to be stress free), ii) an initially isotropic material, and iii) the specialized (decoupled) potential framework discussed in [1]. A Cartesian coordinate reference frame and index notation are utilized (repeated Roman subscripts imply summation).

Given the Gibb's potential in the following form

$$\Phi = \Phi(\sigma_{ij}, \alpha_\gamma, T, \epsilon_{ij}^I) \tag{1}$$

and assuming *a priori* that the inelastic strain is an **independent parameter** (and not an internal state variable), for example

$$\Phi = \bar{E}(\sigma_{ij}, T) - \sigma_{ij}\epsilon_{ij}^I + \bar{H}(\alpha_\gamma, T) - \frac{\sigma_{kk}}{3}\eta(T - T_o), \tag{2}$$

an expression for the total strain rate can be obtained by differentiating, that is,

---

[1]TIMETAL 21S is a registered trademark of TIMET, Titanium Metals Corporation, Toronto, OH.

$$\dot{e}_{ij} = \frac{d}{dt}(\frac{-\partial\Phi}{\partial\sigma_{ij}}) = C_{ijrs}\dot{\sigma}_{rs} + \dot{\epsilon}_{ij}^{I} + \left[\frac{-\partial^2\bar{E}(\sigma_{ij})}{\partial\sigma_{ij}\partial T} + \frac{\delta_{ij}}{3}\left(\eta + \frac{\partial\eta}{\partial T}\Delta T\right)\right]\dot{T} \qquad (3)$$

as well as the rate of change of the conjugate internal variables (A_ξ),

$$\dot{A}_\xi = \frac{d}{dt}(\frac{-\partial\Phi}{\partial\alpha_\xi}) = Q_{\xi\gamma}\dot{\alpha}_\gamma + \theta_\xi\dot{T} \qquad (4)$$

where

$$C_{ijrs} = \frac{-\partial^2\Phi}{\partial\sigma_{ij}\partial\sigma_s} = \frac{-\partial^2\bar{E}(\sigma_{ij},T)}{\partial\sigma_{ij}\partial\sigma_{rs}} \qquad (5)$$

and

$$Q_{\xi\gamma} = \frac{-\partial^2\Phi}{\partial\alpha_\xi\partial\alpha_\gamma} = \frac{-\partial^2\bar{H}(\alpha_\gamma,T)}{\partial\alpha_\xi\partial\alpha_\gamma} \qquad (6)$$

are the external and internal compliance operators, respectively, and

$$\theta_\xi = \frac{-\partial^2\Phi}{\partial\alpha_\xi\partial T} = \frac{-\partial^2\bar{H}(\alpha_\gamma,T)}{\partial\alpha_\xi\partial T} \qquad (7)$$

is the change in the conjugate internal variable (A_ξ) with temperature. Note the three terms in equation (3) may then be identified from left to right as the reversible, irreversible (inelastic), and thermal expansion components of the total strain rate, respectively. Thus,

$$\dot{e}_{ij} = \dot{\epsilon}_{ij}^{R} + \dot{\epsilon}_{ij}^{I} + \dot{\epsilon}_{ij}^{T} \qquad (8)$$

where

$$\dot{\epsilon}_{ij}^{R} = C_{ijrs}\dot{\sigma}_{rs} \qquad (9)$$

and

$$\dot{\epsilon}_{ij}^{T} = M_{ij}\dot{T} \qquad (10)$$

with

$$M_{ij} = \frac{-\partial^2\Phi}{\partial\sigma_{ij}\partial T} = \left[\frac{-\partial^2\bar{E}(\sigma_{ij})}{\partial\sigma_{ij}\partial T} + \frac{\delta_{ij}}{3}\eta_{\text{tan}}\right] \qquad (11)$$

and

$$\eta_{\text{tan}} = \eta + \frac{\partial\eta}{\partial T}\Delta T$$

denoting the instantaneous coefficient of thermal expansion, and $\dot{\epsilon}_{ij}^{I}$ (the inelastic strain rate) is defined separately using the concept of a complementary dissipation potential $\Omega(\sigma_{ij},\alpha_\gamma,T)$.

Given

$$\Omega = \Omega(\sigma_{ij},\alpha_\gamma,T) \qquad (12)$$

and using the Clausius-Duhem inequality; the flow law becomes

$$\dot{\epsilon}_{ij}^{I} = \frac{\partial \Omega}{\partial \sigma_{ij}}$$

(13)

and the evolutionary laws for the thermodynamic conjugate internal state variables are:

$$\dot{\mathcal{A}}_\gamma = -\frac{\partial \Omega}{\partial \alpha_\gamma}$$

(14)

Utilizing equation (4) the internal constitutive rate equations for the internal state variables are obtained,

$$\dot{\alpha}_\gamma = L_{\gamma\xi} \left[\dot{\mathcal{A}}_\xi - \theta_\xi \dot{T} \right]$$

(15)

where

$$L_{\gamma\xi} = [Q_{\gamma\xi}]^{-1} = \left[\frac{-\partial^2 \Phi}{\partial \alpha_\xi \partial \alpha_\gamma} \right]^{-1}$$

(16)

Thus, equations (13) and (14) represent the flow and evolutionary laws, for an assumed $\Omega = \Omega(\sigma_{ij}, \alpha_\gamma, T)$, and equation (15) the internal constitutive rate equations, given a Gibb's potential Φ, wherein both potentials are directly linked through the internal state variables α_γ.

VISCOPLASTIC CONSTUTIVE MODEL

A complete multiaxial statement of a **GVIPS** model can be derived by using the above framework, given a specific form for both the Gibb's potential, Φ, and the complementary dissipation potential, Ω. Form invariance (objectivity) of these potentials, and material symmetry considerations requires that they depend only on certain invariants of their respective tensorial arguments (*i.e.*, an integrity basis [6]). In the spirit of von Mises and because of the deviatoric nature of inelastic deformation, only the quadratic invariant will be considered at this time in specifying the dissipation potential. Similarly, only the linear elastic strain energy contribution will be considered in specifying the Gibb's potential, with the internal state groupings being functions of the respective quadratic invariants. Finally, although equation (12) indicates that an unlimited number of internal state variables can be specified, here our attention will be restricted to a **GVIPS** model with a single *independent*, evolving, internal state variable. This internal state variable, a_{ij}, is taken to be a second order symmetric traceless tensor that represents the internal (or back) stress associated with dislocation substructure. Two additional scalar state variables associated with dislocation density are considered. One representing the drag strength

(κ) is taken to be temperature dependent and yet non-evolving with respect to plastic work; whereas, the other, a yield stress (Y), is uniquely assumed to implicitly evolve with internal stress over a specified temperature range.

Consequently, the Gibb's potential may be written as

$$\Phi = -\frac{1}{2}C_{rskl}(T)\sigma_{rs}\sigma_{kl} - \sigma_{ij}\epsilon_{ij}^{I} + H_1(G,T) - \frac{\sigma_{kk}}{3}\eta(T-T_o) \tag{17}$$

and the complementary dissipation potential as

$$\Omega = \mu(T)\int f(F(T))dF + R_\alpha(T)B_0(T)\int g(G)dG \tag{18}$$

where

$$F(T) = \left\langle \frac{\sqrt{J_2}}{\kappa(T)} - Y(T) \right\rangle \tag{19}$$

$$Y(T) = \left\langle 1 - \beta(T)\sqrt{G} \right\rangle \tag{20}$$

$$G = \frac{I_2}{\kappa_0^2} \tag{21}$$

$$H_1(G) = -B_0(T)(G + B_1 G^p) \tag{22}$$

$$I_2 = \frac{3}{2}a_{ij}a_{ij} \tag{23}$$

$$J_2 = \frac{3}{2}\Sigma_{ij}\Sigma_{ij} \tag{24}$$

$$\Sigma_{ij} = S_{ij} - a_{ij} \tag{25}$$

$$S_{ij} = \sigma_{ij} - \frac{1}{3}\sigma_{kk}\delta_{ij} \tag{26}$$

$$a_{ij} = \alpha_{ij} - \frac{1}{3}\alpha_{kk}\delta_{ij}. \tag{27}$$

and $\kappa_0 = \kappa(T_0)$ where T_0 is the reference temperature.

Note that in the preceding expression for the dissipation potential, the stress dependence, both external and internal, enters through the scalar functions F and G in the form of effective (Σ_{ij}) and internal (a_{ij}) deviatoric stresses, respectively. Furthermore, the function F acts like a threshold surface, since when $F < 0$, no inelastic strain can occur. Clearly, this threshold value is dictated by the magnitude of both the drag strength (κ) and yield stress (Y). A unique aspect of this model is that the internal variable representing the yield stress is specifically taken as a special scaled function of the back stress and drag strength. Consequently, this allows for 1) the model to possess features of a model with three internal variables yet without any additional computational cost, and 2) the presence of an induced strain recovery term (as opposed to the common 'ad-hoc' introduction of such terms) in the evolution of the back strain (i.e., the associated conjugate variable, $\mathcal{A}_\gamma$) at lower temperatures. It is important to realize that the product κY constitutes the radius of the initial threshold surface, thereby dictating from physical arguments, that both Y and κ be always positive valued. Furthermore, given the specified form of Y in equation (20), it is clear that the material parameter ratio (β/κ) will dictate the limit value (i.e., when $Y=0$) of the internal stress (α_{ij}), or the cut-off limit for dynamic recovery.

By selecting the preceding scalar functions, a general yet *complete* potential-based model, with *associated* flow and evolutionary laws, can be constructed. The second invariants , J_2 and I_2 , are also scaled for tension. These invariants could just as easily have been scaled for shear by replacing the coefficient $3/2$ with $1/2$, and modifying the definition of the magnitude of the inelastic strain rate that follows. Also, it should be stated that the linear forms of F and Y in equations (19) and (20) were chosen in order to allow algebraic manipulation and analytical solution of the resulting expressions (e.g., inversion of the flow law), so as to ease the characterization stage of the model, as discussed in [5].

Taking the appropriate derivatives of both Φ and Ω as indicated in equations (3) through (16), one obtains the multiaxial nonisothermal specification particular to the present constitutive model. Here the decomposition of the total strain rate is that of equation (8), where the reversible mechanical and thermal strain rate is given by equations (9) and (10) and the irreversible (or inelastic) component is defined by the following flow law:

$$\dot{\epsilon}_{ij}^I = \frac{3}{2}||\dot{\epsilon}_{ij}^I||\frac{\Sigma_{ij}}{\sqrt{J_2}} \qquad if \quad F \geq 0 \tag{28}$$

or

$$\dot{\epsilon}_{ij}^I = 0 \qquad if \quad F < 0 \tag{29}$$

where

$$||\dot{\epsilon}_{ij}^I|| = \sqrt{\frac{2}{3}\dot{\epsilon}_{ij}^I\dot{\epsilon}_{ij}^I} = \frac{\mu(T)f(F(T))}{\kappa(T)} \tag{30}$$

The internal constitutive rate equation is always given by

$$\dot{a}_{ij} = L_{ijrs}\left[\dot{A}_{rs} - \theta_{rs}\dot{T}\right] \tag{31}$$

while the evolutionary law for the back strain rate is given by :

$$\dot{A}_{kl} = \dot{\epsilon}_{kl}^I - \frac{3}{2}\frac{\beta(T)}{\kappa_o^2}\kappa(T)||\dot{\epsilon}_{ij}^I||\frac{a_{kl}}{\sqrt{G}}Hv[Y] - \frac{3R_\alpha(T)B_0(T)}{\kappa_o^2}\frac{g(G)}{}a_{kl} \qquad if \quad a_{ij}\Sigma_{ij} \geq 0$$

(32)

during internal loading and

$$\dot{A}_{rs} = Q_{rslm}E_{lmnp}\left(\dot{\epsilon}_{np}^I - \frac{3}{2}\frac{\beta(T)}{\kappa_o^2}\kappa(T)||\dot{\epsilon}_{ij}^I||\frac{a_{np}}{\sqrt{G}}Hv[Y] - \frac{3R_\alpha B_0}{\kappa_o^2}\frac{g(G)}{}a_{np}\right)$$

(33)

during internal unloading, i.e., if $a_{ij}\Sigma_{ij} < 0$. The internal stiffness operator is defined as

$$L_{ijrs} = [Q_{ijrs}]^{-1} = \frac{\kappa_o^2}{3B_0(T)(1+B_1pG^{p-1})}\left(I_{ijrs} - \frac{3B_1(p-1)G^{p-2}}{\kappa_o^2(1+B_1pG^{p-1}(6p-5))}a_{rs}a_{ij}\right)$$

(34)

with $I_{ijrs} = \delta_{ir}\delta_{js}$; whereas, the dynamic thermal recovery operator, is defined as:

$$\theta_{rs} = \frac{\partial B_0(T)}{\partial T}(1+B_1pG^{p-1})\frac{3}{\kappa_o^2}a_{rs}$$

(35)

Equation (33) constitutes a new consistent, potential preserving, internal unloading criterion that prevents the classical abnormalities [7] in the cyclic response associated with nonlinear hardening formulations, upon unloading and reversed loading of the *external* variables as described in [5]. Justification for this criterion stems from the work of Orowan [8] and others, and has indicated that upon stress reversals dislocations are remobilized and consequently rapid rearrangement of these dislocations is possible within the wake of the previous load path. To describe this rapid motion of dislocation remobilization (readjustment of internal stress) during material unloading we introduce distinct regions within the state space in which the *rate* of the conjugate internal state variable (internal strain) changes discontinuously. Currently, until more exploratory tests can be performed to fully describe the affected regions of the state space, the internal (back) strain rate during internal unloading is taken to be proportional to the back strain rate during loading, through the product of the external stiffness and internal compliance operators, see eq. (33). Thus, implying that the internal stress and strain during internal unloading are related by the external stiffness tensor. Furthermore, an examination of the evolution of the conjugate of the internal stress (*i.e.*, the internal strain, $\mathcal{A}_{ij}$), clearly shows that equation (32) possesses, as typically assumed in the literature (cf. [9] through [13]), competitive mechanisms consisting of a hardening term (which accounts for strengthening mechanisms) and two state recovery terms (which account for softening mechanisms). The first state recovery term evolves with inelasticity; it is strain induced, and is commonly called dynamic recovery; whereas, the second term, interchangeably

Constants	Units	TIMETAL 21S
Viscoplastic		
κ	MPa	5.86
μ	MPa/sec	5.52×10^{-9}
n	-	3.3
B_0	MPa	5.86×10^{-4}
B_1	-	0.05
p	-	1.8
q	-	1.35
R_α	1/sec	0.1×10^{-5}
β	-	0.01
Elastic		
E	MPa	80,671
ν	-	0.365

Table 1- - Isothermal material parameters for TIMETAL 21S at 650°C

called static or thermal recovery, evolves with time and is thermally induced. The added flexibility provided by the dynamic recovery term is particularly advantageous at low homologous temperatures where thermal recovery is inactive.

Finally, the above expressions are further specialized by assuming the nonlinearity of inelasticity and the thermal recovery process to be represented by power law functions, that is:

$$f(F) = F^n \tag{36}$$

$$g(G) = G^q \tag{37}$$

Characterization

Given the previously characterized isothermal **GVIPS** model [5] with the nine material parameters corresponding to the reference temperature, T_0, of 650°C, the extension of the model to the nonisothermal regime becomes straightforward with the primary task being associated with the determination of the number and functional form of the required temperature dependent parameters. Table 1 lists both the elastic (E and v) and inelastic material parameters utilized in the isothermal model; that is the three parameters associated with the flow law (i.e., κ, μ, and n), the three with the nonlinear hardening operator, (i.e., B_0, B_1, and p), two with the internal thermal recovery term (i.e., R_α and q), and the one parameter with the strain induced (dynamic) recovery term (i.e., β). Considering the physical and mathematical significance (see eqs. (28)-(34)) of the above material parameters one could easily argue that only five of the nine inelastic parameters, in addition to the elastic stiffness and coefficient of thermal expansion, need be

Parameters

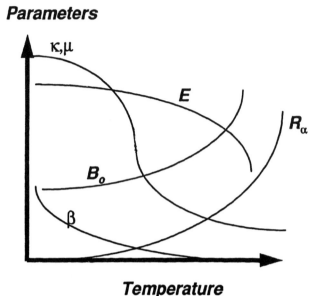

Temperature

FIG. 1- - Schematic illustrating the temperature dependence of the elastic and inelastic material parameters

functions of temperature. Clearly, since the magnitude of κ defines the onset of inelastic (or irreversible) behavior, i.e., through the threshold surface F, one would expect its temperature dependence to be a decreasing function as temperature increases, as depicted schematically in Fig. 1. A similar temperature dependence would be expected for the coefficient μ as it is related to the viscosity of the material, which is known to significantly decrease as the temperature is increased. Two key assumptions that simplify the nonisothermal characterization process considerably, are first, that the internal (L_{ijkl}) and external (E_{ijkl}) stiffness operators have a similar temperature dependence (decreasing with increasing temperature, see Fig. 1). Consequently B_0 is taken initially to have the inverse temperature dependence of the Young's modulus $E(T)$. Second, the exponent n in the flow law is taken to be temperature independent, thus minimizing the non-unique determination of the remaining material parameters μ and κ, and ensuring such theoretical niceties as convexity of the flow surface at all temperatures. Finally, as β scales the dynamic recovery term (which is active at low homologous temperatures and inactive at high homologous temperatures) and R_α scales the static (thermal) recovery term (which is inactive at low homologous temperatures and active at high) each parameter must have the temperature dependence illustrated in Fig. 1.

As previously stated, the material of choice for characterizing the current nonisothermal **GVIPS** formulation is the titanium matrix alloy, TIMETAL 21S. This alloy was selected due to the considerable attention it has received for use in TMCs with application toward advanced airframe structures. TIMETAL 21S is a metastable β-titanium alloy with a nominal composition of Ti-15Mo-3Nb-3Al-0.2Si (wt. %). The coupon specimens used in this study were taken from "fiberless" panels. The fiberless panels were fabricated by hot isostatic pressing 0.13mm thick TIMETAL 21S foils, so as to subject the matrix material to an identical processing history as that seen by the matrix material in the composited form. All specimens were subjected to a pre-test heat treatment, consisting of an 8 hour soak in vacuum at 621°C, to stabilize the $\beta+\alpha$ microstructure of the TIMETAL 21S. For further material, machining, and experimental details see Castelli et al. [14]. All isothermal and nonisothermal tests addressed are uniaxial experiments, thus implying that the multiaxial material constants are typically generalized from their uniaxial counterparts. This need for generalization is precisely why a consistent multiaxial theory, such as that developed from a potential formulation, is imperative. The available tests for use in characterizing the current nonisothermal extension of the **GVIPS** model are, five tensile tests and four creep tests (performed at a specified reference stress level of 103 MPa) spanning the representative domains in temperature. Very few repeats were performed due to the limited amount of available material.

The temperature dependence of the elastic properties were obtained by fitting simple polynomial functions through the experimental data [14]. The resulting continuous functions utilized for the present study are taken to be:

$$E(T) = 114141. \left[1 - 1.071\text{x}10^{-7}T^2 \ln(T)\right] \tag{38}$$

$$\eta_{\text{tan}}(T) = 4.539x10^{-6} + 3.078\text{x}10^{-6} \exp(\frac{T}{719.6}) \tag{39}$$

where the units of E(T) are MPa and the temperature, T, is in degrees Celsius. Note, although the poisson's ratio was shown [14] to have a slight temperature dependence the parameter was taken here, for convenience, to be temperature independent and remain at the constant value given in Table 1.

To obtain the actual temperature dependent functional forms for the five inelastic parameters, $\kappa, \mu, B_0, R_\alpha$, and β the following procedure was followed. The first and most important step was identifying the value of the drag strength as this represents the value of stress below which only elastic (or reversible) behavior is "observed", i.e., no measurable inelastic behavior occurs, and this parameter implicitly influences the determination of all other parameters. The actual numerical value employed is typically non-unique as it is dependent upon the definition of inelasticity employed and thus implicitly dependent upon the sensitivity of the experimental equipment. A typical approach taken to arrive at a value for κ, is to conduct a sequence of creep or relaxation probing tests to determine the maximum value of stress for which no time dependent behavior ($\epsilon^I = 0$) is observed. Such a detailed experimental program was under taken by Castelli et al. [14] to obtain this time dependent threshold. In the process, however, it was discovered that TIMETAL 21S exhibits both a time and temperature dependent reversible (linear viscoelastic) and

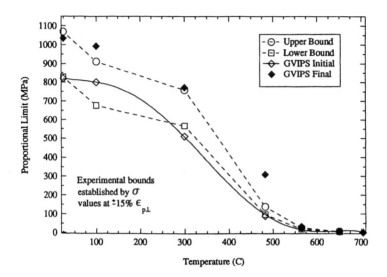

FIG. 2- - The upper and lower bounds on the experimentally measured drag stress and the initial and final correlation values employed.

irreversible (viscoplastic) domain[2]. Consequently, an alternative experimental procedure was conducted to identify the temperature dependent, irreversible domain (F>0) wherein the viscoplastic inelastic strain component defined in eq. (28) is active. The procedure explained in [14], consists essentially in conducting a number of extremely slow (e = 0.5x10^{-6}), strain controlled tensile tests at various temperatures and looking for the stress value, above which, the stress strain curve deviates from proportionality. In this way the material itself provides the definition of inelasticity, independent of any *a priori* assumption regarding the accumulation of inelastic strain over a specified time. Given the results of these tests (relative to a conservative and non-conservative definition, i.e., the stress at ±15% of the proportional limit strain) the drag stress is seen to possesses a sigmoidal like functional dependence on temperature, see Fig. 2. The initial temperature dependent function for κ is taken to be:

$$\kappa(T) = -0.55 + \frac{121.2}{\left(1 + \exp\left(\frac{(T-333.)}{71.2}\right)\right)} \tag{40}$$

and is shown, relative to the experimentally obtained values in Fig. 2.

Given the drag stress as a function of temperature, and the initial temperature depen-

---

[2]Inclusion of this significant time dependent reversible domain along with the irreversible (viscoplastic) domain will be reserved for a later study.

dence of B_0 taken to be;

$$B_0(T) = B_0(T_0) \frac{1.41}{[1 - 1.071 \times 10^{-7} T^2 \ln(T)]}$$

and taking the initial temperature dependence of μ to be[3]

$$\mu(T) = \frac{\mu(T_0)}{\kappa_0} \kappa(T) z(T)$$

one can then proceed to determine the required μ (or function $z(T)$) and β to simulate the experimental tensile response at the given temperatures. Obtaining these parameters, one would turn to the creep response at the given temperatures to obtain the magnitude of thermal recovery (R_α) required to simulate these histories. Given the assumed temperature independence of the exponent n and the elected use of continuous functions, the effort and compromises required to establish the present set of five temperature dependent inelastic parameters[4] is minimal in comparison to the effort required to establish the nine inelastic isothermal parameters at the reference temperature of 650°C. Note that in the above format, $z(T)$ represents the familiar diffusivity parameter commonly employed in other viscoplastic models in the literature, e.g., see [7],[10] and [11].

This original set of parameters with the single power-law format for rate sensitivity as used here, was found to lead to significant rate sensitivity even at low temperatures. This behavior was not found to be characteristic of this material, however. Therefore, a final recalibration of κ, μ and B_0 was undertaken; such that κ was increased to approximately its upper bound throughout the temperature range, μ was increase by at least three orders of magnitude at the lower temperatures and B_0 was similarly reduced by a factor of 2 to 6 over the lower temperature regime. The final set of temperature dependent inelastic parameters for the present nonisothermal extension of the **GVIPS** model are given in Table 2, and also shown in Fig. 2.

The corresponding correlation for the tensile response of TIMETAL 21S (at 25, 300, 482, 565 and 650°C given a total strain rate of 8.33×10^{-5}/sec) and short term creep response (at 482, 565, 650 and 704°C at the reference stress level of 103 MPa) are shown respectively, in Figs. 3 and 4. As an aside, note that the way in which one chooses to interpolate between the discrete temperatures given in Table 2 is somewhat arbitrary.For example, one might select the simplest approach of a piecewise linear interpolation, particularly in the case of a significant number of known discrete points, or elect to use continuous functions whenever possible (as done in this study). Note, applying either approach will achieve similar predictive behavior, however, the use of continuous functions allows one to analytically determine the temperature dependent parameters and slopes

---

[3]This form for μ is assumed as the magnitude of inelastic strain is scaled by the ratio $\frac{\mu}{\kappa}$ and the viscosity has a similar temperature dependence as does κ.

[4]Which accurately simulated the available tensile and creep responses over the given temperature range.

(see eq. (35)) at all temperatures. The specific material parameter interpolation functions selected for this study are given in appendix A.

TIMETAL 21S		Temperature, $^\circ$ C					
Constants	Units	25	300	482	565	650	704
κ	MPa	1034.	772.	310.	33.	5.86	0.75
μ	MPa/sec	689.	138.	6.89×10^{-3}	5.86×10^{-7}	5.52×10^{-9}	7.10×10^{-12}
B_0	MPa	6.89×10^{-5}	1.03×10^{-4}	1.72×10^{-4}	4.86×10^{-4}	5.86×10^{-4}	6.36×10^{-4}
R_α	1/sec	0.	0.	1.679×10^{-7}	1.685×10^{-7}	1.0×10^{-6}	6.0×10^{-5}
β	-	0.001	0.	0.	0.	0.	0.0

Table 2- - Nonisothermal material parameters for TIMETAL 21S.

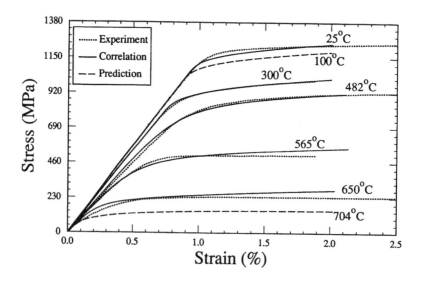

FIG. 3- - **GVIPS** correlation with experimental tensile data at 25, 300, 482, 565 and 650°C, given a total strain rate of 8.33×10^{-5}/ sec.

The previously [5] discussed rate dependent tensile response at 650°C over three orders of magnitude in total strain rate (i.e., $\dot{e} = 8.33 \times 10^{-4}, 8.33 \times 10^{-5}$ and 8.33×10^{-6}/sec, compliments of Ashbaugh and Khobaib [15]) , short term creep at three stress levels (i.e., $\sigma = 72$, 110, and 128 MPa), and relaxation at 103, 238 and 345 MPa, are repeated here in Figs. 5, 6 and 7, respectively, for completeness. In particular, Fig. 5 illustrates the present model's capability to capture the significant rate dependence of this material

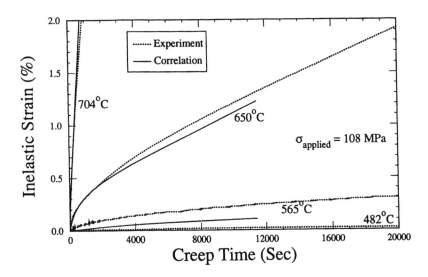

FIG. 4- - **GVIPS** correlation with experimental creep data, at the reference stress level of 15 ksi and three temperatures, 565, 650, and 704°C.

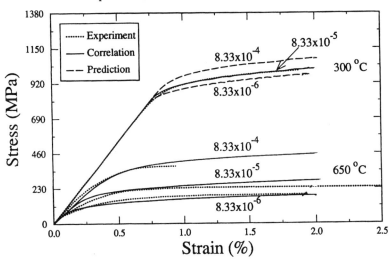

FIG. 5- - **GVIPS** correlation and prediction of strain rate sensitivity at 300 and 650°C; 650 data taken from [13].

at 650°C, whereas Figs. 6 and 7 demonstrate the accurate stress-level sensitivity of the model under both creep and relaxation conditions. In addition, note the ability of the model, in Fig. 6, to accurately represent both the primary and secondary creep regions. Clearly, the present model does an excellent job, given the wide variety of loading conditions and temperature range examined. This is particularly true when one considers the relatively small number of temperature dependent and independent material constants required to simulate these histories.

GVIPS PREDICTIONS

The present nonisothermal **GVIPS** model, characterized to represent the behavior of TIMETAL 21S from 23 to 704°C, is now exercised and its predictive capability assessed.

<u>Relaxation Behavior</u>

The assessment begins by considering relaxation tests performed at stress levels above and below that used (i.e., $\sigma = 238$ MPa) in the characterization of the model at the reference temperature of 650°C, see Fig. 7. Clearly, the overall agreement with short term behavior is quite good with the initial stress rates being accurately predicted at the higher stress levels and yet diminishing more rapidly than do the actual experimental values at the lower stress levels. Note that the three starting stress levels in Fig. 7 represent stress levels[5] ; i) below the apparent yield, ii) within the knee (e.g. 115 to 330 MPa), and iii) near the ultimate stress level of the material (approximately 385 MPa), see Fig. 5. As mentioned earlier, the significant reversible time dependent domain observed in TIMETAL 21S has not been included in the present model; therefore, it is not surprising that the relaxation response at the lower stress levels within the apparent elastic regime are not as accurately represented. Consequently, attention was focused on the relaxation response of TIMETAL 21S at 300 and 565°C given a considerable prior total strain history of 1.9%. The corresponding stress-strain and stress-time test and model prediction results for these two temperatures are shown in Figs. 8 and 9 respectively. Clearly, both short and mid term behaviors are accurately predicted.

---

[5]Given a 5.0 x 10^{-4} total strain rate ramp-up history, see Fig. 5.

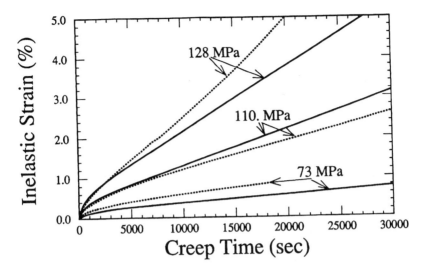

FIG. 6- - **GVIPS** correlation (solid lines) of experimental creep data (dotted lines) under various stress levels at the reference temperature, 650°C.

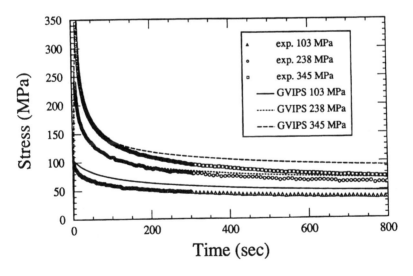

FIG. 7- - Short term relaxation response of TIMETAL 21S at 650°C: **GVIPS** simulation (lines) versus experimental results (symbols). Initial loading total strain rate of 5.0 x 10^{-4}/sec.

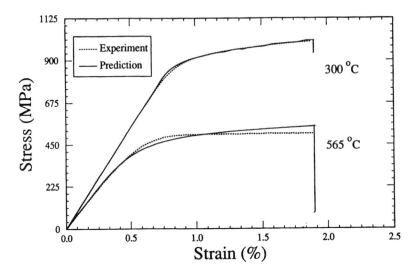

FIG. 8- - The experimental and **GVIPS** stress-strain response resulting from relaxation tests at 300 and 565°C.

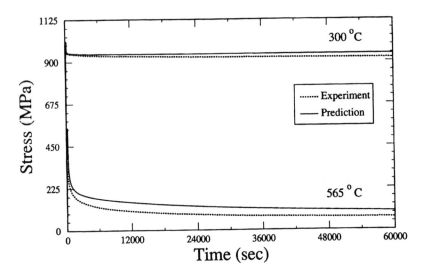

FIG. 9- - The experimental and **GVIPS** stress-time response resulting from two relaxation tests at 300 and 565°C.

Rate Sensitivity

Figure 5 illustrates the rate sensitivity of the present model at 300°C and at the reference temperature 650°C, given a variation in the applied strain rate of three orders of magnitude. The accurate correlation of the rate sensitivity at 650°C is evident and, although the rate sensitivity is reduced as the temperature is decreased, the present model is still considered to over-predict the rate sensitivity of TIMETAL 21S at lower temperatures. The difficulties associated with predicting a highly rate sensitive material at elevated temperatures, as compared with a rate insensitive material at lower temperatures are believed to be enhanced because of neglecting the significant reversible, time dependent domain present at elevated temperatures in this material. As stated previously, inclusion of this feature will be the focus of our future research.

Cyclic Behavior

Given a total strain rate of $8.33 \times 10^{-5}/\text{sec}$, the cyclic stress-strain behavior of the present nonisothermal model is demonstrated for three temperatures, i.e., 25, 482, and 650°C as illustrated in Fig. 10 [15]. The cyclic response at 650 is considered to be a correlation with experiment whereas those at 482 and 25°C are actual predictions. Obviously, the room temperature cyclic behavior agrees quite well with experimental data. Furthermore, these cyclic results illustrate that the consistent, potential preserving, internal unloading criterion prevents, over a wide temperature range, the classical abnormalities in the cyclic stress-strain response associated with nonlinear hardening formulations, even with the exclusion of any isotropic hardening component.

Temperature Step Test

In order to examine the significance of the new dynamic thermal recovery term, resulting from the nonisothermal extension of the **GVIPS** model, a multiple-step temperature creep test was conducted. This test began with a 103 MPa creep test at 565°C, followed by a temperature step, under zero load, to 650°C; whereupon a 103 MPa creep test is once again conducted. This history was then followed by another temperature step, under zero load, to 704°C and a subsequent 103 MPa creep test. The resulting experimentally measured creep strain versus creep time response (dotted line) is shown in Fig. 11 along with the **GVIPS** prediction (solid line). Comparing the model simulation to the experimental results it is apparent that the model under-predicts the material creep response at 565°C by approximately 35%. This is not surprising, as the correlation of the 565°C creep response was similarly under predicted, see Fig. 6. However, several key qualitative features of the temperature step test are represented with reasonable accuracy, including the re-initiation of primary creep subsequent to the steps, and the overall total strain accumulation.

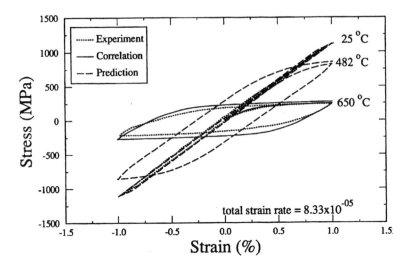

FIG. 10- - **GVIPS** simulation of cyclic stress-strain behavior at 25, 482 and 650°C [13].

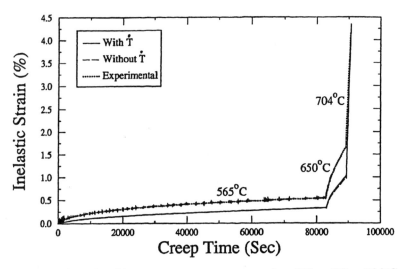

FIG. 11- - Prediction of multi-step temperature creep test, i.e., 565 to 650 to 704°C, with and without the new dyanmic thermal recovery term.

One interesting aspect of this step test as discussed in Castelli et al. [14] is the increased primary creep rate, over that of the virgin sample, subsequent to the jump in temperature to 650°C, thus suggesting the need for a dynamic thermal recovery mechanism. Just such a mechanism is naturally provided in the present nonisothermal extension of the complete potential framework discussed above. To illustrate the impact of the dynamic thermal recovery mechanism, the identical history as that described above was conducted (and is shown by a dashed line in Fig. 11) but with the removal of the $\theta_{ij}\dot{T}$ term in the internal constitutive rate equation, i.e., eq. (31). Although, for the present characterization and rate of thermal loading the influence was minimal, the inclusion of such a mechanism (as required by the potential structure) does provide an increase in the subsequent creep response. An entire study devoted to examining the importance of this new dynamic thermal recovery mechanism and its sensitivity to characterization and thermal loading rate will be the focus of future research.

TMD Test

As a final verification test a thermomechanical deformation (TMD) test was conducted, which involved a strain controlled (i.e., $\dot{e} = 8.33 \times 10^{-4}/$ sec) tensile test at 300°C to 2.0% total strain, followed by an unload to zero load and then a reload under stress control (i.e., $\dot{\sigma} = 68.95$ MPa / sec) to 103 MPa. Holding the load at 103 MPa, the temperature was then increased at a rate of 1 °C/sec to 650°C; whereupon the load and temperature were held fixed at their respective magnitudes for 8 hours. The resulting experimentally obtained stress-total strain and total strain versus time responses and model simulations are illustrated in Figs. 12, 13 and 14. Here, the thermal strain component was zeroed at the initial test temperature of 300°C. Examining Fig. 12, it is obvious that the prior tensile overload is over-predicted, thus one might question the ability of the model to accurately predict the subsequent creep behavior. Part of the discrepancy in the initial tensile response can be attributed to the fact that the model exhibits a notable positive strain rate sensitivity at 300°C whereas the experimental data suggests a minimal strain rate dependence. Consequently, given that the characterization of the present model was performed at a strain rate of $8.333 \times 10^{-5}/$ sec and the verification test was conducted at $8.33 \times 10^{-4}/$ sec, one would anticipate the present over-prediction of the tensile response. Further discrepancies can also be attributed to material variability[6] and experimental scatter. Figures 13 and 14 show the comparison of the experimental and **GVIPS** simulation of the subsequent 103 MPa TMD response. Again, the simulation under predicts the experimental results. Realizing that the experiment performed is actually a constant load test and not a constant stress creep test, as is the simulation, another simulation was conducted wherein the applied stress was increased by approximately 7% to account for the maximum change in cross sectional area of the test coupon. As a

---

[6]The current TMD test coupon consisted of 12 plies, whereas all previous tensile test data were obtained from coupons constructed with 5 plies.

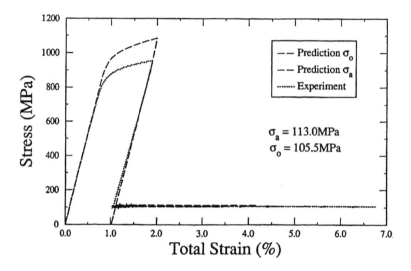

FIG. 12- - Prediction of stress-strain response for a thermomechanical deformation test.

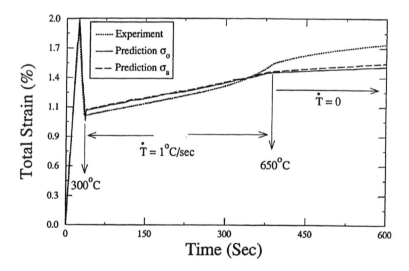

FIG. 13- - Prediction of short term total strain versus time response for a thermomechanical deformation test.

result, the accumulation of predicted inelastic strain was increased, thus providing an upper bound on the prediction, as shown by the dashed line in Fig. 14. Clearly, even with the inclusion of this additional increase in stress, the present characterization of the model still significantly under predicted the experimental results.

Examination of Fig. 13, which is an amplification of the transient temperature regime occurring within the first 600 seconds of the test provides some additional insight into the possible cause for the under-prediction of the present **GVIPS** model. For example, focusing upon the transient temperature region, one observes a nonlinear accumulation of strain starting at the 200 second mark and continuing on (but at a different rate) past the point at which the temperature becomes constant at 650°C. This accumulation equals approximately 0.2% strain, which is precisely the amount of deviation between the simulation and the experimental observation at 600 seconds into the test; this suggests that the inelasticity predicted by the model at low stress levels and temperatures below 650°C is insufficient. This conclusion is further supported by the previous correlations shown in Fig. 4 where the model under-predicted the creep strain at 565 and 482°C by an increasing amount, as the temperature was decreased. Furthermore, the neglecting of the time dependent, reversible domain, as discussed previously may also contribute significantly to this nonlinear accumulation of strain during the temperature transient. Such areas will be the focus of future research.

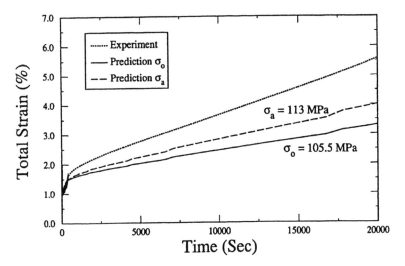

FIG. 14- - Prediction of long term total strain versus time response for a thermomechanical deformation test.

SUMMARY

A fully associative, multiaxial, nonisothermal, nonlinear kinematic hardening viscoplastic model has been presented. It contains three internal state variables (two scalars and one tensor) and both thermal and strain-induced recovery mechanisms. The two, non-evolving (yet temperature dependent), scalar internal state variables are associated with the dislocation density and are defined as the drag and yield stress. The evolving tensorial variable known as the internal (or back) stress is a second order, traceless, symmetric tensor and is associated with the dislocation substructure. A unique aspect of the present model is the inclusion of nonlinear hardening through the use of a compliance operator, derived from the Gibb's potential, in the evolution law for the back stress. This nonlinear tensorial operator is significant in that it allows both the flow and evolutionary laws to be fully associative (and therefore easily integrated), greatly influences the multiaxial response under non-proportional loading paths and in the case of nonisothermal histories, introduces an instantaneous thermal softening mechanism proportional to the rate of change in temperature. The resulting nonisothermal unified viscoplastic model was then characterized for the titanium based alloy TIMETAL 21S from room temperature to 704°C. Results illustrated the good overall correlation and predictive capabilities of the model for a wide range of mechanical and thermal loading conditions.

ACKNOWLEDGEMENT

Thanks be to God. We would also like to acknowledge the efforts of Dr. T.E. Wilt in the development of IDAC and implementation of this nonisothermal **GVIPS** model, and the reviewers for their helpful comments.

REFERENCES

[1] Arnold, S.M.; and Saleeb, A.F.: On the Thermodynamic Framework of Generalized Coupled Thermoelastic Viscoplastic - Damage Modeling, *Int. Jnl. of Plasticity*, Vol. 10, No. 3, 1994, pp 263-278.

[2] Saleeb, A.F. and Wilt, T.E.: Analysis of the Anisotropic Viscoplastic-Damage Response of Composite Laminates-Continuum Basis and Computational Algorithms, *Int. Jnl. Num. Meth. Engng.*, Vol. 36, 1993, pp. 1629-1660.

[3] Arnold, S.M., Saleeb, A.F, and Wilt, T.E.: A Modeling Investigation of Thermal and Strain Induced Recovery and Nonlinear Hardening in Potential Based Viscoplasticity, *Jnl. of Engng. Materials and Technology*, Vol. 117, No. 2, 1995, pp. 157-167.

[4] Saleeb,A.F., Seif, Y., and Arnold, S.M.: Fully-Associative Viscoplasticity with Anisotropic and Nonlinear Kinematic Hardening, submitted *Int. Jnl. of Plasticity*, 1995.

[5] Arnold, S.M.; Saleeb, A.F., and Castelli, M.G.: A Fully Associative, Nonlinear Kinematic, Unified Viscoplastic Model for Titanium Based Matrices, *Life Prediction Methodology for Titanium Matrix Composites, ASTM STP 1253*, Johnson, W.S., Larsen, J. M., and Cox, B.N. Eds., American Society for Testing and Materials, Philadelphia, 1995. NASA TM-106609, 1994.

[6] Spencer, A.J.M.: Continuum Physics, Vol. 1, A.C. Eringen, Ed., Academic Press., London, 1971, p. 240.

[7] Lemaitre, J.; and Chaboche, J.L.: Mechanics of Solid Materials, Cambridge University. Press, New York, 1990.

[8] Orowan, E.: Causes and Effects of Internal Stresses, Internal Stresses and Fatigue of Metals, *Proceedings* of the Symposium on Internal Stresses and Fatigue in Metals, Detroit Mich., Eds., G.M. Rassweiler and W.L. Grvise, Elsevier Publishing, 1959.

[9] Robinson, D.N; and Swindeman, R.W.: Unified Creep-Plasticity Constitutive Equations for 2 1/4 Cr-1Mo Steel at Elevated Temperature, ORNL TM-8444, 1982.

[10] Miller, A.K., Ed.:Unified Constitutive Equations for Plastic Deformation and Creep of Engineering Alloys, Elsevier Applied Science, New York, 1987.

[11] Freed, A.D.; Chaboche, J.L.; and Walker, K.P.: A Viscoplastic Theory with Thermodynamic Considerations, *Acta Mech*, Vol. 90, 1991, pp. 155-174.

[12] Neu, R.W.; Nonisothermal Material Parameters for the Bodner-Partom Model, MD-Vol. 43, Material Parameter Estimation for Modern Constitutive Equations, Eds, L.A. Bertram, S. B. Brown, and A. D. Freed, 1993, pp. 211-226.

[13] Sherwood, J. A. and Quimby, H.M.; "Micromechanical Modeling of Damage Growth in Titanium Based Metal-Matrix Composites", to appear *Comp. Struc.*, 1995.

[14] Castelli, M.G., Arnold, S.M., and Saleeb, A.F.:Specialized Deformation Tests for the Characterization of a Viscoplastic Model: Application to a Titanium Alloy, NASA TM-106268, 1995.

[15] Ashbaugh, N.E., and Khobaib, M.: Unpublished Data, University of Dayton Research Institute, Dayton, Ohio.

APPENDIX A: Interpolation Functions

Here, we list the functions and applicable temperature ranges used in this study to describe the five temperature dependent inelastic material parameters identified at discrete temperatures in Table 2. Note, although the magnitudes of the functions describing any one parameter are continuous across temperature boundaries, no attempt was made to ensure continuity of the slopes. This may be an extremely important fact depending upon the integration scheme utilized, and should be cautiously considered.

Constants	Interpolation Functions	Temperature Range, °C
κ	$a_1 + \dfrac{b_1}{\left(1+\exp\left(-\frac{(T-c_1)}{d_1}\right)\right)}$	$T < 565$
	$a_2 + \frac{b_2}{T^{1.5}} + \frac{c_2}{T^2}$	$T \geq 565$
μ	$\frac{\mu_0}{\kappa_0}\kappa(T)\left[\exp\left(a_3 + b_3 T \ln(T) + c_3 T^{2.5}\right)\right]$	$T < 565$
	$\frac{\mu_0}{\kappa_0}\kappa(T)\left[a_4 + \dfrac{b_4}{\left(1+\exp\left(-\frac{(T-c_4)}{d_4}\right)\right)}\right]$	$T \geq 565$
B_0	$a_5 + b_5 \exp\left(-\frac{T}{c_5}\right)$	$T \leq 482$
	$a_6 + b_6 T + c_6 T^2 \ln(T) + d_6 T^{2.5} + e_6 T^3$	$T > 482$
R_α	$a_7 + b_7 \left[\dfrac{4\exp\left(\frac{-(T-c_7)}{d_7}\right)}{\left(1+\exp\left(\frac{-(T-c_7)}{d_7}\right)\right)^2}\right]$	$T > 482$
	$a_8 - b_8 (482 - T)^2$	$300 \leq T \leq 482$
	0	$T < 300$
β	0.001	$T \leq 100$
	$0.001 - 1.0 \times 10^{-4} \langle T - 100 \rangle$	$100 < T \leq 200$
	0	$T > 200$

where $\mu_0 = 5.52 x 10^{-9}$ and $\kappa_0 = 5.86$. The above interpolation constants are:

$a_1 = -1027.4$ $b_1 = 2118.5$ $c_1 = 566.04057$ $d_1 = -154.9935$

$a_2 = 254.76$ $b_2 = -1.997\text{x}10^7$ $c_2 = 4.04\text{x}10^8$

$a_3 = 20.197653$ $b_3 = 2.4519076\text{x}10^{-3}$ $c_3 = -3.41797\text{x}10^{-6}$

$a_4 = -0.22876195$ $b_4 = 3961.1386$ $c_4 = 397.89176$ $d_4 = -31.582681$

$a_5 = 5.782\text{x}10^{-5}$ $b_5 = 1.003\text{x}10^{-5}$ $c_5 = -197.95617$

$a_6 = -5.430\text{x}10^{-2}$ $b_6 = 3.603\text{x}10^{-4}$ $c_6 = -4.006\text{x}10^{-7}$ $d_6 = 1.156\text{x}10^{-7}$

$\qquad\qquad\qquad\qquad\qquad\qquad\qquad\qquad\qquad\qquad\quad e_6 = -1.196\text{x}10^{-9}$

$a_7 = 1.670\text{x}10^{-7}$ $b_7 = 1.0381\text{x}10^{-4}$ $c_7 = 722.0$ $d_7 = 11.597$

$a_8 = 1.670\text{x}10^{-7}$ $b_8 = 5.068\text{x}10^{-12}$

Ming Gao[1], William Dunfee[2], Carl Miller[3], Robert P. Wei[4], William Wei[5]

THERMAL FATIGUE TESTING SYSTEM FOR THE STUDY OF GAMMA TITANIUM ALUMINIDES IN GASEOUS ENVIRONMENTS

REFERENCE: Gao, M., Dunfee, W., Miller, C., Wei, R. P., Wei, W., "**Thermal Fatigue Testing System for the Study of Gamma Titanium Aluminides in Gaseous Environments,**" Thermomechanical Fatigue Behavior of Materials: Second Volume, ASTM STP 1263, Michael J. Verrilli and Michael G. Castelli, Eds., American Society for Testing and Materials, 1996.

ABSTRACT: To critically assess the thermal fatigue resistance of γ-titanium aluminides in hydrogen and other gases, a thermal fatigue test system and associated procedures were developed. The test equipment consisted of an environmental chamber, a rigid test fixture with fixed grips, and a thermal controller. Direct electrical resistance heating was used to heat the specimen, and cooling was accomplished by a chilled gas jet. Characterization of the thermal-mechanical features of the equipment showed that the system allows for rapid heating and cooling rates with acceptable control of the thermal stresses and has excellent repeatability between cycles.

Proof tests were performed on a Ti-48Al-2Cr alloy in helium, hydrogen, and air, with temperature cycling between 25 and 900°C and a preload equal to 50% of the material's yield strength. The results showed that the equipment and test method developed here are an effective tool for material evaluation, specifically for the critical assessment of materials for high temperature applications in hydrogen and hydrogenous gases.

KEYWORDS: thermo-mechanical fatigue, thermal cycling test, environmentally-assisted cracking, hydrogen embrittlement, gamma titanium aluminides, intermetallic compounds

Due to their low density and attractive high-temperature characteristics, titanium aluminide intermetallics are among the most promising new materials for use in intermediate and high temperature applications in future aircraft engines and hypersonic vehicles [1-3]. However, a major concern with these materials is their potential for embrittlement from hydrogen exposure and degradation by the earth's atmosphere [1-8].

[1]Principal Research Scientist [2]Graduate Student [3]Senior Technician [4]Professor and Chairman, Department of Mechanical Engineering and Mechanics, Lehigh University, Bethlehem, PA 18015.
[5]Professor, Universiteit Twente, Enschede, The Netherlands; formerly MTU Motoren-und-Turbinen Union, Münich, Germany.

Also, when considering the intended applications for titanium aluminides, it becomes apparent that the materials will be subjected to these deleterious environments as well as thermal cycling. It is therefore important to assess their ability to resist degradation through the combined actions of environment and thermal fatigue.

In an effort to determine the material's thermal fatigue response, a test system and associated procedures, described herein, were developed to perform environmentally-assisted thermal fatigue using a fixed-grip condition on specimens of a γ-titanium aluminide alloy. The test system used direct electrical resistance heating to rapidly heat the specimen and a chilled gas jet to provide swift cooling. During testing, the specimen was enclosed within an environmental chamber to explore the effect of various gaseous environments: namely, helium, hydrogen, and air. The features of the system were characterized in terms of thermal profile, heating and cooling rates, and thermal stresses developed during thermal cycling. A long focal length traveling microscope was used for visual inspection of the specimen at 20X during the test to monitor the development of fatigue cracks.

SPECIMEN

A two-phase $(\alpha_2 + \gamma)$ gamma titanium aluminide, with nominal composition Ti=50, Al=48, Cr=2 in atomic percent was used. The thermal fatigue specimens were designed to be thin and flat, with a 1.3 mm thick by 6.3 mm wide by 12.7 mm long gage length, to account for special environmental, thermal, and stress considerations. This geometry provides a high surface area-to-volume ratio to maximize the environmental exposure area for reaction, allows for rapid forced cooling, and minimizes the thermal gradient through-out the sample thickness. The specimen also has a sharp, wedge-shaped edge (< 0.025 mm edge tip thickness) on one side to improve rapid cooling from a chilled gas jet. The details of the specimen geometry are shown in Fig. 1.

A minimal cross-sectional temperature gradient in the sample is desirable to prevent the development of lateral thermal stresses and keep the material in pure tension loading. Therefore, the thermal fatigue specimen was designed to be thin enough to prevent significant lateral thermal stresses from developing within the specimen [9,10]. It should be noted, however, that a significant temperature gradient is expected to build up in the cross sections at and near the knife-edge of the sample during forced cooling. This gradient will produce higher strains and stresses, thus enhancing crack initiation and growth.

Buckling was of concern [11], and was tested experimentally. The results showed that the specimen was able to withstand compressive loads of at least 1 000 N without buckling; a value well above that generated by thermal cycling between 25 and 900°C [10].

Specimen Preparation

Electro-discharge machining (EDM) was used to cut the specimens from the hot-isostatically pressed ingot of material. Mechanical grinding was then used to remove the damaged layer induced by EDM at the surface. The wedge was ground on a computerized grinding machine and was manually polished with diamond paste down to 6 μm to remove tool marks and further minimize the influence of surface defects on crack initiation and growth. The remainder of the specimen surfaces was manually polished with 600 grade silicon-carbide paper. The polishing was along the longitudinal direction of the specimen so that any remaining scratches would be parallel to the loading direction. After polishing, each specimen was carefully inspected for defects using an optical stereo microscope at 50X. A K-type thermocouple (0.381 mm diameter wires) was then spot welded to the

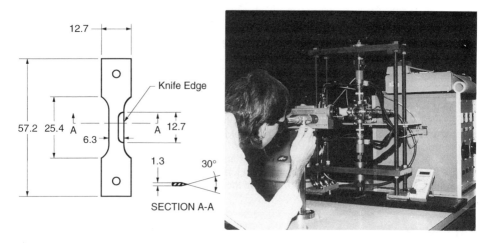

FIG. 1. Details of thermal fatigue test specimen design. Dimensions in millimeters

FIG. 2. A photograph of the test equipment.

center of the specimen along the edge opposite to the wedge, Fig. 1. Prior to mounting in the test fixture, each specimen was ultrasonically cleaned in acetone and then in methanol for 15 minutes each.

Although each specimen's location in the original ingot is known, the selection of specimens for testing was randomized.

TEST SYSTEM DESCRIPTION

The test system was designed to perform thermal fatigue under an essentially fixed-grip condition with an applied preload. The system consists of a test fixture, an environmental chamber, gas supply/exhaust apparatus, and a thermal cycling controller. The thermal and load cycling profiles are monitored using a two-pen chart recorder. A long focal length traveling microscope is used to visually monitor the crack initiation and growth. Figures 2 and 3 are a photograph and a schematic of the test system, respectively.

Test Fixture

The test fixture is a rigid box frame which provides sufficient support to the grip arms to ensure an essentially fixed-grip condition, Fig. 4. Total motion of the grips out of a fixed-grip condition during thermal cycling is estimated to be under 0.02 mm, producing a strain of less than 0.08%. The grip arm assemblies are fabricated of 25 mm diameter brass rods (to provide good electrical conduction). The grip at the end of each grip arm uses a pin to position the specimen and a clamping plate to lock the specimen in place. The grip arms are electrically insulated from the remainder of the test fixture frame. Each grip arm is water cooled by a coil of copper tubing soldered to it.

At the base of the lower grip arm is a screw-driven mechanism which allows for the manual application of a preload to the test specimen. At the base of the upper grip arm are both a 6.7 kN (± 0.5% full scale accuracy) load cell and a grip arm alignment mechanism which allows for the upper grip arm to be aligned with the fixed lower one. The load cell was calibrated *on-site* by using both dead-weights and another certified load cell. The alignment was carried out with the aid of a strain-gauged sample.

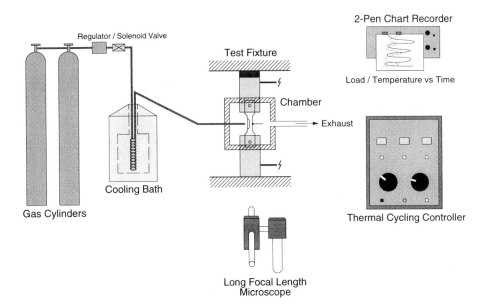

FIG. 3. Schematic of the thermal fatigue test system

Environmental Chamber

The environmental chamber has four feed-throughs: two for the grip arms, one for the inlet gas nozzle, and one for the exhaust gas exit and thermocouple. The front of the chamber has a window to allow for visual monitoring of the specimen during testing. The rear wall of the chamber is internally water-cooled. All chamber seals for the feed-throughs and the window are vacuum-tight. The seal between the chamber and the grip arms is provided by a set of clamped O-rings, which also serve to electrically insulate the chamber from the grip arms. The thermocouple wires were fed through the exhaust flange of the chamber.

Cooling Gas Apparatus

The cooling gas system supplies chilled gas to rapidly cool the specimen, Fig. 3. Up to three gas cylinders provided the high purity gas (99.995% for He, 99.99% for H_2, and air which contained less than 50 ppm water vapor and 19.5 - 23.5% oxygen) for cooling. The gas flow is controlled by a solenoid valve, and the flow rate is set by regulating the output pressure of the gas cylinders. A coil in the gas line is immersed in a dewar full of either liquid nitrogen (-196°C) or an ethyl alcohol/dry ice mixture (-83°C), depending on the gases used. Liquid nitrogen is used during hydrogen and helium testing, while the dry ice solution must be used when testing with air to prevent liquidation of the air within the coil. This cooling produced a gas jet temperature at the nozzle exit of approximately -26 °C with hydrogen, -7 °C with helium, and 0 °C with air.

The gas enters the chamber through a nozzle formed of 0.8 mm ID stainless steel tubing aimed at the center of the knife edge and located 0.6 mm away from the edge. When cooling the specimen from 900 to 25 °C in ten seconds, the gas velocity at the nozzle exit was calculated to be 56 m/s with air, 45 m/s with helium, and 30 m/s with hydrogen.

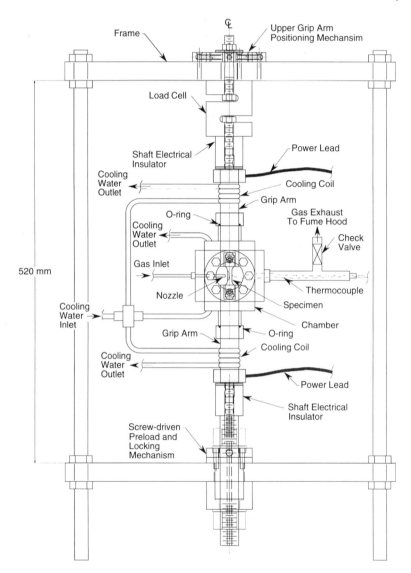

FIG. 4. Schematic of the thermal fatigue test fixture

Safety Features

Safety, especially during testing with hydrogen gas, is an important issue. Thus, several precautions were taken to ensure that air cannot enter the system to cause an explosive hazard: (1) all of the fittings were leak tested both during an argon flush and also during the initial hydrogen introduction, (2) check valves prevented air from back flowing into the system, (3) as a result of the check valves, the entire test system was maintained under a positive pressure of 9 kPa above atmospheric pressure, (4) the exhaust gas line vents into a fume hood rated for hydrogen use, and (5) the entire gas system was flushed with argon at least 500 times its volume prior to hydrogen introduction.

The above steps ensure that air is nearly completely removed from the chamber and cannot re-enter the system to create an explosion hazard. Note that even if a leak allowed air into the system, it would take 25% by volume of air to enter and mix with the hydrogen to reach the upper limit of inflammability [12]. Also, a small amount of hydrogen leaking into the laboratory would quickly dissipate through adequate ventilation. The high flow rate of the hood (20 m^3/min) compared to the relatively small flow of exhaust hydrogen (<0.03 m^3/min for ten seconds at thirty second intervals) rapidly diffuses the exhaust hydrogen to well below the lower limit of inflammability (4% hydrogen in air) [12] as it is released into the atmosphere.

Thermal Cycling Controller

The thermal cycling controller utilizes three electronic programmable timers to control the timing of the three phases of the thermal cycle: heating, holding at temperature, and cooling. The rapid heating of the specimen is achieved by direct electrical resistance heating, i.e., by passing a high electrical current through the specimen. Current is delivered to the specimen via the grip arm/grip assemblies. A manually-set variable transformer is used to adjust the voltage (0 to 4.91 V) across the specimen which provides the proper amount of current to achieve the desired temperature profile in the heating phase of the cycle. Thus the heating profile can be adjusted by varying the current, or the timer setting, or both. Approximately 290 W were required for heating to 900°C in 10 seconds. A second transformer (0 to 4.91 V) controls the specimen temperature during the holding phase of the cycle.

A constant voltage transformer feeds the thermal cycling controller with a voltage of 120 V AC ± 3% for a 15% line voltage fluctuation. This permits excellent repeatability (to within ±0.5°C) between succeeding heating cycles.

During the cooling phase, the electrical current is cut off, and the third timer opens the solenoid valve in the gas line, allowing the cooling gas to flow. The cooling profile can be adjusted by varying, independently or in combination, the gas flow rate, the gas temperature (via the cooling bath), or the timer setting.

Data Acquisition

A two-pen chart recorder recorded the output from both the load cell and the thermocouple for the duration of the test. During typical testing, the chart speed was set at 5 mm/min. The 6.7 kN (± 0.5% full scale accuracy) load cell output was recorded at the 5 V range. The K-type thermocouple output was recorded at the 50 mV range, and had an accuracy of ± 2.2°C or 0.75% of the measured temperature, whichever was greater. The cycles are counted on an electromechanical rotary type counter on the controller.

TEST SYSTEM CHARACTERISTICS

Thermal Profile

A test was conducted to characterize the thermal profiles of the specimen during the heating, holding and cooling phases of the cycle. Naturally, since the specimen was heated by internal resistance and the grips were cooled, the temperature profile along the gage length was not constant. Five K-type thermocouples (0.381 mm gage chromel and alumel wires) were spot welded along the gage length of the specimen on the edge opposite the knife edge. Thermocouple #3 was mounted in the center (comparable with a standard test), with two thermocouples spaced at 4 mm intervals on either side. The center thermocouple was used to set the temperature, again as is the case during standard testing. The output of each thermocouple was recorded on one of three calibrated chart recorders set in the 50 mV range at a chart speed of 15 cm/min.

The cycling parameters chosen for the proof test were the same as those to be used later in actual testing, that is, a thirty second cycle period with ten seconds for heating, ten

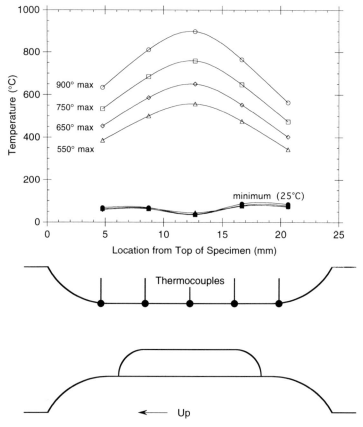

FIG. 5. Thermal profiles along the thermal fatigue specimen gage length at maximum and minimum temperatures for thermal cycling ranges between 25 °C and 900, 750, 650, and 550 °C. Accuracy: ± 2.2 °C or 0.75% , whichever is greater.

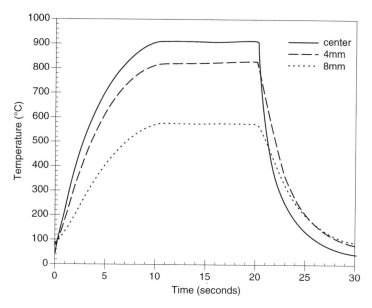

FIG. 6. Thermal cycle temperature profile during cycling between 25 and 900 °C as measured by thermocouples located at 4 mm intervals above the center of the specimen (see Fig. 5). Accuracy: ± 2.2 °C or 0.75% , whichever is greater.

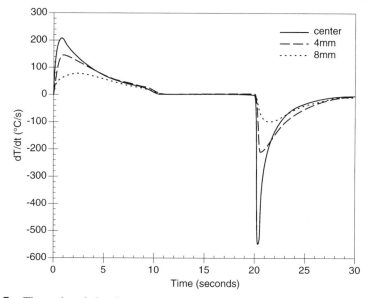

FIG. 7. Thermal cycle heating and cooling rates during cycling between 25 and 900 °C as measured by thermocouples located at 4 mm intervals above the center of the specimen (see Fig. 2.7). Accuracy: ± 10 °C/s or 17% , whichever is greater.

for holding, and ten for cooling. The lower temperature in the cycle was always 25 °C ± 10°C. Several upper temperatures, namely, 900, 750, 650, and 550 °C, were used. These temperatures would be used in later tests. Helium gas was used for the test with the chamber window open to allow multiple thermocouple access. The test system was run for over 150 cycles prior to recording any data, to allow the grips and chamber to warm up to operating temperatures and the output to become stabilized.

Figure 5 shows the thermal profile along the gage length of the specimen at both maximum and minimum temperatures for all four temperature ranges. Although the specimen is symmetrical, the temperatures on the upper portion are about 20 - 50 °C hotter than on the lower portion at maximum temperatures due to convection inside the chamber, while they are slightly colder (about 5°C) at minimum temperatures due to a slight difference in the cooling conditions between the upper and lower arms.

Figure 6 shows the thermal profile of the upper half of the specimen during cycling to 900°C, and Fig. 7 is the corresponding heating and cooling rates. Similar data were obtained for the other temperature ranges [10]. It should be noted that the cooling rates at the knife edge are significantly higher than those measured by the thermocouples at the rear of the specimen.

Thermal Stress Cycle

The thermal stress cycle is 180° out-of-phase with respect to the temperature cycle due to the fixed-grip condition, thus the lowest stress occurs at the highest temperature, and vice-versa. The load ratio (R) is determined by the combination of the preload level and the range of the thermal cycle, for example, with a preload of 1.57 kN at room temperature and cycling between 25 and 900°C, R = 0.14.

Temperature and Load Drift

As stated earlier, the specimen's upper and lower temperatures were adjusted whenever they drifted out of a ±10 °C range. The load was not adjusted after twenty cycles.

Cycle to cycle temperature repeatability was excellent with any deviations usually below the resolution limits of the thermocouples. However, there was a tendency for temperature to drift, especially during the first 1 000 cycles. After the initial 50 cycle stabilization phase, the drift rate for 900 °C cycling was typically less than 3 °C/hr downwards. This drift was most likely caused by an increase in the resistance of the specimen. After the first 1 000 cycles, typical drift rates were less than 2 °C/hr, and could be in either direction. Drift rates were lower at lower temperatures, but followed the same patterns.

The load on the specimen also tended to drift downward during testing, while the load range remained constant. Since all specimens were tested in a tension-tension condition, this load drift is probably due to creep of the test specimens. A typical example of this load drift is shown in Fig. 8. Note that most of the drift occurs during the first 500 to 1 000 cycles, reducing the maximum load from 1.57 kN to 1.37 kN by the 1 000th cycle, with further reduction to 1.35 kN by the 4 500th cycle. The minimum load follows the same pattern; however, the load range (i.e., maximum minus minimum) remains constant.

TEST PROCEDURES

Since environmental thermal fatigue testing is not a "conventional" test, the test

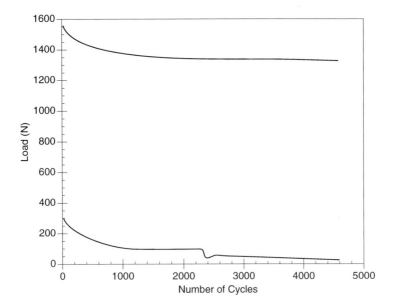

FIG. 8. Typical load drift of the maximum and minimum loads over the course of a test cycling between 25 and 900 °C. Jumps in the load are due to adjustments in temperature as it drifted out of the ± 10 °C tolerance. Accuracy: ±0.5%.

procedures needed to be evolved along with the development of the test apparatus. These procedures were designed to ensure accuracy of the data and consistency between tests. The procedures consisted of specimen mounting, test initiation, monitoring the specimen and test conditions, and test termination.

Care was taken during specimen mounting to ensure that the upper grip arm and grip were perfectly aligned with the specimen. After the chamber was sealed, a digital voltmeter was utilized to test for electrical continuity between the chamber and the grip arms. If continuity existed, the chamber mounts were adjusted to tilt the chamber slightly in order to bring it parallel to the grip arms.

The system was then flushed with the test gas more than 500 times its volume. When testing with hydrogen, argon was used for the initial flush to remove the air from the system prior to flushing with hydrogen.

After application of the preload, the controller was activated, beginning the thermal cycling, and the maximum and minimum temperatures were adjusted to the required levels. During the first twenty cycles as the grip arms heated up, the load was also readjusted upwards to account for the expansion of the grip arms. By the twentieth cycle, the vast majority of this thermal expansion was completed, and the load was not adjusted for the remainder of the test in order to maintain a fixed-grip condition. The temperature (either maximum or minimum) was adjusted whenever it drifted out of a range of ±10 °C for the duration of the test.

The chart recorder registered the output from both the thermocouple and the load cell for the entire test, enabling any deviations to be detected. The specimen was periodically examined using the long-focal length traveling microscope to detect fatigue cracks and surface damage. However, only those cracks that were on the window-side of the specimen and not covered by oxidation could be detected.

The thermal fatigue test was terminated for one of two reasons: fracture of the specimen or the attainment of a predetermined number of cycles. Upon the completion of the test, the load was immediately released (if the specimen was not fractured); otherwise, the cooling of the grip arms would increase the load well beyond the preload level. The specimen was then carefully removed after cool-down from the chamber for later analysis.

PROOF TESTS

The proof tests were performed under a fixed fixture condition with a preload (1.57 kN) equal to a tensile stress of 241 MPa (50% of room temperature yield strength). Details of the material microstructure, the cyclic thermal-loading profiles and fractographic analyses are reported elsewhere [10,13-15]. Results were obtained for the thermal fatigue response of γ-titanium aluminide in helium (as an inert reference), hydrogen, and air. These results demonstrate the adequacy of the system for the desired performance and the importance of environmentally assisted thermal fatigue testing.

The thermal fatigue testing of γ-TiAl revealed a significant effect of environment on the material's thermal fatigue lifespan [10,13-15]. In particular, hydrogen was found to severely attack the material during thermal fatigue, resulting in a lifetime as low as three (3) cycles, while no failures were observed in helium for test durations of over 4 100 cycles, Fig. 9. However, the hydrogen sensitivity of the material is strongly dependent on the partial pressure of residual oxygen [10,13-15]. The impurity levels of oxygen in the thermal fatigue test system are dependent on the length of the gas flush prior to the test and on the specimen cleanliness. In order to fully quantify the effect of residual oxygen, more

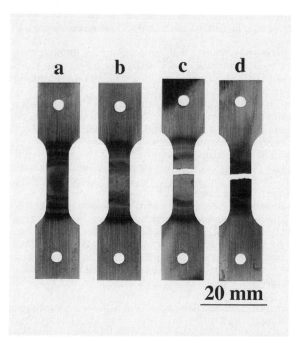

FIG. 9. Specimens cycled to 900 °C in (a) helium, 4 626 cycles, (b) air, 4 100 cycles, (c) air, lifetime: 2 674 cycles, and (d) hydrogen, lifetime: 3 cycles.

control over the impurity oxygen level is required. This can be accomplished by modifying the equipment to permit evacuation of the system to remove impurities, and by adding further purification of the source gas.

The optical system used for crack monitoring proved to be adequate for monitoring a single, slow-growing crack, but monitoring the fast-growing multiple cracks produced by a strong hydrogen attack was difficult. Thus, alternatives need to be explored to improve the monitoring of crack initiation and growth.

Thermal fatigue testing in gaseous environments appears to be a critical tool for evaluating γ-titanium aluminide and other advanced materials for use in engineering applications in deleterious environments at high temperatures. Hydrogen attack on γ-titanium aluminide has been shown to be tremendously enhanced by thermal cycling at high temperatures, an effect that no other constant-temperature test in a hydrogen atmosphere revealed [1-8]. The results showed that the equipment and test method developed here are an effective tool for material evaluation, specifically for the critical assessment of materials for high temperature applications in hydrogen and hydrogenous gases.

SUMMARY

A test system and associated procedures were developed to conduct studies on the combined effects of thermal fatigue and gaseous environments on a gamma titanium aluminide alloy, with the goals of assessing the alloy's resistance to environmentally enhanced thermal fatigue, and developing a mechanistic understanding of the processes involved. Proof tests demonstrated the system's ability to produce repeatable rapid thermal cycling with good thermal stress control, as well as the ability to safely maintain a gaseous hydrogen environment. The results indicate that thermal fatigue testing in gaseous environments is a critical tool for developing and evaluating advanced materials for use in engineering applications in deleterious environments at high temperatures.

REFERENCES

[1]   Lipsitt, H. A.,"Titanium Aluminides - An Overview", Proceedings of Materials Research Society Symposium on High Temperature Ordered Intermetallic Alloys, Vol. 39, Materials Research Society, Pittsburgh, PA, 1985, pp. 351-364.

[2]   Froes, F. H., Suryanarayana, C., and Eliezer, D., "Review of Synthesis, Properties, and Applications of Titanium Aluminides", Journal of Materials Science, Vol. 27, No. 19, 1992, pp. 5113-5134.

[3]   Kim, Y-W., "Intermetallic Alloys Based on Gamma Titanium Aluminide", Journal of Metals, Vol. 41, No. 7, 1989, pp. 24-30.

[4]   Chan, K. S. and Kim, Y-W., "Influence of Microstructure on Crack Tip Micromechanisms and Fracture Behavior of a Two-Phase TiAl Alloy", Metallurgical Transactions, Vol. 23A, 1992, pp. 1663-1677.

[5]   Nakamura, M., Hashimoto, K., and Tsujimoto, T., "Environmental Effect on Mechanical Properties on TiAl base Alloys", Journal of Materials Research, Vol. 8, no. 1, 1993, pp. 68-77.

[6]   Chan, K. S. and Kim, Y-W., "Rate and Environmental Effects on Fracture of a Two-Phase TiAl Alloy", Metallurgical Transactions, Vol. 24A, 1993, pp. 113-125.

[7]   Liu, C. T. and Kim, Y-W., "Room-Temperature Environmental Embrittlement in a TiAl Alloy", Scripta Metallurgica et Materialia, Vol. 27, 1992, pp. 599-603.

[8]   Takasugi, T., Hanada, S., and Yoshida, M., "Environmental Embrittlement of Gamma Titanium Aluminide", Journal of Materials Research, Vol. 7, 1992, pp. 2739-2745.

[9]   Incropera, F. P. and DeWitt, D. P., Fundamentals of Heat and Mass Transfer, 3rd
      ed., John Wiley & Sons, New York:, 1981.
[10] Dunfee, W., "Environmentally Enhanced Thermal Fatigue and Cracking of a Gamma-
      based Titanium Aluminide Alloy", Master of Science Thesis, Lehigh University,
      1994.
[11] Shigley, J. E. and Mischke, C. R., Mechanical Engineering Design, 5th ed.,
      McGraw-Hill, New York, 1989.
[12] CRC Handbook of Chemistry and Physics, ed.,Weast R. C., 66th edition., CRC
      Press, Boca Raton, 1985.
[13] Gao, M., Dunfee, W., Wei, R. P., and Wei, W., "Thermal Fatigue of Gamma
      Titanium Aluminide in Hydrogen", Proceedings of TMS Symposium on Fatigue and
      Fracture of Ordered Intermetallic Materials, The Minerals, Metals and Materials
      Society, Pittsburgh, 1993, pp. 225-237.
[14] Gao, M., Dunfee, W., Wei, R. P., and Wei, W., "Thermal Fatigue of Gamma
      Titanium Aluminide in Hydrogen and Air", submitted to TMS Symposium on Fatigue
      and Fracture of Ordered Intermetallic Materials II, Chicago, 1994.
[15] Dunfee, W., Gao, M., Wei, R.P., and Wei, W., "Hydrogen Enhanced Thermal
      Fatigue of γ-Titanium Aluminide", accepted for publication in Scripta Metallurgica et
      Materialia, 1995.

J. Dai[1], N.J. Marchand[2] and M. Hongoh[3]

THERMAL MECHANICAL FATIGUE CRACK GROWTH IN TITANIUM ALLOYS: EXPERIMENTS AND MODELLING

REFERENCE: Dai, J., Marchand, N. J., and Hongoh, M., ''Thermal Mechanical Fatigue Crack Growth in Titanium Alloys: Experiments and Modelling,'' Thermomechanical Fatigue Behavior of Materials: Second Volume, ASTM STP 1263, Michael J. Verrilli and Michael G. Castelli, Eds., American Society for Testing and Materials, 1996.

ABSTRACT: Strain controlled thermal–mechanical fatigue crack growth (TMFCG) tests were conducted on two titanium alloys, namely Ti–6Al–4V and Ti–6Al–2Sn–4Zr–6Mo, to evaluate the effect of phase angle between strain and temperature on the TMFCG rates. Three fracture mechanics parameters were used to correlate the data: the ΔK, ΔK_ϵ and ΔK_{eff}. A fractographic study of the specimens tested under TMF was carried-out to identify the mechanisms responsible for cracking in these two titanium alloys. Hence, specimens tested under in–phase (ϵ_{max} at T_{max}), out–of–phase (ϵ_{min} at T_{max}) and counter–clockwise diamond (90° out–of–phase) conditions were compared to specimens tested under isothermal conditions (T_{min} and T_{max}) for different ΔK_{eff} levels. The dominant TMF cracking mechanisms were mechanical fatigue (crack tip plasticity) and oxygen–induced embrittlement. The ΔK_{eff} was found to be the only parameter to properly correlate all the data obtained under various testing conditions. A model is developed to predict the TMFCG rates based solely on isothermal data. The model uses a linear summation of the contributions to crack growth of the two dominant mechanisms which are active at the minimum and maximum temperature of the cycle. A discussion on the applicability of the model to predict the fatigue lives of actual components is discussed.

KEYWORDS: Titanium alloys, crack growth, oxygen embrittlement, fatigue life prediction, TMF.

INTRODUCTION

Titanium alloy forgings are extensively used in gas turbine engines for applications ranging from impellers, compressor discs, sealings, cases and

[1]Formerly Graduate Student, Dep. of Materials Engineering, École Polytechnique, Montréal, Canada, H3C 3A7. Now Stress Engineer, Bombardier/Canadair, Ville St–Laurent, Québec, Canada H4R 1K2.

[2]Formerly Associate Professor, Dep. of Materials Engineering, École Polytechnique, Montréal, Canada, H3C 3A7. Now R & D Engineer AMRA Technologies, 4700 de la Savane, Montreal (Quebec), Canada, H4P 1T7.

[3]Project Engineer, Stress Analysis Group, Pratt and Whitney Canada, 1000 Marie–Victorin, Longueuil, Canada, J4G 1A1.

fan blades just to name a few. In recent years, the demand toward achieving higher thrust–to–weight ratios has led to increasing the temperature of all stages of the engines. As a result of operating at higher temperatures, compounded by the introduction of new regulations for safety, the accurate predictions of the crack initiation lives ($a_o \approx 1/32$"– 800 μm) and crack propagating lives (da/dN) have become paramount for Ti– based components [1–2].

In most engine applications and in some advanced supersonic air– frame designs, the service conditions includes severe temperature excursions combined with simultaneous cycling of mechanical loads, commonly termed thermal–mechanical fatigue (TMF) [3]. The influence of TMF cycles on the material's fatigue crack growth resistance (da/dN) is not well understood at present. A review of the available data pertaining to thermal mechanical fatigue crack growth (TMFCG) can be found elsewhere [4– 5]. At present, the common procedure to predict the LCF lives is to use isothermal data at the maximum temperature of a cycle or spectrum block to calculate the TMF crack growth rates. This procedure can often be extremely conservative leading to unrealistic crack propagation life predictions. Moreover, there are instances where this procedure is non– conservative, i.e. the TMFCG rates exceed those at the maximum temperature (isothermal case) [4–5]. In either situation, these predictive methodologies based on isothermal data have not addressed the fatigue mechanisms involved. At present, physically and mechanistically based predictive schemes for TMFCG are almost non–existent in the literature [4].

TMF cycling is expected to introduce a multitude of cyclic deformation and damage mechanisms. In the higher temperature portion of the TMF cycle, plastic deformation and various time–dependent creep mechanisms may operate, along with aggressive environmental attack and microstructural changes. In the lower temperature portion, some of these mechanisms become inoperative because of insufficient thermal activation. This alternate operation of high and low temperature mechanisms differs considerably from the situation encountered in isothermal fatigue.

The purpose of this paper is to report on the fatigue behaviours and micromechanisms involved in TMFCG of two titanium alloys: Namely Ti–6Al– 4V (Ti64) and Ti–6Al–2Sn–4Zr–6Mo (Ti6246). The fractographic observations are focused on the relationship between the fracture morphology and the α and β phase microstructural features, including the α/β interfaces. In turn, this information was used to develop a crack growth model for predicting TMFCG rates. Hence an engineering procedure for TMFCG rate predictions is presented which is based solely on isothermal data and developed considering the physical mechanisms responsible for crack growth over the range of temperatures of the TMF cycles.

MATERIALS AND EXPERIMENTAL PROCEDURES

Two titanium alloys were investigated, respectively Ti 64 (Ti–6Al– 4V) and Ti6246 (Ti–6Al–2Sn–4Zr–6Mo). All the specimens were taken from compressor discs forgings [5–6]. The initial microstructure of the α/β forged Ti64 consisted of equiaxed α particles evenly distributed in the α + β matrix. The primary α grain size was approximately 32 μm [7]. The microstructure of the β–forged Ti6246 consisted of coarse acicular α–phase laths randomly distributed in the transformed β matrix [7]. The average size of the α laths was about 60 μm x 2 μm. The volume fraction in both alloys was approximately 50% for each phase. The microstructure of each alloy is shown in Figure 1.

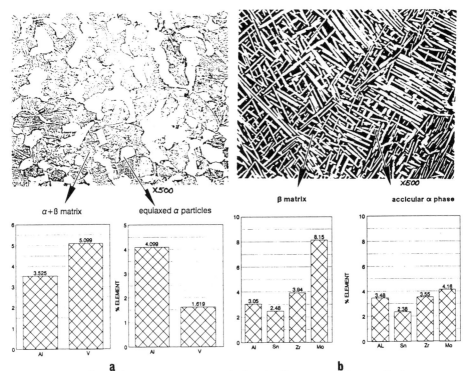

a   b

FIG. 1— Microstructures and chemical compositions of the
respective phases in (a) Ti64 and (b) Ti6246

The specimens used in this fatigue study were single edge notched
(SEN) specimens with rectangular cross section (W = 10.67 mm, B = 5.33 mm)
and effective gauge section (for the extensometer) of 25.4 mm (see Figure
2). The isothermal tests and the LCF crack initiation tests ($a_o \approx$ 1/32" –
800 μm) were carried out on specimens having a semi-circular notch which
radius was ρ = 0.042" (1.06 mm) thus entailing ρ/W = 0.1. The specimens
used in the TMFCG program were pre-flawed by EDM with the initial notches
having 0.254 mm width and 0.5 mm depth. Before TMF testing the specimens
were fatigued (pre-cracked) at room temperature. The pre-cracking was
carried-out at a frequency of 10 Hz, with the R-ratio (σ min/σ max) equal
to zero. The maximum load for precracking was set to one-third of the
expected TMF testing load level. In all cases the crack length after pre-
cracking was less than 1.05 ± 0.1 mm (a/W ≃ 0.1).

All the TMFCG tests were performed under fully reversed (R = −1)
mechanical strain controlled conditions with a cyclic period of 100 sec
(0.6 cpm or 0.01 Hz). The cyclic stress-strain histories (hysteresis
loops) were recorded continuously. The TMFCG data were obtained for a
temperature cycle of 150 to 400°C for Ti64 and of 200 to 480°C for Ti6246.

Three types of TMF cycles were investigated. These included in-
phase (IP – ϵmax at Tmax), out-of-phase (OP – ϵmax at Tmin) and counter-
clockwise diamond (CCD) or faithful cycling (90° out-of-phase). Isothermal
FCG tests at Tmin and Tmax were also carried-out for the sake of
comparison and for identifying the parameters used to characterize the
cracking mechanisms. All the details pertaining to the testing procedures

(temperature calibration, alignment procedures, etc.) can be found elsewhere [5–6, 9–10]. The testing matrices for both alloys are shown in Table 1. Because little scatter was found between specimens of Ti6246 tested under the same loading conditions (strain range, R-ratio, frequency and temperature) [5–6, 9], fewer TMFCG experiments were carried out on Ti6246 as compared to Ti64

TABLE 1—Thermal mechanical fatigue crack growth test matrices

(a) Ti464 ($R_\epsilon = -1$)

TS_NAME	$\Delta\epsilon$ %	$\dot{\epsilon}$ (sec^{-1})	waveform	Temp. (oC)	Remarks
TMF 51	0.40	1.6E–4	triangular	400 ↔ 150	in–phase TMFCG
TMF 52	0.40	1.6E–4	triangular	400 ↔ 150	in–phase TMFCG
TMF 53	0.35	1.6E–4	triangular	400 ↔ 150	in–phase TMFCG
TMF 54	0.35	1.4E–4	triangular	400 ↔ 150	in–phase TMFCG
TMF 58	0.35	1.4E–4	triangular	400 ↔ 150	in–phase TMFCG
TMF 55	0.35	1.4E–4	triangular	400 ↔ 150	out–of–phase TMFCG
TMF 56	0.40	1.6E–4	triangular	400 ↔ 150	out–of–phase TMFCG
TMF 81	0.35	1.4E–4	triangular	400 ↔ 150	out–of–phase TMFCG
TMF 57	0.35	1.4E–4	triangular	400 ↔ 150	diamond TMFCG
FCG 58	0.35	1.4E–4	triangular	150	isothermal
FCG 83	0.35	1.4E–4	triangular	150	isothermal
FCG 59	0.35	1.4E–4	triangular	400	isothermal
FCG 84	0.35	1.4E–4	triangular	400	isothermal

(b)   Ti6246 ($R_\epsilon = -1$)

TS_NAME	$\Delta\epsilon$ (%)	$\dot{\epsilon}$ (sec^{-1})	waveform	Temp. ($^\circ$C)	Remarks
TMF 61	0.300	1.2E–4	triangular	480 ↔ 200	in–phase TMFCG
TMF 62	0.350	1.4E–4	triangular	480 ↔ 200	in–phase TMFCG
TMF 65	0.375	1.5E–4	triangular	480 ↔ 200	in–phase TMFCG
TMF 63	0.350	1.4E–4	triangular	480 ↔ 200	out–of–phase TMFCG
TMF 66	0.375	1.5E–4	triangular	480 ↔ 200	out–of–phase TMFCG
TMF 64	0.375	1.5E–4	triangular	480 ↔ 200	diamond TMFCG
TMF 61	0.350	1.4E–4	triangular	200	isothermal
TMF 67	0.375	1.5E–4	triangular	480	isothermal

The electrical potential drop (EPD) technique was employed to monitor crack initiation and growth. This techniques make use of an increase in the electrical resistance of a conducting material due to crack initiation and growth. The well-established DC potential drop methods use constant DC current passing through the specimen and measure the potential drop between the probes on either side of the crack or notch. However, DCPD requires large currents to produce an adequately measurable potential drop, which is problematic for measuring small cracks with high resolution.

The AC potential drop (ACPD) methods use high frequency AC current of constant amplitude passing through the specimen and measure the potential drop between the probes. As compared to DCPD systems, the sensitivity and linearity of ACPD system are enhanced by the skin effect resulting from the high frequency current passing through the specimen. Using the new generation of Lock-in amplifiers and advanced phase shift detection (PSD) circuits, the accuracy and sensitivity of AC potential measurement systems have been greatly improved in the recent years [9].

ACPD Probe Set-up

The schematic for the ACPD probe set-up is illustrated in Figure 2. It can be seen that ceramic spacers were used to protect and control the probe configurations during the test. Two sets of probes, namely the working (PD) and reference probes, were used to sense the AC potential drop signals. For the TMFCG tests, the probe vertical spacing was about 0.38 mm. The AC current leads used were 0.5 mm (0.020″) in diameter while

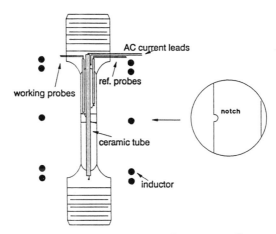

FIG. 2—Schematic of the ACPD probe set-up for crack
initiation and propagation monitoring

the PD probes were 0.127 mm (0.005″) in diameter. Both types of wires were made of pure titanium to minimize any junction effects. Spot welding of the probes were performed using a specially designed holder and an UNITEK welding system mounted on a microscope. With this system the welding position and pressure can be easily controlled to guarantee reproducibility of the spot welds. Since the specimen was heated by an RF inductive heating coil, the ACPD probes and current leads were set in parallel with the RF magnetic field to minimize undesired emf noise pick-up from the induction heating coil.

The ACPD signals were amplified by a newly developed system CGM5[a] which consists of an advanced two-channel pre-amplifier, a multi high frequency current source and an automatic AC signal phase shift control circuit (PSD). A similar system is described in detail elsewhere [9]. In this study, an AC current of 30 KHz with an amplitude 1000 mA was used for the both titanium alloys. Such a current was insufficient to generate any notable heating of the specimen. Details of the experimental system to measure the TMFCG rates, including thermal compensation associated with thermal cycling and thermal fluctuations, are provided elsewhere [9].

Fractographic Observations

The fractured specimens were observed in a JEOL 840 scanning electron microscope equipped with a LINK energy dispersive X-ray analysis system. The relative concentrations of the beta stabilizers (V in Ti64 and Mo in Ti6246) were employed to identifiy the phases of the different microregions on the fractured surfaces (see Figure 1). For example, the primary α-phase in Ti64 contains approximately 1.5% V while the $\alpha + \beta$ matrix contains approximately 3.2% V. The microfractographic features at different stages (corresponding to different values of ΔK_{eff}) of fatigue crack growth were studied. The fractographic results pertaining to the isothermal tests at T_{min} and T_{max} are reported elsewhere [7].

TMFCG CORRELATING PARAMETERS

Three types of fracture mechanics parameters were investigated in the da/dN data correlation procedures; namely the stress intensity factor range (ΔK), the maximum stress intensity factor (K_{max}) and the strain intensity factor (ΔK_ϵ). The ΔK's, which are commonly used for FCG data correlation, were computed using

$$\Delta K = \left(\frac{\Delta N}{BW}\right) \cdot \sqrt{\pi a} \cdot H\left(\xi, \eta, \eta_b\right) \quad . \tag{1}$$

Here ΔN is the **measured** load range during the strain controlled cycling, B is the thickness, W is the width, "a" is the crack length measured from the specimen edge, and H (ξ, η, η_b) is a geometrical function which characterizes the SEN geometry as well as the boundary conditions imposed during the testing [5-6, 9-10].

The K_{max}'s were computed using

$$K_{max} = \left(\frac{N_{max}}{BW}\right) \cdot \sqrt{\pi a} \cdot H\left(\xi, \eta, \eta_b\right) \tag{2}$$

where N_{max} is the peak value of the **measured** load during cycling. The ΔK_ϵ were computed using

$$\Delta K_\epsilon = \Delta \epsilon \sqrt{\pi a} \cdot H\left(\xi, \eta, \eta_b\right) \tag{3}$$

---

[a]Trademark Matelect/AMRA Technologies

where $\Delta\epsilon$ is the far field mechanical strain which is controlled during the test. Although this parameter does not have a clear physical meaning it was used in this study because it has been employed in many reported studies [4].

It should be emphasized here that the tests were carried-out under strain controlled conditions, and that under strain controlled testing, the minimum and maximum stresses in the cycle are not controlled and achieve values depending on the inelastic deformation and hardening or softening characteristics of the material. The average of these two, the mean stress, is also free to stabilize to whatever value is dictated by the flow behaviour of the material. Therefore, the stress R-ratio is not necessarily constant throughout a test and the determination of the effective stress intensity factor range (ΔK_{eff}) defined as

$$\Delta K_{eff} = \frac{(N_{max} - N_{op})}{BW} \sqrt{\pi a}\ H\left(\xi,\ \eta,\ \eta_b\right) \qquad (4)$$

is difficult because it is difficult to measure (N_{op}) the opening load level. However it is assumed that the cracks were opened (when the loads were tensile even if the imposed strains were compressive. Thus the ΔK_{eff} is equal to K_{max}. The measured a vs N curves, for all testing conditions, have been analysed in terms of ΔK, ΔK_{eff}, (K_{max}) and ΔK_{ϵ}.

RESULTS AND DISCUSSION

ACPD Response to Fatigue Crack Initiation and Growth

Although the main purpose of this paper is to report about TMFCG rates, the corresponding isothermal tests were also carried out. The most interesting features pertaining to the isothermal testing are reported here to emphasize the capabilities of the crack monitoring system. Typical ACPD responses (working probes) as a function of time are displayed in Figure 3. It can be seen that the ACPD signals continuously change during each fatigue cycle, as a result of straining or crack opening/closure and for complex mechanical-magnetic interactions. As the damage cracking at the root of the notch grows with increasing number of cycles, the AC potential time curves gradually shifts upward.

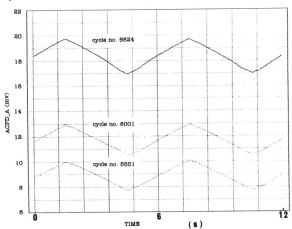

FIG. 3—ACPD versus time (two cycle periods) curves
at three cyclic stages.

By plotting the ACPD changes as a function of the far field stress (see Figure 4) a hysteresis loop behaviour can be observed, which is associated with the stress–strain behaviour of the material and crack closure effets [9]. Note that stress ratio (R_σ) employed to generate the date shown in Figure 4, was equal to zero. On the other hand the TMFGC data were generated under strain controlled R_ϵ with varying R_σ. In this particular testing program, no attempt was made to determine N_{op} since the opening stresses for real components in service are almost impossible to measure or compute.

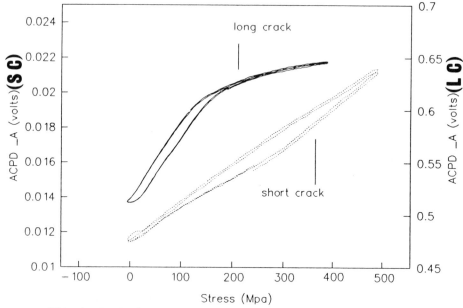

FIG. 4—Typical ACPD signal plotted as a function of the applied stress for short (SC) and long (LC) cracks (Ti6246).
Tests performed at RT, stress controlled and R=0.

As shown in Figure 5 (for Ti64 tested at 400°C), the peak values of the ACPD loops were plotted as a function of the number of fatigue cycles. From this plot, a clear picture of the crack initiation and growth processes can be obtained. The reference ACPD signal is constant which indicates that the testing temperature and the far-field strain remained stable throughout the test. The gradual increase of the ACPD signal of the working probes is due to crack initiation and/or growth at the root of the notch. At the end of this particular test, the crack length was measured to be 1.07 mm (0.042″). Thus the ACPD crack growth monitoring resolution is about 18 μm per 1 mv ACPD change (at 60dB). The crack growth measurements were performed with a resolution better than 2 μm per 1mv at low temperature (< 550oC).

A linear relationship was found between the ACPD responses and the (physically) measured crack lengths as shown in Figure 6. Although the slope of the curve was slightly affected by the testing temperature and probe positioning, this feature (linearity) is another advantage of ACPD technique over other crack detection techniques. Because the initial readings of the ACPD signal are sensitive to factors such as probe lengths, positioning, spacing and connections etc. unlike the signal processing procedures used by Hwang and Ballinger [11], normalization of the ACPD signals was not used in this study.

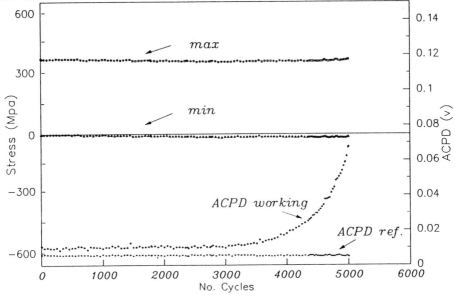

FIG. 5—Typical ACPD signal (maximum value within each cycle) as a function of number of cycles (Ti64, 400°C).

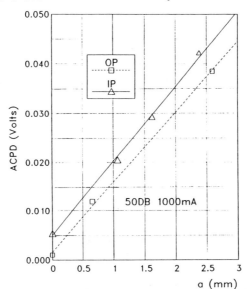

FIG. 6—ACPD signal as a function of crack length (Ti6246).

Fatigue crack growth analysis—A typical crack growth rate vs. crack length curve (Ti64 tested at 400°C) is shown in Figure 7. When the cracks were short with their average depth from the root of the notch less than 60 μm, which is on the order of the microstructural characteristic features (i.e. the equiaxed α-grains) of the material, the da/dN varies erratically with increasing crack length. In this region most of da/dN data span two orders of magnitude, which is similar to the data reported by Miller [12] and Newman [13] for aluminum alloys. This implies that these growth rates (for a <60μm) are controlled by local microstructural

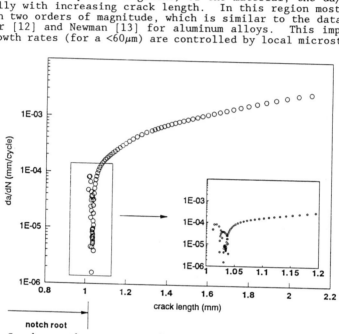

FIG. 7—Crack growth rates as a function of crack length (Ti64, 400°C)

events and thus cannot be rationalized by a fracture mechanics parameter alone. On the other hand, when the cracks are longer than 60μm, their growth rates start increasing monotonically with increasing crack length and thus can be correlated if a proper fracture mechanics parameter is used. This suggest that the da/dN's are controlled by a fracture mechanics parameter soon after the average crack length (depth) has reached 60μm for Ti64. A similar notch crack behaviour was also found with Ti6246.

The TMFCG behaviour of Ti64—Figure 8 shows the crack growth rates (da/dN) plotted as a function of ΔK, ΔK_ϵ and ΔK_{eff} (i.e. K_{max}). These results are obtained using the da/dn vs a curves along with the definition of ΔK, ΔK_ϵ and ΔK_{eff} defined earlier. The following conclusions can be drawn from these plots. First it is obvious that ΔK and ΔK_ϵ are not able to correlate the data obtained from the various testing conditions. Furthermore, the scatter increases with increasing ΔK or ΔK_ϵ which indicates that the compressive part of the load cycle, which depends on the temperature waveform, does not contribute significantly to crack growth. However a good correlation is obtained using ΔK_{eff}. All the test data fell in a narrow scatterband. A close look at this plot (Figure 8.c) indicates that the out-of-phase data are similar to the isothermal data at Tmin and that the in-phase data are similar to the Tmax results. All the da/dN vs ΔK_{eff} curves are basically parallel to each other indicating

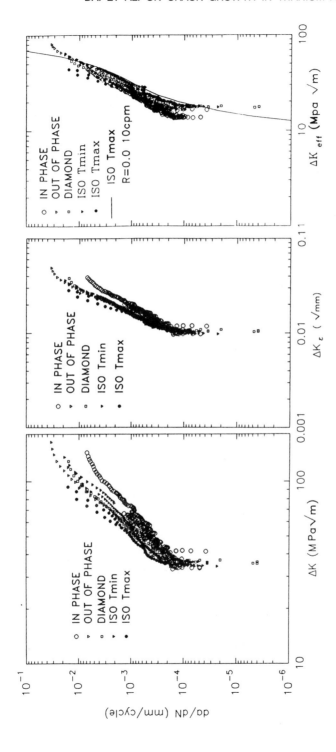

FIG. 8—TMFCG data of Ti64 plotted as a function of (a) ΔK, (b) ΔK_ϵ and (c) ΔK_{eff}

similar operative crack growth mechanisms for all testing conditions. This conclusion is supported by fractographic analyses which showed similar features for all testing conditions.

The results obtained at Tmax but at a frequency of 10 cpm (16.7 times faster than the TMF cyclic frequency) are also shown in Figure 8.c [14]. These data indicates that the effect of temperature decreases with increasing frequency suggesting that fatigue cracking in Ti64 is affected by time—dependent mechanisms. The FCG data available from PWA [6, 14], which were obtained using CT specimens cycled at 10 cpm, show no temperature effect on the crack growth rates from room temperature up to 400°C.

The TMFCG behaviour of Ti6246—The results (da/dN curves) pertaining to Ti6246 are shown in Figure 8. As can be seen the same conclusions drawn for Ti64 can be applied to Ti6246; namely that ΔK and ΔK_ϵ are not the proper driving force for fatigue cracking whereas ΔK_{eff} correlates the various data much better. However, contrary to the data of Ti64, a pronounced effect of environment can be observed in Ti6246 tested at high temperatures (isothermal Tmax and in—phase cycling). Note the shape of the da/dN vs ΔK_{eff} curves obtained at Tmax (isothermal) and for in—phase cycling. Here a plateau—like shape curve, typical of corrosion—fatigue experiments is observed. The conclusion that the time—dependent damage mechanisms in Ti64 and Ti6246 are associated with environmental attack (oxygen and oxidation embrittlement) is supported by the microfractographic study of the fractured surfaces. As expected, with increasing crack growth rates, the environmental damage becomes less important and all the da/dN data merge into a narrow scatterband.

Fractographic Studies

Fractography of Ti64—Figure 10 shows a montage of the fracture surfaces obtained under in—phase (IP) condition ($\Delta K_{eff} \simeq 22MPa \sqrt{m}$). The overall fracture surface is quite crystallographic and the various microstructural features of Ti64 (primary α and the $\alpha + \beta$ matrix) can be easily identified. Evidence of cleavage—like facets associates with the α—phase can be identified (see arrow A). The $\alpha + \beta$ matrix appears to fracture in a more ductile manner and tearing—like features can be found between the α—phase grains (see arrow B). Well—defined fatigue striations are often found in the equiaxed α—grains (see arrow C). Fractographic features of the cleavage—like facets showed that fracture began near or at the α/β interface and proceeded into the grain. These observations indicated that the α—phase islands failed first and that the surrounding $\alpha + \beta$ matrix fractured under considerably higher cyclic strains as confirmed by the presence of coarse striations and faint crystallographic features in the $\alpha + \beta$ matrix. Some secondary cracks were also found at the α/β interfaces [5].

Figure 11 shows the corresponding fracture surface ($\Delta K_{eff} \simeq 22$ MPa $\sqrt{m}$) obtained under out—of—phase (OP) conditions. Pronounced rubbing lines parallel to the crack growth direction are seen. Depressions which present faint fatigue striations are also present (see arrow A). The overall fractographic features are flat and less crystallographic than for the IP tests. There are indications of crack surface rubbing and thus, of plasticity—induced crack closure effects (see arrow B). Note that for the OP condition, the maximum compressive strain is applied at the maximum temperature when the yield strength is relatively low.

Additionnally, secondary cracking along the α/β interfaces was found to be less prominent than for the IP tests.

In the counter—clockwise diamond (CCD) conditions, the fractographic features were intermediate between those observed for the IP and OP tests, as shown in Figure 12. There are less α—phase islands which appear as depressions and there are less indications of fracture surface rubbing than for the IP tests. Also, well—defined striations and some heavily depressed areas as for the IP tests can be observed. No continuous

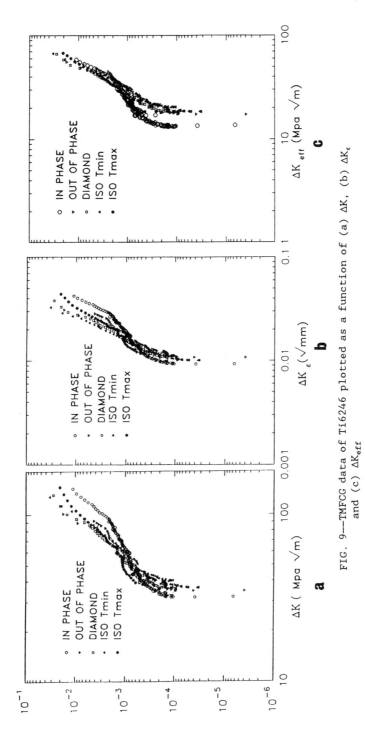

FIG. 9—TMFCG data of Ti6246 plotted as a function of (a) ΔK, (b) ΔK_ε and (c) ΔK_{eff}

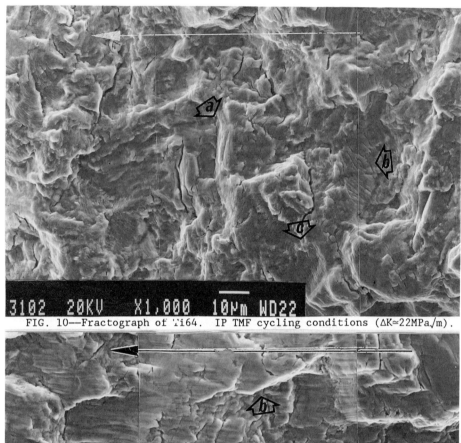

FIG. 10—Fractograph of Ti64. IP TMF cycling conditions ($\Delta K \approx 22 MPa\sqrt{m}$).

FIG. 11—Fractograph of Ti64. OP TMF cycling conditions ($\Delta K \approx 22 MPa\sqrt{m}$).

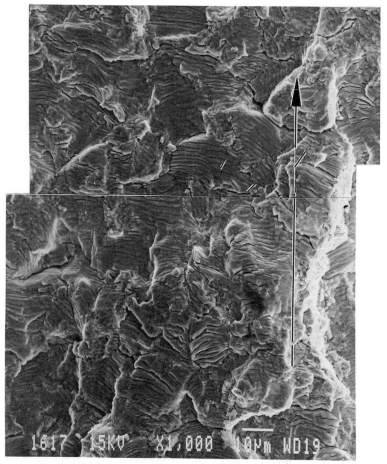

FIG. 12—Fractograph of Ti64.   CCD TMF cycling conditions
(ΔK≈22MPa√m).

rubbing lines are found as in the case of OP. The fracture surfaces are
less crystallographic and secondary cracking is less pronounced than for
the IP tests.   The same fractographic differences between the IP, OP and
CCD tests were also found at higher ΔK_{eff} values.

The fracture surfaces of the isothermal FCG tests at T_{max} and T_{min}
were also studied [7,10].   The overall fractographic features presented
more secondary cracking but otherwise appeared similar to those for the IP
tests.   Furthermore the effect of increasing temperature was to accentuate
the features describes above.

Fractography of Ti6246—Fractographic features observed for an
intermediate ΔK_{eff} ($\approx$ 20 MPa √m) are presented in Figures 13, 14 and 15 for
IP, OP and CCD cycling conditions respectively. The overall fractographic
features associated with the IP cycling (see Figure 13) appear relatively
crystallographic and can be easily associated with the microstructural

FIG. 13—Fractograph of Ti6246. IP TMF cycling conditions ($\Delta K \approx 20 MPa\sqrt{m}$).

features of this alloy. Many acicular α-phase packets were found to form
flat facets which presented fine, well-defined striations (see arrow A).
Acicular α phase laths with other orientations with respect to the loading
axis were delaminated from the β matrix leaving parallel channel-like
traces on the fracture surface (see arrow B).   These features indicate
preferential cracking paths along the α/β interfaces.   As a result,
depending on the relative orientation of the α/β interfaces with respect

to the loading axis, either striation–like lines were found along the delaminated areas or secondary cracking occurred along the α/β interfaces (see arrow C).   With increasing ΔK, the amount of secondary cracking increases. Rather different fractographic features were obtained for the OP conditions (at similar ΔK_{eff}'s).   As seen in Figure 14, the overall fractographic aspect is eminently flat and less crystallophic than for the IP cycling.   Furthermore, the relationships between the fractographic and metallographic features of the material are much less evident.

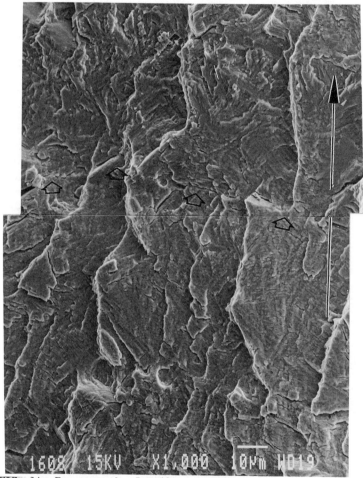

FIG. 14—Fractograph of Ti6246. OP TMF cycling conditions ($\Delta K \approx 20 MPa\sqrt{m}$).

Nevertheless there is evidence that the crack was locally stopped by microstructural obstacles, such as grain boundaries, thus producing clear crack arrest lines (see arrow A).

The fracture surface obtained under CCD conditions present fractographic features which are intermediate between those found for IP and OP cycling (see Figure 15).   There is less evidence for interfacial crack growth than for IP cycling and the overall morphology is much less crystallographic.   These observations are in agreement with the result that the macroscopic da/dN vs ΔK_{eff} behaviour for the CCD cycling was more similar to that for OP than for IP cycling.

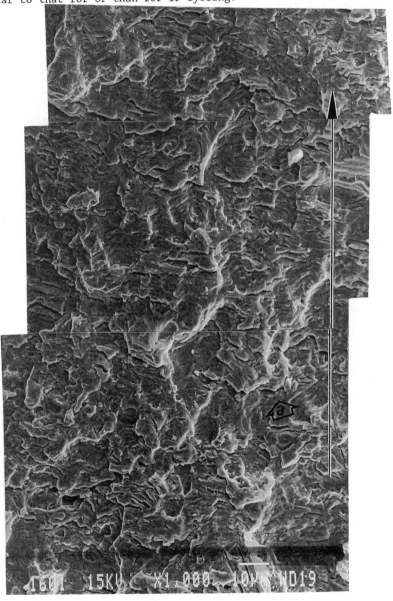

FIG. 15—Micrograph of Ti6246.   CCD TMF cycling conditions ($\Delta K \approx 20 MPa \sqrt{m}$).

With increasing ΔK_{eff} the differences between the fracture surfaces of the IP, OP and CCD became small. Again, this is consistent with the result that a high ΔK_{eff} all the da/dN vs ΔK_{eff} curves converge to form a single line. For all three TMF conditions the amount of secondary cracking was found to increase with increasing ΔK_{eff}.

At low ΔK_{eff}'s ($\Delta K_{eff} < 15$ MPa $\sqrt{m}$), the effects of environment became more pronounced. Hence oxidation deposites, brittle fracture facets and secondary cracking along α/β interfaces (which can be attributed to the α/β interfaces weakened by oxidation) become more important and fatigue striations become much less visible. This indicates that environmentally-assisted cracking becomes the dominant mechanism.

The α/β interfaces, especially for Ti6246, strongly influence the local crack propagation. Interfaces perpendicular to the stress axis provide easy cracking paths; interfaces steeply inclined to this plane favour the formation of steeply inclined secondary cracks (for examples see arrow C Figure 13 and arrow A Figure 15). As suggested by Eylon [15], surface–connected interfacial cracking is an indication of environmentally–assisted fatigue crack initiation or crack–tip re-initiation. Diffusion of oxygen (at elevated temperatures) from the surface into the material is enhanced along the α/β interfaces either by lattice mismatch between the hexagonal α and cubic β phases or by the interface stresses developed during the load cycles. Interfacial diffusion is also promoted by the relatively high surface/volume ratio associated with the acicular α–phase and higher dislocation density [15]. These resulting higher concentration of interstitial solutes embrittle interfacial regions [16] and reduce their ability to deform plastically. The net result is the resistance to fatigue crack growth in decreased.

MODELLING TMFCG IN Ti64 AND Ti6246

The FCG rates are modelled by assuming that the observed da/dN is a summation of contributions to crack growth of the two dominant mechanisms which are active at the minimum and maximum temperature of the cycle: namely the mechanical fatigue and environmentally assisted crack growth. Thus we can write

$$\left(\frac{da}{dN}\right)_{tot} = \left(\frac{da}{dN}\right)_{fat} + \left(\frac{da}{dN}\right)_{env} \tag{5}$$

where (da/dN) $_{fat}$, the mechanical fatigue contribution, is correlated by a K–based fracture mechanics parameter. Here (da/dN) $_{env}$ is the time – dependent contribution related to the kinetics of oxygen induced embrittlement. However, it should be pointed out that the environmentally driven crack growth rates (da/dN) $_{env}$ are assumed governed by rate controlling processes **in conjunction** with a mechanical driving force characterized by the stress intensity factor (K_{max}). Since the transport of oxygen (and its reaction products) to the newly produced crack surfaces is sufficently rapid, the crack growth is assumed to be determined by the rate of diffusion of oxygen in the material and from the crack tip to the fracture process zone [17]. Thus the (da/dN) $_{env}$ can be written as:

$$\left(\frac{da}{dN}\right)_{env} = \left(\frac{da}{dt}\right)_{ox} \cdot t_{eff} \tag{6}$$

with t_{eff} (unit of s/cycle) the effective time period within one cycle in which the oxygen–embrittlement assisted cracking mechanisms operate. The (da/dt) $_{ox}$ is assumed to be controlled by the diffusion of oxygen along the α/β interfaces and grain boundaries of the material. In particular,

$(da/dt)_{ox}$ can be written as:

$$\left(\frac{da}{dt}\right)_{ox} = \left(\frac{da}{dt}\right)_{ox}^{o} \exp\left(\frac{-Q}{RT}\right) \tag{7}$$

where $(da/dt)_{ox}^{o}$ is a material constant (intrinsic embrittlement of the material [4, 18]), Q is the apparent activation energy for oxygen transport, R is the universal gas constant, and T is the temperature in ^{o}K. Here a growth rate dependence upon K_{max} is implictly assumed for $(da/dt)_{ox}$ to reflect the expected proportionnality between the size of the oxygen induced damage zone and the crack tip plastic zone. Combining eqs 6 and 7 yields:

$$\left(\frac{da}{dN}\right)_{env} = t_{eff}\left(\frac{da}{dt}\right)_{ox}^{o} \exp\left(\frac{-Q}{RT}\right) , \tag{8}$$

or for a given specified effective period (t_{eff}):

$$\left(\frac{da}{dN}\right)_{env} = \left(\frac{da}{dN}\right)_{ox}^{o} \exp\left(\frac{-Q}{RT}\right) . \tag{9}$$

Using the da/dN data obtained at T_{min} and T_{max} the constants in equations 8 and 9 were obtained, i.e. $(da/dt)_{ox}^{o} = 2.36 \times 10^{-4}$ mm/sec and Q=16.9 KJ/mole. Here t_{eff} was taken as the time interval where the **measured** load is tensile.

It is interesting to note that the low frequency of loading (0.6 cpm) used in the present study might also be associated with creep deformation mechanisms occuring at the crack tip. However, the measured apparent activation energy (16.9 KJ/mole) of the time dependent contribution to cracking is much lower than the 300 KJ/mole observed in creep deformation experiments [19] where bulk diffusion is believed to be the rate controlling mechanism in these creep experiments. Note that, the 16.9 KJ/mole obtained from the isothermal tests is akin to the 33.8 KJ/mole found from oxidation testing of Ti-alloys [21]. However, oxidation behaviour is most likely limited by diffusion through the oxide scale while oxygen embrittlement of the α/β interface is controlled by interfacial diffusion. The low value of activation energy (16.9 KJ/mole) further supports the conclusion that **oxygen–induced embrittlement**, which kinetics is controlled by interfacial α/β diffusion, is the dominant time dependent damage mechanism in Ti6246. Also, the coarse acicular α–phase microstructure with straight and continuous α/β interface favours oxygen diffusion thus rendering Ti 6246 **prone** to environmental attack.

Using equations 5 and 8, the time–dependent contribution as well as the mechanical fatigue contribution were obtained. The mechanical fatigue contributions pertaining to Ti6246 are shown in Figure 16.a where, as expected, there is no dependence of the testing temperature in agreement with other reported data [14]. For the TMF data shown in figure 16.b the time–dependent contributions were computed using

$$\left(\frac{da}{dN}\right)_{ox} = \int_{t_1}^{t_2} \left(\frac{da}{dt}\right)_{ox}^{o} \exp\left(\frac{-Q}{RT(t)}\right) dt \tag{10}$$

with t_1 and t_2 pertaining to the starting and finishing time within one cycle where oxygen induced embrittlement operate. Assuming that t_1 and t_2

are related to the opening and closure processes of the crack, the time dependent contributions (eq. 10) were estimated by integrating over the tensile load part of the cycle.   In turn, the mechanical fatigue contributions were computed by subtracting $(da/dN)_{env}$ from $(da/dN)_{tot}$. This result, i.e. $(da/dN)_{fat}$ as a function of ΔK_{eff} for the various TMF cycle types, is shown in figure 16.b.

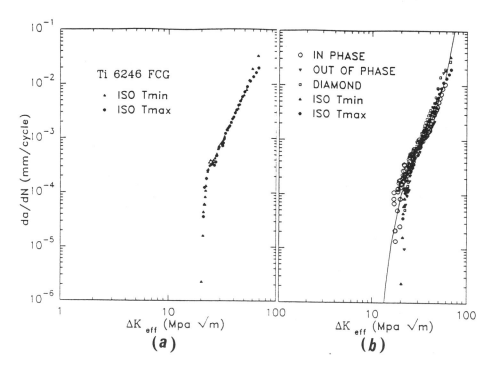

FIG. 16—Mechanical fatigue contribution to crack growth after partitionning the environmental assisted cracking for Ti6246.(a) Isothermal tests only, (b) all testing conditions

As can be seen the mechanical fatigue contribution is very well correlated using ΔK_{eff} (K_{max}).   It should be emphasized that figure 16.b was obtained using the constants determined from isothermal tests only.   The practical implication of the results presented in this paper is that the taking into account of environment is paramount to correctly predicting the fatigue lives of Ti6246 components.   Hence, if environment is not compounded for properly, the data produced under isothermal conditions (<400°C) might lead to a signifiant underestimation of the crack growth rates in components experiencing subtained loads periods or low frequency TMF cycle types.   In order words, combining the data shown in figure 16.b (mechanical fatigue) with equation 10 (contribution of oxygen–induced embrittlement) allows the accurate prediction of the actual crack growth rates $(da/dN)_{tot}$.

The same equations and reasoning developed for Ti6246 can be applied to Ti64.   However the fine equiaxed α–phase microstructure with curved α/β

interfaces is such that there is no continuous path for oxygen embrittlement ahead of the crack tip, thus rendering Ti64 more resistant to environmental damage. In first approximation the activation energy for oxygen transport in Ti64 should be similar to that of the activation energy for **bulk diffusion** of oxygen in Ti-α alloys, which is about 213 KJ/mole [22]. Furthermore, the fractographic observations pertaining to Ti64 have shown evidence of surface rubbing. Thus closure mechanisms associated with crack tip plasticity were also important in determining the growth rates than the environmental damage. These conclusions pertaining to oxygen embrittlement and closure effect as being the two dominant mechanisms to control crack growth in Ti64 have also been drawn elsewhere [23]. For long exposure time at elevated temperatures, equations 5 to 10 should be used for predicting (da/dN) with Q taken as 213 KJ/mole. Meanwhile for short t_{eff} (< 1 hour) the use of ΔK_{eff} with the data displayed in Figure 8 should be sufficient to predict $(da/dN)_{tot}$ and thus the fatigue lives.

CONCLUSIONS

Strain controlled thermal-mechanical fatigue crack growth rates tests were carried-out on Ti64 and Ti6246 respectively. The main conclusions drawn from this study can be summarized as follow:

-    The ΔK and ΔK_ϵ are not the proper fracture mechanics parameters to correlate the TMFCG rates data.

-    The K_{max}, taken as ΔK_{eff}, was found to correlate all the various isothermal and TMF crack growth rates data.

-    The main time-dependent contribution to cracking was found to be oxygen-induced embrittlement in Ti6246.

-    A model is developed to predict the TMF crack growth rates based solely on isothermal data. This model sums the contribution of mechanical fatigue to the contribution of oxygen-induced embrittlement.

-    The measured activation energy for oxygen-induced embrittlement in Ti6246 is consistent with the energy for interfacial diffusion of oxygen along α/β interfaces.

ACKNOWLEDGEMENTS

It is a pleasure to acknowledge the support for this work provided by the National Science and Engineering Research Council of Canada (NSERC) and by Pratt and Whitney Canada. The authors are also indebted to Dr. B.A. Unvala (Imperial College, London, UK) for his enlightening comments and fruitful discussions.

REFERENCES

[1]   King, T. T., Cowie, N. D. and Reiman, W. M. in "Damage Tolerance Concepts for Critical Engine Components", AGARD Conference Proceedings No. 393, 1985, pp. 31 – 35, 1987.

[2]   "Engine Structural Integrity Program (ENSIP)", MIL-STD-1783, Nov. 1984.

[3]   Halford, G. R. in <u>Thermal Stresses</u>, ed. by R.B. Metmarki, Elsevier Science Publishers, Amsterdam, pp. 329–428, 1987.

[4]   Nicholas, T., Heil., M. L. and Haritos, G. K., *Int. J. of Fract.*, vol. 41, pp. 157–176, 1989.

[5]   Dai, Y., "Thermal–Mechanical Fatigue in Aircraft Engine Materials", Ph. D. Thesis, École Polytechnique de Montréal, 1993.

[6]   Marchand, N. J., Dai, Y. and Hongoh, M., *P&WC, Report WY – 910069*, 102 pages, 1991.

[7]   Dai, Y., Marchand, N. J., Hongoh, M. and Dickson, J.I., *Titanium–92*, ed. by F. M. Froes and I. L. Caplan, pub. by TMS, vol. II, pp. 11803–11810, 1993.

[8]   Marchand, N.J., Dorner, W. and Ilschner, B., in *Surface Crack Growth : Models, Experiments and Structures*, ASTM, STP 1060, ed. by W. G. Reuter, J.H. Underwood and J.C. Newman, ASTM, pp. 237–259, 1990.

[9]   Dai, J., Marchand, N.J. and Hongoh, M. in *Applications of Advanced Techniques for Crack Size Determination*, ASTM STP 1251, ed. by J.J. Ruschau and J.K. Donald, ASTM, pp. 22–47, 1995.

[10] Dai, Y., Marchand, N.J. and Hongoh, M., Proc. of the Int. Conf. on *Low Cycle Fatigue and Elasto–Plastic Behaviours of Solids*, ed. by K.–T. Rie, Elsevier pub., 1992, pp. 594–600.

[11] Hwang, I. S. and Ballinger, R. G., *Mea. Sci Technol.*, vol. 3, pp. 62–74, 1992.

[12] Miller, K. J., *Fract. of Engng. Mat. et Struc.*, vol. 5, pp. 223–232., 1982.

[13] Newman, J. C., in "Short Crack Growth Behaviour in an Aluminium Alloy–An AGARD Cooperative Test Program", AGARD R–732, 1988.

[14] Private Communication PWA, PWA Data Bank, Hartford, Connecticutt, 1990.

[15] Eylon, D. E., in "Titanium–1980", ed. by H. Kimura and O. Izumi pub. by TMS, pp. 1845–1854, 1980.

[16] Gray, G.T., *Met Trans.*, vol. 21A, pp. 95–105, 1990.

[17] Wei, R.P. and Gangloff, R.P., in *Fracture Mechanics: Perspectives and Directions*, ASTM STP 1020, ed. by R.P. Wei and R.P. Gangloff, ASTM, pp. 233–264, 1989.

[18] Marchand, N. J., Baïlon, J.–P. and Dickson, J.I., *Met Trans.*, vol 19A, pp. 2575–2587, 1988.

[19] *The Elsevier Materials Selector*, ed. by N.A. Waterman and M.F. Ashby, CRC Press, New York, NY, vol. 2, 1991.

[20] Hauffe, K., "Oxidation of Metals", Plenum Press, 1965.

[21] Liu, Z. and Welsch, G. H., *Met. Trans.*, vol. 19A, pp. 1121–1125, 1988.

[22] Petit, J., Berata, W. and Bouchet, B., in *Titanium–1992*, ed. by Froes, F.M. and Caplan I. L., pub. by TMS, vol. II., pp. 1819–1826, 1993.

Titanium Matrix Composites

Demirkan Coker[1], Richard W. Neu[2] and Theodore Nicholas[3]

ANALYSIS OF THE THERMOVISCOPLASTIC BEHAVIOR OF [0/90] SCS-6/TIMETAL®21S COMPOSITES

REFERENCE: Coker, D., Neu, R. W., and Nicholas, T., **"Analysis of the Thermoviscoplastic Behavior of [0/90] SCS-6/Timetal®21S Composites,"** Thermomechanical Fatigue Behavior of Materials: Second Volume, ASTM STP 1263, M. J. Verrilli and M. G. Castelli, Eds., American Society for Testing and Materials, 1996.

ABSTRACT: Micromechanical modeling is used to determine the stresses and strains due to both mechanical and thermal loads in [0/90] titanium matrix composites (TMCs) subjected to cooldown from the processing temperature and subsequent thermomechanical fatigue (TMF) loading conditions. The [0/90] composite is treated as a material system composed of three constituents: fiber and matrix in the [0] ply, and a [90] ply. The [0/90]$_S$ layup is modeled by a uniaxial stress rule of mixtures model for the [0] ply and adding a parallel element to the [0] model and invoking strain compatibility and stress equilibrium in the loading direction. The fiber in the [0] ply is treated as elastic and the matrix in the [0] ply is treated as viscoplastic with temperature dependent mechanical properties. The [90] ply is characterized as a viscoplastic material including damage from fiber/matrix interface separation. Computations are made for isothermal fatigue as well as in-phase and out-of-phase TMF conditions for the crossply SCS-6/Timetal®21S composite. Effects of frequency and maximum temperature on the composite and constituent stress-strain behavior are evaluated. Fiber/matrix separation and strain ratchetting are found to be important factors in describing the response. Fiber stresses are shown to be dominant in isothermal fatigue at low frequencies as well as under in-phase TMF conditions. Matrix stresses dominate the behavior under high frequency isothermal fatigue and out-of-phase TMF. The use of tenth cycle constituent stresses is shown to be a good compromise between capturing the fully relaxed behavior and computational efficiency.

KEYWORDS: titanium matrix composite, fatigue, thermomechanical fatigue, analysis, micromechanics, plasticity, damage.

Titanium and titanium aluminide matrix composites reinforced with silicon carbide fibers are attractive candidate structural materials for several high temperature aerospace applications. In both hypersonic aircraft and turbine engine components, the composites

[1]Research engineer, University of Dayton Research Institute, Dayton, OH 45429

[2]Formerly, NRC associate, Wright Laboratory Materials Directorate, Wright-Patterson AFB, OH 45433; presently, assistant professor, Woodruff School of Mechanical Engineering, Georgia Institute of Technology, Atlanta, GA 30332-0405

[3]Senior scientist, Wright Laboratory Materials Directorate, Wright-Patterson AFB, OH 45433

may be subjected to combinations of monotonic, cyclic, or sustained loading as well as large temperature excursions. In addition to the range of maximum temperatures that has to be considered for design, a wide range of vibratory frequencies must also be taken into account in assessing the capability of this class of materials. An adequate life prediction methodology must address all of these issues, as well as consider the onset and propagation of service induced damage such as fiber/matrix interfacial sliding or separation or, fiber or matrix cracking. Analyses should have the capability to predict service life, depending on the definition of life, as well as determine the progressive deformation, stiffness, and residual strength characteristics during the entire service life of a component. Noting that the behavior of titanium matrix composites (TMCs) at elevated temperatures is a complex combination of matrix creep, fatigue, environmental degradation processes, and damage accumulation, the task of describing the micro and macro response of these materials is a challenging one indeed.

To describe composite behavior for design and analysis purposes, analytical tools have been under development for several years. The capabilities of several of these procedures or computer codes to reproduce experimental data on [0/90] cross-ply TMCs at room and elevated temperatures have been compared recently [1]. Limited comparisons of stress-strain, cyclic fatigue, and thermomechanical fatigue behavior show, in general, that several methodologies exist which can provide reasonable and efficient numerical simulations of composite behavior as well as provide information on the micromechanical stress and strain states in the fiber or matrix. One of the methods which was demonstrated to provide both accuracy and numerical efficiency in the computation of response of a unidirectional composite is the concentric cylinder model (CCM) coded into the program FIDEP, a finite difference elastic-plastic analysis code [2]. Various modifications have been made to the computer code, including modifying the material model for the matrix from elastic-plastic to thermoviscoplastic [3], introducing a parallel element to represent a [90] ply in a [0/90] lay-up [3], and introducing damage in the form of fiber/matrix separation in the [90] ply [4]. The code has been validated through comparisons with experimental data as well as finite element computations [5] and has provided reasonable agreement with the results of a number of other computational procedures in a recent comparative study [1]. Further, the uniaxial stress model (USM) with a [90] ply parallel element has been shown to produce macro stress-strain results which are nearly equivalent to those obtained with the concentric cylinder model with a parallel element [4]. For this reason, coupled with computational simplicity, the USM was chosen for this investigation.

Having a model to represent a specific composite architecture allows the calculation of material response such as stress-strain behavior under monotonic or cyclic conditions and comparisons of the response with experimental data. In addition, micromechanical stresses in the fiber and the matrix arising from both processing and service conditions can be determined and used in life prediction modeling and in the development of failure criteria. Nicholas et al. [6] performed calculations of fiber, matrix, and ply stresses for SCS-6/Ti-24-11 including damage using an elastic-plastic material model in a FEM analyis. Results obtained were used to identify regions where matrix failure might occur and demonstrated that using an imperfect interface provided better correlation with experimental data than those obtained using perfect bonding. The FIDEP model, in particular, has been used in a series of investigations to analyze both macromechanical and micromechanical composite behavior. For unidirectional SCS-6/Ti-24Al-11Nb, thermomechanical fatigue (TMF) behavior was evaluated using the elastic-plastic matrix formulation and found to provide adequate correlation with experimentally observed stress-strain behavior [2]. The model using a viscoplastic model for the [0] plies and elastic-plastic model for the [90] plies was applied to SCS-6/Timetal®21S [0/90]$_S$ under tension as well as in-phase (IP) and out-of-phase (OP) TMF and provided good agreement overall with experimental data [3]. Finally, the model has been applied to the same composite using a viscoplastic model with damage for the [90] ply under isothermal fatigue and TMF at different frequencies to determine the macro response of a [0/90]$_S$ lay-up as well as to a number of conditions for the [0]$_4$ lay-up [4]. Overall, the model has been found to provide good agreement with experimentally observed behavior.

It is the objective of this investigation to review the observations to date and to conduct additional calculations of the micromechanical and macromechanical response of a [0/90] TMC composite to evaluate effects of temperature, temperature range, and frequency under isothermal fatigue and TMF loading conditions. In particular, the accumulation of strain leading to ratchetting is evaluated in detail in order to determine the time or number of cycles necessary to achieve stress stability after shakedown. Comparisons of the computed stresses are made with experimental data on fatigue life to deduce the governing failure criteria over a wide variety of loading conditions.

MATERIAL AND MODEL

The material used in this study was a titanium matrix composite (TMC) comprised of silicon carbide fibers (SCS-6) and a beta-titanium matrix (Timetal®21S) with a composition of Ti-15Mo-3Nb-3Al-0.2Si wt %. The composite panels were assembled by alternating layers of rolled titanium foils with fiber mats of SCS-6 woven with a Ti-Nb ribbon, and then consolidated by hot isostatic pressing. The fibers were 142 μm in diameter and were spaced at 126±3 fibers per inch of fiber mat, the foil thickness was 114 μm, and the thickness of the resulting 4-ply composite panels nominally averaged 0.9 mm. Specimens were generally taken from more than one panel. The fiber volume fraction (V_f) was generally dependent on the panel and varied from 0.32 to 0.41. Specimens were heat treated in vacuum at 621°C for 8 hours prior to testing to help stabilize the Ti matrix.

The basic formulation for a unidirectional composite is given in Coker et al. [2] where an elastic-plastic material model is used to represent the matrix. Subsequent modifications of FIDEP included incorporation of a thermoviscoplastic model for the matrix based on the equations of Bodner and Partom [7], with modifications to reflect non-isothermal material behavior as developed by Neu [8]. Material parameters to describe Timetal®21S, the matrix material considered in this investigation, are presented in Kroupa and Neu [5]. The ability to analyze a [0/90] composite with FIDEP was added by Coker et al. [3] by introducing a parallel element which represents the [90] ply. Stress equilibrium and axial strain compatibility with the concentric cylinder representing the [0] ply are invoked, and the [90] ply was represented by an elastic-plastic material model fit to numerical results from finite element method (FEM) computations. The representation of the [90] ply has been modified subsequently by treating it as a viscoplastic material, incorporating damage in the form of fiber/matrix separation, and calibrating the model with detailed FEM computations [4]. For this same representation of the [90] ply, the uniaxial stress model (USM) produces similar response as the concentric cylinder model (CCM) under fatigue loading [4]. It was noted in the comparisons that the most significant features in reproducing experimental fatigue stress-strain response accurately are describing the [0] ply matrix and [90] ply with an accurate thermoviscoplastic model as well as including damage in the form of fiber/matrix separation in the model formulation. For these reasons, the USM and CCM produce similar results since they incorporate all these features. The main difference between the USM and CCM is the magnitude of the calculated thermal residual stress upon cooldown from composite processing temperature. This difference for a [0/90] lay-up, which amounts to 11% for the fiber stress (-847 MPa for the USM, -953 MPa for the CCM), can be attributed to the difference between a three-dimensional stress state in the [0] ply for the CCM compared to one-dimensional stress in the USM. This difference was not considered to be significant for the computations reported herein which used the USM.

The model of the [0/90] laminate using the USM consists of three parallel elements representing the 0° fiber, [0] ply matrix, and [90] ply. Compatibility and equilibrium are imposed in the direction parallel to the [0] ply fiber. The constituent strain is equal to the composite strain in this direction,

$$\varepsilon_f = \varepsilon_m = \varepsilon_{90} = \varepsilon_c \tag{1}$$

where f is the fiber, m is the matrix in the [0] ply, 90 is the [90] ply, and c is the composite. The stress-strain relations are

$$\sigma_i = E_i(\varepsilon_c - \varepsilon_i^{th} - \varepsilon_i^{in}), \quad i = f, m, 90 \tag{2}$$

where ε_i^{th} is the thermal strain and ε_i^{in} is the inelastic strain. Load equilibrium in the axial direction is

$$\sigma_{app} = \frac{1}{2}(\sigma_f V_f + \sigma_m(1 - V_f) + \sigma_{90}) \tag{3}$$

where V_f is the fiber volume fraction assuming a symmetric layup. From these equations the expression for total strain assuming the fiber is thermoelastic becomes

$$\varepsilon_c = \frac{1}{E_c}\left(\sigma_{app} + \alpha_c E_c \Delta T + \frac{1}{2}(1 - V_f)E_m \varepsilon_m^{in} + \frac{1}{2}E_{90}\varepsilon_{90}^{in}\right) \tag{4}$$

where subscript c refers to the composite. The stresses in the constituents are obtained from equations (2) and (4), with E_c determined by rule of mixtures.

For the SCS-6/Timetal 21S composite, the fiber is considered elastic with temperature-dependent elastic modulus, Poisson's ratio, and CTE. The matrix is treated as thermo-viscoplastic and is described using the Bodner-Partom model with directional hardening formulation [7]. The nonisothermal forms of these equations are described in Ref. 9 with modifications described in Ref. 8.

The constitutive model describing the [90] ply response, which treats the material as an equivalent homogeneous media, incorporates both matrix viscoplasticity and fiber/matrix separation [4]. Matrix viscoplasticity is described with the Bodner-Partom model while continuum damage mechanics concepts are used to describe the voids created when the fiber and matrix separate.

Damage (i.e., the voids) alters the stress in the ply and is described using the idea of an effective stress, $\tilde{\sigma}$, [10,11] defined by

$$\tilde{\sigma} = \frac{\sigma}{1 - \eta D} \tag{5}$$

where σ is the average transverse stress on the lamina, D is the damage where $D = 1$ indicates the matrix is cracked through the entire ply, and η is a measure of the fraction of damage described through a sigmoidal function. Damage is fully active when $\eta = 1$ and is fully passive when $\eta = 0$ [4,12].

The effective stress is used in the constitutive relationships,

$$\tilde{\sigma} = E \, \varepsilon^e \qquad (6)$$

Taking the derivative with respect to time and writing the stress in terms of the [90] ply stress,

$$\dot{\sigma} = E\left[(1 - \eta D)(\dot{\varepsilon} - \dot{\varepsilon}^{in} - \dot{\varepsilon}^{th}) - (\dot{\eta} D + \eta \dot{D}) \frac{\sigma}{E(1 - \eta D)} \right] + \frac{\partial E}{\partial T} \dot{T} \frac{\sigma}{E} \qquad (7)$$

The inelastic strain is defined by the Bodner-Partom flow rule using the effective stress.

Only one mode of damage, fiber/matrix separation, is considered. The damage evolution is dependent on the strength of the bond, the number of debonds that have already occurred, and the magnitude of the mechanical bond, which is dependent on the residual stress from the difference in the fiber and matrix coefficient of thermal expansion (CTE). As the ply stress overcomes both the mechanical and chemical bond strength, denoted by σ_m and σ_{ch}, respectively, the debond damage rate increases rapidly. The rate eventually tails off as more fibers become debonded and only a small portion of the fiber diameter remains in contact with the matrix. This process is represented with a Weibull function,

$$\frac{D}{D^*} = 1 - \exp\left[-\left(\frac{\sigma - \sigma_m - \sigma_{ch}}{\theta} \right)^m \right] \quad \text{updated when } \sigma - \sigma_m - \sigma_{ch} > \sigma_p \qquad (8)$$

where m is the shape parameter (also called the Weibull slope), and θ is the scale parameter. Both are taken as independent of temperature, while σ_p is the peak stress ever reached. D is a monotonically increasing function and is updated when the damage rate is positive; therefore, no healing of damage occurs. D^* represents the maximum damage due to complete fiber/matrix separation [4].

Both σ_m and σ_{ch} represent the effect mechanical and chemical bonding have on the remote stress, rather than the local stress. The initial value of σ_{ch} is about the same at both 23°C and 650°C and is assumed to be 80 MPa for all temperatures [4]. On the other hand, the mechanical bond strength, σ_m, is a function of temperature [4]. Even though σ_{ch} is assumed to be independent of temperature, it is a function of D. A simple approach is taken by assuming σ_{ch} degrades linearly with D,

$$\sigma_{ch} = \sigma_{ch\,o} \left(1 - \frac{D}{D_{ch}} \right), \quad D < D_{ch}$$

$$= 0, \qquad\qquad\qquad D \geq D_{ch} \qquad (9)$$

D_{ch} is designated by Weibull statistics.

Material parameters for both fiber and matrix are given in Ref. [5]. Material parameters for the [90] ply model are given in Ref. [4]. These models have been incorporated into the FIDEP2 code [13]. All simulations were conducted using this code. In addition to the features outlined above, the model treats the CTE of the composite in a different manner than most codes. As introduced by Neu et al. [4], the CTE of the [90] ply is based on FEM analysis under no external mechanical load when damage is passive. Castelli [14] showed experimentally that the transverse CTE may be dependent on active damage, though CTE is considered independent of damage in this study. If the [90] ply CTE increases with active damage and approaches the matrix CTE, the slope of the composite response at the lower stresses is captured a little better [4]. This observation can be explained by the fact that the [90] ply acts more like a matrix with a hole after the initial fiber/matrix separation.

In the application of the model described above to TMF simulations, it should be noted that experiments performed by Castelli [15] showed that IP TMF produces a small degradation in CTE early, but stabilizes and remains constant for most of life. OP TMF, on the other hand, shows significant degradation of CTE due to matrix cracking. This can be attributed to the extensive amount of matrix cracking observed. The CTE tends towards the lower CTE dictated by the [0] fibers. No attempt was made in the present investigation to capture the complex CTE behavior of a real composite. Further, no damage other than fiber/matrix separation was modeled. It has been hypothesized and observed that fiber fracture is a dominant mechanism leading to strain accumulation under in-phase TMF conditions [15-19]. Analytical modeling to support this hypothesis has also been performed [20], but no attempt was made to include this damage mechanism in the analysis used in this investigation.

A number of isothermal and TMF test conditions were simulated to study the composite response as well as the micromechanical stresses as a function of number of applied cycles. Two methods are used throughout to illustrate the behavior. First, plots are made of stress as a function of time for typically ten cycles for the stress in the fiber, the average matrix stress in the [0] ply, and the average stress in the [90] ply. Second, these same stresses are plotted against mechanical strain to illustrate the stress-strain behavior for each of the three stresses indicated. For all of the simulations, a stress ratio, R=0.1 is assumed. In all of the cases studied, cooldown from a processing temperature is simulated assuming an unstressed condition at that temperature. The residual thermal stresses resulting from cooldown during processing are included in all the computations. The constants and material properties were obtained solely from isothermal tension and fatigue tests on [0] and [90] lamina.

RESULTS AND DISCUSSION

The first test condition simulated is an in-phase TMF test utilizing a temperature range of 150-650°C at the baseline frequency of 5.6 x 10^{-3} Hz, and a maximum stress of 550 MPa. For this particular condition, the numerical simulation has been shown to represent the experimentally observed stress-strain behavior up to ten cycles quite well [4]. The stress-time and stress-strain behavior for this condition are shown in Figs. 1(a) and 1(b), respectively. The upsidedown triangle in Fig. 1(b) and subsequent fiber stress-strain curves indicates the stress at maximum load during the first TMF cycle. The fiber stress is seen to increase from a maximum of 1973 MPa after the first cycle to 2453 MPa after 10 cycles where the stress level appears to be approaching a steady state value. While the fiber maximum stress and stress range are both large under in-phase TMF, the [0] matrix stress and [90] ply stress are both small in magnitude as well as stress range. From Fig. 1(b) it can be seen that the major change in these latter stresses occurs during the first cycle when fiber/matrix separation occurs in the [90] ply. From an examination of the numerical output (which is not plotted), it was noted in all cases reported that the first point of deviation from linearity in the [90] ply response during the first cycle is due to fiber/matrix separation as opposed to matrix yielding. For this first condition studied, Fig. 1(b), changes in the [90] ply stress and [0] matrix stress subsequent to the first cycle are small. To evaluate the long term stability of the stresses after the first few cycles, the simulation was performed again for a total of 100 cycles and the results are presented in Fig. 1(c). It can be seen that the

An out-of-phase TMF condition is simulated next at the same frequency as the baseline in-phase TMF of 5.6×10^{-3} Hz which represents the condition used in most of the experiments reported [4]. At a maximum stress of 500 MPa, the referenced experiments display a larger hysteresis loop than produced by the numerical simulation after either one or ten cycles. It should be noted, however, that this is probably due to an artifact in the experiment, perhaps a phase lag between load and temperature, because the hysteresis loop observed experimentally disappears as frequency is decreased. The stress-time and stress-strain results are presented in Figs. 3(a) and 3(b), respectively. Some significant differences from the in-phase results, Fig. 1, can be immediately noted. First, the fiber stress and stress range are considerably lower in the out-of-phase condition than in the in-phase condition for a similar maximum applied load. This is attributed to the contribution of the thermally induced streses which produce compression in the fiber and tension in the [0] ply matrix at the low temperature of the TMF cycle [6]. The second observation is that the stress range in the [0] ply matrix is now significant, while the average stress in the [90] ply is somewhat lower. This is due to a combination of thermally induced stresses which produce tension in the matrix at low temperature (where applied load is maximum under out-of-phase TMF) as well as the existence of fiber/matrix interface damage which increases the compliance of the [90] ply. The third observation is that there is essentially no strain ratchetting under out-of-phase TMF as seen clearly in Fig. 3(b). With the exception of the damage induced in the [90] ply during the first cycle, the stress-strain behavior of the fiber, [0] ply matrix, and [90] ply are all nearly linear elastic during the first ten cycles. The main reason for this is that maximum stresses in the matrix material occur at the low temperature of the cycle where no creep occurs in this matrix material. For out-of-phase TMF, therefore, it can be concluded that the matrix carries a substantial portion of the load at the low temperature when the applied stress is high.

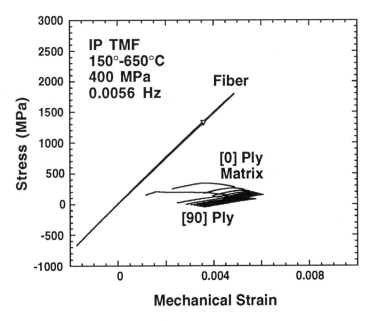

FIG. 4--Stress-strain response of fiber, [0] ply matrix, and [90] ply over 10 cycles to in-phase TMF during 180 s cycle, σ_{max}=400 MPa.

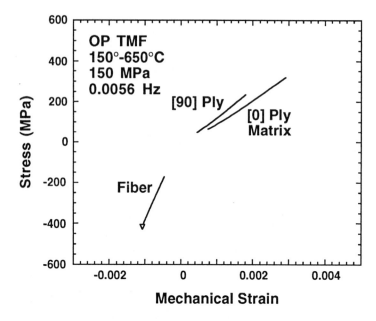

FIG. 5--Stress-strain response of fiber, [0] ply matrix, and [90] ply over 10 cycles to out-of-phase TMF during 180 s cycle, σ_{max}=150 MPa.

The effect of maximum composite stress on the in-phase and out-of-phase TMF behavior of a composite can be studied by comparing the results for the baseline conditions presented above with those obtained at a lower maximum stress. For the in-phase case, a maximum stress of 400 MPa was simulated to compare with the results at 550 MPa. The results are presented in Fig. 4 which can be compared with similar data at 550 MPa shown in Fig. 1(b). The reduction in applied stress reduces the maximum fiber stress at cycle 10 from 2453 MPa to 1798 MPa. It is of interest to note that this reduction in composite stress results in an increase in fatigue life from 36 to in excess of 5700 cycles to failure [21]. The maximum stresses and stress or strain range in the [0] ply matrix and [90] ply are still small under these conditions. For the out-of-phase simulation, the maximum composite stress is decreased from 500 to 150 MPa, resulting in an increase in fatigue life from 435 to 6264 cycles to failure [21]. The results, shown in Fig. 5, can be compared with the baseline results at 500 MPa shown in Fig. 3(b). The reduction in stress results in a significant reduction in all of the stresses being compared. The fiber now remains in compression over the entire cycle and its behavior is dominated by the thermal excursions. The [0] ply matrix and [90] ply stresses are also much lower than before and, further, there is no evidence of ratchetting or damage even on the first cycle. Examination of the numerical files generated during this simulation indicates that no fiber/matrix separation occurs at this low applied stress level. This is the only case simulated during this investigation where separation did not occur. The material behavior is effectively elastic for this case as seen from Fig. 5. The effect of maximum applied stress on the maximum constituent stress was evaluated by performing simulations at a number of different stress levels for both in-phase and out-of-phase TMF at the baseline frequency of 5.6 x 10^{-3} Hz and over the baseline temperature range of 150-650°C. The results of these calculations are summarized for the

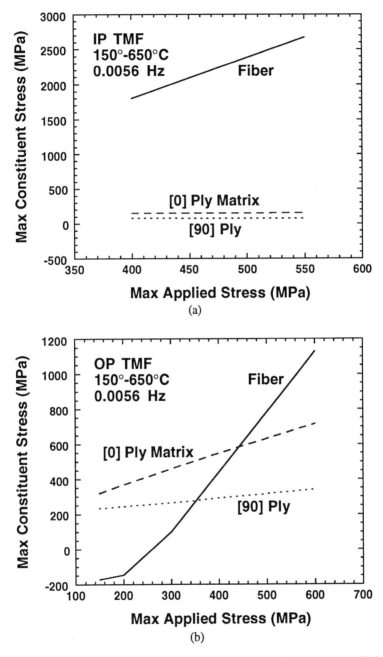

FIG. 6--Maximum constituent stresses as a function of maximum applied stress for 150=650°C TMF: (a) in-phase, (b) out-of-phase.

in-phase and out-of-phase cases in Figs. 6(a) and 6(b), respectively. As noted earlier, the reduction of applied stress results in a reduction in fiber stress for the in-phase case, Fig. 6(a), but has little or no effect on the stresses in the [0] ply matrix or [90] ply. On the other hand, reduction of applied stress for the out-of-phase case results in a reduction of all three computed stresses as shown in Fig. 6(b). The amount of this reduction is greatest in the fiber stress, though this reduction does not affect fatigue life because the out-of-phase fatigue life is not affected by fiber stresses of these magnitudes which are well below those of the in-phase case as shown in Fig. 6(a). The fiber stresses are also well below a level of approximately 2000 MPa, a value reported in micromechanics computations where fibers appear to start failing [22,23]. The nonlinearity in the fiber stress curve in Fig. 6(b) at low applied stress is due to the fact that no fiber/matrix separation occurs at this level, but separation occurs at higher stresses. For the range of conditions simulated, the in-phase TMF life is governed primarily by fiber stresses while out-of-phase TMF life is dominated by the stresses in the matrix [15,19,21].

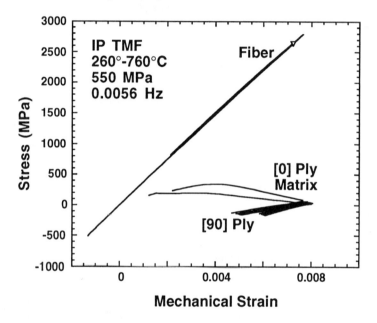

FIG. 7--Stress-strain response of fiber, [0] ply matrix, and [90] ply over 10 cycles to in-phase TMF, 260-760°C, σ_{max}=550 MPa.

Of added interest in these simulations is the effect of maximum and minimum temperature on the constituent stresses. For the baseline frequency of 5.6 x 10^{-3} Hz, and maintaining a temperature range of 500°C and maximum composite stress of 550 MPa, in-phase TMF simulations were carried out for a higher maximum temperature of 760°C and a lower maximum of 593°C to compare with the baseline maximum of 650°C for which the data are presented in Fig. 1(b). For comparison purposes, these correspond to experiments where the observed lives were 31, 1382, and 5046 cycles to failure for maximum temperatures of 760, 650, and 593°C, respectively. (These data have not been reported previously). The results for the higher and lower maximum temperature are presented in Figs. 7 and 8, respectively, and can be compared directly with the baseline results for a maximum temperature of 650°C presented in Fig. 1(b). Such comparison shows that increasing the maximum temperature results in an increase in the fiber maximum stress (Fig. 7). Conversely, lowering the maximum temperature lowers the fiber maximum stress (Fig. 8). The primary reason for this trend is the decrease in strength of the matrix material with increase in temperature. From Fig. 7 it can be seen that the [0] ply matrix carries very little stress at the maximum temperature of the cycle where the load and mechanical strain are both at maximum. There is also a considerable strain accumulation during the first cycle in the composite due to both fiber/matrix separation in the [90] ply as well as viscoplastic flow in the matrix material. This results in most of the load being transferred to the 0° fibers. On the other hand, at a lower maximum temperature of 593°C, Fig. 8, the matrix flows much less during the first cycle, the stress carried by the matrix is higher, and the 0° fibers carry less of the total load. Since the matrix still displays viscoplastic behavior at this maximum temperature and remains under load, a larger amount of strain ratchetting is observed over the first ten cycles. Nevertheless, the [0] ply matrix and [90] ply still share in the load carrying capacity of the composite compared to the condition of 760°C maximum temperature where the 0° fibers carry almost all of the load. Since the in-phase TMF life of the composite is governed primarily by the stresses in the 0° fibers, the fatigue life is observed to decrease with increasing maximum temperature as noted above. In a companion paper [22], the in-phase TMF life is attributed to no more than the time it takes for the matrix to relax to the point where the 0° fiber stress reaches the strength of a fiber bundle. Thus, the matrix stress relaxation continues well beyond the ten cycles simulated here (Fig. 8).

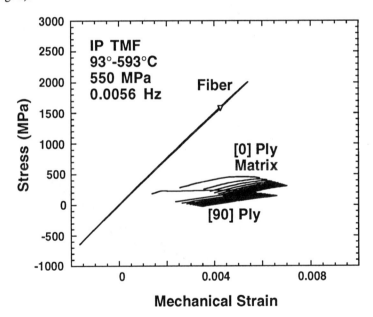

FIG. 8--Stress-strain response of fiber, [0] ply matrix, and [90] ply over 10 cycles to in-phase TMF, 93-593°C, σ_{max}=550 MPa.

Similar comparisons of the effect of maximum temperature can be made for out-of-phase TMF. Here, the simulations were carried out for a maximum stress of 400 MPa where no computations were carried out for the baseline maximum temperature of 650°C. For reference purposes, the fatigue life for the 150-650°C baseline out-of-phase TMF condition increases from 435 to 953 cycles to failure when the maximum load is decreased from 500 to 400 MPa. At 400 MPa, the out-of-phase fatigue lives are 420, 953, and 1499 cycles to failure for maximum temperatures of 760, 650, and 593°C, respectively [24]. The computed results of the stress-strain behavior for maximum temperatures of 760 and 593°C are presented in Figs. 9 and 10, respectively. As noted above, the fiber stresses are low compared to in-phase TMF and are not important in the lifing of the composite. Nevertheless, the fibers carry a higher load at the higher maximum temperature, even though the maximum load occurs at minimum temperature. The lower load carried by the matrix at the higher maximum temperature results in a higher load being carried by the fibers. Several reasons cause this stress redistribution in the composite. First, a higher maximum temperature is accompanied by a higher minimum temperature for a fixed value of ΔT. The higher minimum temperature results in a lower thermal residual stress (tension in matrix, compression in fiber). Second, even though maximum load occurs at minimum temperature, the combination of loads and temperatures experienced during a cycle is sufficient to cause viscoplastic flow in the matrix material as evidenced by the ratchetting in Fig. 9. By comparison, there is no strain ratchetting at the lower maximum temperature of 593°C as seen in Fig. 10. There, the same combination of loads with lower temperatures does not produce any measurable viscoplatic flow. The only stress redistribution in the lower temperature case is due to fiber/matrix separation in the [90] ply during the first cycle. It is of interest to note that the differences in [0] ply matrix stresses between the two cases shown in Figs. 9 and 10 are small, and the differences in cycles to failure are only a factor of 3.6. By contrast, for the same temperature range change under in-phase TMF, where fatigue life is governed by fiber stress, the life decreases by a factor of 163 for a maximum composite stress of 550 MPa.

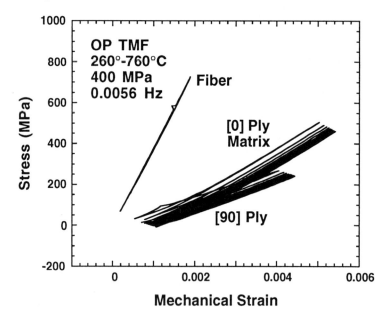

FIG. 9--Stress-strain response of fiber, [0] ply matrix, and [90] ply over 10 cycles to out-of-phase TMF, 260-760°C, σ_{max}=400 MPa.

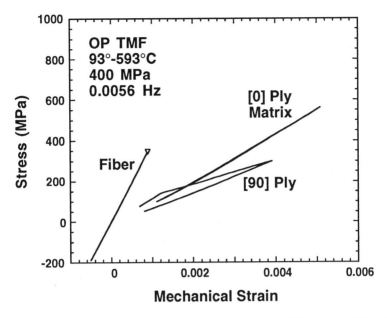

FIG. 10--Stress-strain response of fiber, [0] ply matrix, and [90] ply over 10 cycles to out-of-phase TMF, 93-593°C, σ_{max}=400 MPa.

The final comparison involves the effect of frequency on isothermal fatigue at 650°C and a maximum stress of 550 MPa. Two frequencies were simulated, 0.01 and 200 Hz, which cover the range of frequencies which have been evaluated experimentally in several investigations [21,25]. There are no experimental data available at a maximum stress of 550 MPa although it may be noted that for these two frequencies, at stresses around 400 MPa, the experimentally observed cycles to failure is typically 10^4 for 0.01 Hz and 10^5 for 200 Hz tests. The results for the stress-time and stress-strain behavior are presented in Figs. 11 and 12 for the 0.01 and 200 Hz cases, respectively. The low frequency condition is simulated over 10 cycles while the high frequency case covers 100 cycles which amounts to only 0.5 s. It is immediately apparent that the fiber stresses over the first several cycles are significantly different in the two cases, with the low frequency case having the higher fiber stress. The primary reason for this difference is the strain rate sensitivity of the matrix material at the test temperature of 650°C. Since the frequency (or strain rate) difference is 2 x 10^4, the flow stress difference is quite high. At the low strain rate, the matrix carries little stress, but at the strain rate in a 200 Hz test, the matrix is considerably stronger as illustrated in Fig. 12 where it can be seen that the matrix carries over 400 MPa stress. At both frequencies, the computations show that there is strain ratchetting taking place over the entire time covered in the analysis. Comparison of Fig. 11(b) for 0.01 Hz isothermal fatigue at 550 MPa with Fig. 1(a) for in-phase TMF at the same maximum stress shows that the maximum fiber stress for the in-phase case is slightly higher than for isothermal fatigue, even when the comparison is made for the same time as opposed to the same number of cycles. Since the isothermal test spends all the time at maximum temperature compared to the TMF test where the temperature and the load are concurrently low, for the same load profile one would expect more stress relaxation in the

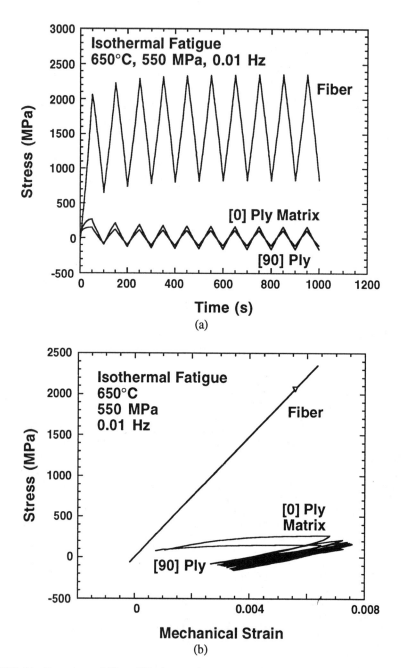

FIG. 11--Response of fiber, [0] ply matrix, and [90] ply over 10 cycles to 0.01 Hz isothermal fatigue, σ_{max}=550 MPa: (a) stress-time profiles, (b) stress-strain curves.

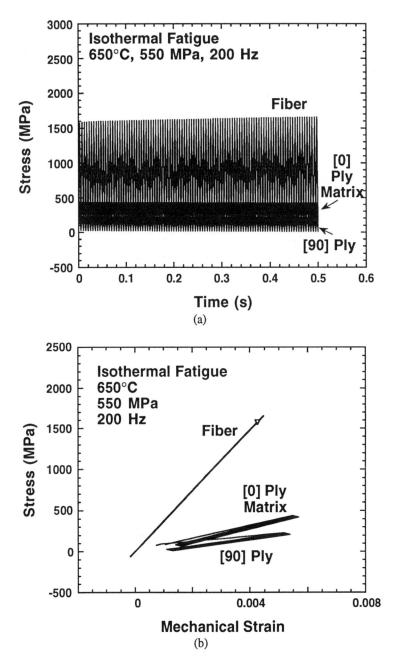

FIG. 12--Response of fiber, [0] ply matrix, and [90] ply over 100 cycles to 200 Hz isothermal fatigue, σ_{max}=550 MPa: (a) stress-time profiles, (b) stress-strain curves.

isothermal case and, therefore, higher fiber stresses. This is opposite to the calculated behavior. The explanation is that during the first loading cycle, the matrix material in the isothermal case undergoes more inelastic deformation, for the same reason as above because it is always at high temperature. Consequently, if the total strain at maximum load, which is governed primarily by the 0° fibers, is the same in both cases, the TMF case will produce a larger fiber/matrix separation. From the damage model which represents the [90] ply, the stiffness of the ply decreases with increasing damage in the form of fiber/matrix separation. Thus, the TMF case should develop less stress in the [90] ply compared to the isothermal case. A comparison of Figs. 1(a) and 11(b) shows this clearly by observing the difference between the profiles of the [0] ply matrix stress, which is about the same in both cases, and the [90] ply stress which is about 100 MPa lower in the TMF case, Fig. 1(a). This difference, which accounts for stresses across half of the composite cross section, produces a significant difference in fiber stress for the same maximum applied load and explains the observed increase in fiber stress in the TMF case.

Some final comments on the use of these simulations in life prediction of TMC's are in order. Recent papers on life prediction have used the micromechanical stress values obtained after ten cycles to provide a better estimate of life than those after the first half cycle [21,26]. The rationale was that strain ratchetting changes the stresses which, in turn, affect the predicted fatigue life. Since the correlation of fatigue life with computed stresses is highly empirical, the specific cycle taken does not have a major effect on the ability to correlate stress with fatigue life. Nonetheless, the damage induced in the form of fiber/matrix separation, and the extensive redistribution of stress during the first few cycles in some cases, makes the use of a near equilibrium stress more desirable. Mirdamadi et al. [23], in the numerical simulation of TMF and nonisothermal fatigue tests, used VISCOPLY which incorporates thermoviscoplastic matrix material behavior to determine 0° fiber stresses. The calculations were carried out until the fiber stress range did not change with further cycling, which was usually within five load cycles. From the results presented herein, it can be seen that the effect of fiber/matrix separation occurs primarily on the first cycle. For all cases where strain ratchetting takes place, a quasi steady state condition is achieved within the first ten cycles. Note that for the case where the computations were carried out for 100 cycles, Fig. 1(c), the amount of change in fiber stress during the last 90 cycles is small. For these reasons, it can be concluded that use of tenth cycle stresses is a reasonable compromise between accurate representation of steady state stresses for life prediction purposes and computational efficiency.

CONCLUSIONS

The following conclusions are drawn from the results of this investigation:

1.   Micromechanical stress computations aid in understanding the fatigue process in MMCs at elevated temperatures where inelastic behavior of the matrix material and fiber/matrix separation can take place.

2.   Strain ratchetting occurs in isothermal fatigue as well as under in-phase TMF. The resultant stress redistribution is an important part of the fatigue failure process.

3.   Differences in behavior due to small changes in frequency can be attributed to the difference in time available (per cycle) for stress relaxation to occur in the matrix. Differences due to large changes in frequency are a result, primarily, of the strain rate sensitivity of the matrix material and the resulting differences in load carrying capability.

4.   Composite and constituent response under in-phase TMF and low frequency isothermal fatigue are dominated by fiber stress. Stress-strain behavior under out-of-phase TMF is dominated by matrix stresses.

5.   Accurate representation of fiber/matrix separation and resultant [90] ply stiffness along with realistic modeling of the thermoviscoplastic behavior of the matrix are important in capturing the cyclic response of a [0/90] composite.

6.   The use of tenth cycle stresses in life prediction appears to be a reasonable compromise between accuracy and computational efficiency.

REFERENCES

[1]   Kroupa, J.L., Neu, R.W., Nicholas, T., Coker, D., Robertson, D.D., and Mall, S., "A Comparison of Analysis Tools for Predicting the Inelastic Cyclic Response of Cross-Ply Titanium Matrix Composites," Life Prediction Methodology for Titanium Matrix Composites, ASTM STP 1253, W. S. Johnson, J.M. Larsen and B.N. Cox, Eds., American Society for Testing and Materials, Philadelphia, 1995.

[2]   Coker, D., Ashbaugh, N.E. and Nicholas, T., "Analysis of Thermomechanical Cyclic Behavior of Unidirectional Metal Matrix Composites," Thermomechanical Fatigue Behavior of Materials, ASTM STP 1186, H. Sehitoglu, Ed., American Society for Testing and Materials, Philadelphia, 1993, pp. 50-69.

[3]   Coker, D., Ashbaugh, N.E., and Nicholas, T., "Analysis of the Thermomechanical Behavior of [0] and [0/90] SCS-6/Timetal 21S Composites," Thermomechanical Behavior of Advanced Structural Materials, W. F. Jones, Ed., AD-Vol. 34/AMD-Vol. 173, ASME, 1993, pp. 1-16.

[4]   Neu, R.W., Coker, D., and Nicholas, T., "Cyclic Behavior of Unidirectional and Cross-Ply Titanium Matrix Composites," International Journal of Plasticity, 1995 (accepted for publication).

[5]   Kroupa, J.L. and Neu, R.W., "The Nonisothermal Viscoplastic Behavior of a Titanium Matrix Composite," Composites Engineering, Vol. 4, 1994, pp. 965-977.

[6]   Nicholas, T., Kroupa, J.L. and Neu, R.W., "Analysis of a [0°/90°] Metal Matrix Composite Under Thermomechanical Fatigue Loading," Composites Engineering, Vol. 3, 1993, pp. 675-689.

[7]   Chan, K.S., Bodner, S.R. and Lindholm, U.S., "Phenomenological Modeling of Hardening and Thermal Recovery in Metals," Journal of Engineering Materials and Technology, Vol. 110, 1988, pp. 1-8.

[8]   Neu, R.W., "Nonisothermal Material Parameters for the Bodner-Partom Model," Material Parameter Estimation for Modern Constitutive Equations, L.A. Bertram, S.B. Brown, and A.D. Freed, Eds., MD-Vol. 43/AMD-Vol. 168, ASME, 1993, pp. 211-226.

[9]   Chan, K.S. and Lindholm, U.S., "Inelastic Deformation Under Nonisothermal Loading," Journal of Engineering Materials and Technology, Vol. 112, 1990, pp. 15-25.

[10] Kachanov, L.M., "On Creep Rupture Time," Izv. Acad. Nauk SSSR, Otd. Techn. Nauk, No. 8, 1958, p. 26.

[11] Rabotnov, Y.N., Creep Problems in Structural Members, North-Holland, Amsterdam, 1969.

[12] Krajcinovic, D. and Fonseka, G.U., "The Continuous Damage Theory of Brittle Materials, Part 1: General Theory," Journal of Applied Mechanics, Vol. 48, December 1981, pp. 809-815.

[13] Coker, D. and Ashbaugh, N.E., "FIDEP2 User's Manual," University of Dayton Research Institute, Dayton, OH (to be published).

[14] Castelli, M.G., "Thermomechanical and Isothermal Fatigue Behavior of a [90]s Titanium Matrix Composite," Proceedings of the American Society for Composites, 8th Tech. Conf. on Composite Materials, Mechanics, and Processing, Technomic Publishing, Lancaster, PA, 1993, pp. 884-892.

[15] Castelli, M.G., "Thermomechanical Fatigue Damage/Failure Mechanisms in SCS-6/Timetal 21S [0/90]s Composite," Composites Engineering, Vol. 4, 1994, pp. 931-946.

[16] Neu, R.W. and Nicholas, T., "Effect of Laminate Orientation on the Thermomechanical Fatigue Behavior of a Titanium Matrix Composite," Journal of Composites Technology & Research, Vol. 16, 1994, pp. 214-224.

[17] Castelli, M.G., Bartolotta, P.A. and Ellis, J.R., "Thermomechanical Fatigue Behavior of SiC(SCS-6)/Ti-15-3,"Composite Materials: Testing and Design (Tenth Volume), ASTM STP 1120, Glenn C. Grimes, Ed., American Society for Testing and Materials, Philadelphia, 1992, pp. 70-86.

[18] Russ, S.M., Nicholas, T., Hanson, D.G. and Mall, S., "Isothermal and Thermomechanical Fatigue of Cross-Ply SCS-6/β21-S," Science and Engineering of Composite Materials, Vol. 3, 1994, pp. 177-189.

[19] Neu, R.W. and Roman, I., "Acoustic Emission Monitoring of Damage in Metal Matrix Composites Subjected to Thermomechanical Fatigue," Composites Science and Technology, Vol. 52, 1994, pp. 1-8.

[20] Nicholas, T. and Ahmad, J., "Modeling Fiber Breakage in a Metal Matrix Composite," Composites Science and Technology, Vol. 52, 1994, pp. 29-38.

[21] Nicholas, T., Russ, S.M., Neu, R.W. and Schehl, N., "Life Prediction of a [0/90] Metal Matrix Composite Under Isothermal and Thermomechanical Fatigue," Life Prediction Methodology for Titanium Matrix Composites, ASTM STP 1253, W. S. Johnson, J.M. Larsen and B.N. Cox, Eds., American Society for Testing and Materials, Philadelphia, 1995.

[22] Nicholas, T. and Johnson, D.A., "Time- and Cycle-Dependent Aspects of Thermal and Mechanical Fatigue in a Titanium Matrix Composite," Thermo-Mechanical Fatigue Behavior of Materials: Second Symposium, ASTM STP 1263, M.J. Verrilli and M.G. Castelli, Eds., American Society for Testing and Materials, Philadelphia, 199#.

[23]   Mirdamadi, M., Johnson, W.S., Bahei-El-Din, Y.A. and Castelli, M.G., Analysis of Thermomechanical Fatigue of Unidirectional Titanium Metal Matrix Composites," Composite Materials: Fatigue and Fracture, Fourth Volume, ASTM STP 1156, W.W. Stinchcomb and N.E. Ashbaugh, Eds., American Society for Testing and Materials, Philadelphia, 1993, pp. 591-607.

[24]   Neu, R.W. and Nicholas, T., "Thermomechanical Fatigue of SCS-6/TIMETAL®21S Under Out-of-Phase Loading", Thermomechanical Behavior of Advanced Structural Materials, AD-Vol. 34, AMD-Vol. 173, W.F. Jones, Ed., American Society of Mechanical Engineers, New York 1993, pp. 97-111.

[25]   Nicholas, T., Russ, S., Schehl, N. and Cheney, A., "Frequency and Stress Ratio Effects on Fatigue of Unidirectional SCS-6/Ti-24Al-11Nb Composite At 650°C," FATIGUE 93, Volume II, J.-P. Bailon and J.I. Dickson, Eds., Chameleon Press, London, U.K., 1993, pp. 995-1000.

[26]   Nicholas, T., "An Approach to Fatigue Life Modeling in Titanium Matrix Composites," Materials Science and Engineering, 1995, (in press).

David D. Robertson[1] and Shankar Mall[1]

ANALYSIS OF THE THERMO-MECHANICAL FATIGUE RESPONSE OF
METAL MATRIX COMPOSITE LAMINATES WITH INTERFACIAL NORMAL
AND SHEAR FAILURE

REFERENCE: Robertson, D. D. and Mall, S., **"Analysis of the Thermo-Mechanical Fatigue Response of Metal Matrix Composite Laminates with Interfacial Normal and Shear Failure,"** Thermomechanical Fatigue Behavior of Materials: Second Volume, ASTM STP 1263, Michael J. Verrilli and Michael G. Castelli, Eds., American Society for Testing and Materials, 1996.

ABSTRACT: In this study, an analysis of the thermo-mechanical fatigue (TMF) behavior of titanium-based metal matrix composites (MMCs) is accomplished by employing a modified method of cells micromechanics model coupled with a fiber/matrix interfacial failure scheme and the Bodner-Partom unified constitutive theory. Interfacial failure is based on a probabilistic failure criterion which considers the equivalent interfacial compliance to be a function of the interfacial stress. In this approach, the effects of both normal and shear failure of the interface have been included. MMC laminates with angle-ply and quasi-isotropic layups are analyzed for monotonic, cyclic, and thermo-mechanical fatigue loads, and where it is appropriate, comparisons with experimental data are made.

KEYWORDS: thermo-mechanical fatigue, metal matrix composite, micromechanics, fiber/matrix interface, interfacial damage, titanium composites, statistical modeling

INTRODUCTION

The importance of fiber/matrix interfacial characteristics on the behavior of titanium-based metal matrix composites is well documented [1]. In particular, such composites exhibit interfacial failure and debond at relatively low applied stress fields (i.e. 150 to

---

[1] Assistant Professor and Professor & Head, respectively, Department of Aeronautics and Astronautics, Air Force Institute of Technology, Wright-Patterson AFB, OH, 45344

250 MPa at room temperature) [2]. Hence, the interface for these materials is normally modeled as an unbonded fiber/matrix contact [3]. This approach has been relatively successful, but it fails to account for any finite interfacial strength. However, a recent study examined the effects of finite interfacial failure strengths on the composite behavior through a statistical representation of the interfacial stresses [4]. The study examined the effects of interfacial failure and debond normal to the interface and its impact on both unidirectional and crossply laminates of the SCS6/Ti-15-3 composite system. The present study extends this analysis by incorporating tangential or shear failure along the interface. The approach outlined here parallels the previous formulation where a statistical representation is employed to create the mathematical model for the interface. Once the mathematical model for the interface is obtained, it is embedded in a general purpose micromechanics approach for analysis of laminated composites.

The micromechanics technique chosen for this study is a modified method of cells approach [5,6] used in the computer program LISOL (Laminate Inelastic SOLver) which was developed at the Air Force Institute of Technology [7]. This micromechanics technique partitions a unit fiber/matrix cell into various regions of constant stress which can be related through an elementary mechanics of materials analysis by choosing appropriate equilibrium and compatibility relations. These local or micromechanical relations are then assembled for the global response of the laminate through the classical laminated plate theory.

Therefore, the objective of this paper is to develop a failure scheme to model both normal as well as axial shear failure of the interface and incorporate it into the micromechanics model. The effects of this approach on the characteristics of various layups for the SCS6/Ti-15-3 composite system for monotonic and thermomechanical fatigue (TMF) loading will then be examined. To this end, the formulation of the interface model is presented first, followed by a brief explanation of the micromechanics model and results.

MODEL FOR INTERFACIAL FAILURE

Consider a transverse applied load to a unidirectional composite. Due to the variations in fiber packing and interfacial properties, all interfaces do not fail at the same moment. If it is assumed that all interfacial failure strengths are the same and that the random failure of the interfaces may be described in terms of the interfacial stresses, then the progressive failure of all interfaces may be depicted by Figure 1. In this figure, the distribution of interfacial stresses, σ_I, is presented where all interfaces with stresses above the interfacial failure strength, σ_{Iult}, represent the fraction of interfaces that have failed under the current loading. If a Gaussian distribution for the interfacial stresses is assumed, then [4]

$$p(\sigma_I) = \frac{1}{S\sqrt{2\pi}} \exp\left[-\frac{(\sigma_I - \sigma_m)}{2S^2}\right] \qquad \text{for} \quad -\infty < \sigma_I < \infty \qquad (1)$$

where S and σ_m are the standard deviation and mean of the distribution, respectively.

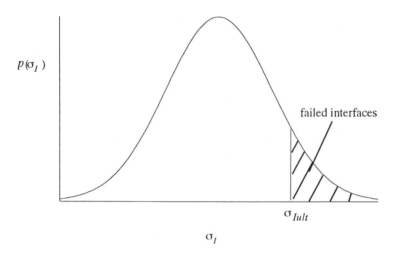

Figure 1. Distribution of Interfacial Stresses in the Composite

Equation (1) applies for a composite with all intact or undamaged interfaces. For a composite possessing failed interfaces, Eq (1) applies for the intact interfaces while all failed interfaces possess zero stress. As mentioned in the introduction, the formulation for the behavior of the interface in the normal or perpendicular direction from the fiber has previously been presented [4]. The present study performs the same analysis but adds the effects of shear failure. Therefore, the equation formulation will be presented in terms of the interfacial shear stress, τ_I. The average interfacial shear stress may be given by

$$\tau_{I\,avg} = \int_{-\tau_{I\,ult}}^{\tau_{I\,ult}} \tau_I p(\tau_I) d\tau_I \tag{2}$$

which may be solved to produce

$$\tau_{I\,avg} = \frac{S}{\sqrt{2\pi}}\left(\exp\left[\frac{-(\tau_{I\,ult}+\tau_m)^2}{2S^2}\right] - \exp\left[\frac{-(\tau_{I\,ult}-\tau_m)^2}{2S^2}\right]\right)$$
$$+ \frac{\tau_m}{2}\left(erf\left[\frac{\tau_{I\,ult}+\tau_m}{S\sqrt{2}}\right] + erf\left[\frac{\tau_{I\,ult}-\tau_m}{S\sqrt{2}}\right]\right) \tag{3}$$

where S and τ_m are the standard deviation and mean of the distribution of interfacial shear stresses for a composite possessing all intact interfaces. Note that $\tau_m \neq \tau_{I\,avg}$ because $\tau_{I\,avg}$ contains the additional effect of the failed interfaces.

The relative displacement of all intact interfaces are zero while the relative displacement between fiber and matrix at each failed interface is assumed to be proportional to the stress on that interface if it had not failed, and hence, may be related by a constant, C_v. Therefore, the relative displacement of each interface, v_I, and average displacement of all interfaces, $v_{I avg}$, is given by

$$v_{I avg} = \int_{-\infty}^{\infty} v_I p(\tau_I) d\tau_I \qquad \text{where} \qquad v_I = \begin{cases} 0 & \text{for } |\tau_I| \leq \tau_{I ult} \\ C_v \tau_I & \text{for } |\tau_I| \geq \tau_{I ult} \end{cases} \qquad (4)$$

which results in

$$v_{I avg} = C_v \left(\tau_m - \tau_{I avg} \right) \qquad (5)$$

By utilizing the above equations, the general behavior of the average interfacial stress and average interfacial displacement may be determined for a monotonically increasing load as shown in Figure 2. Also, by recognizing that during unloading the percentage of failed interfaces remains constant as the stress is decreased, the behavior of the average interfacial stress and displacement during unloading may also be determined [4]. By combining the conditions for interfacial failure during loading and unloading, the general behavior of the average interfacial stress and displacement throughout the composite for

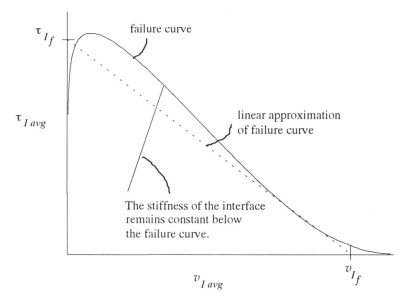

Figure 2. Average Interfacial Stress and Displacement Using a Statistical Representation for Interfacial Failure

any loading condition can be evaluated. For instance, the increasing applied stress on the composite produces additional interfacial failures which eventually results in a drop in the average interfacial stress while the average interfacial displacement increases. Hence, the average stiffness of the interface decreases during this process, but when the applied stress is decreased, the average stiffness of the interface remains constant until the material is again loaded to the point where more interfacial failures occur. This general behavior of the interface as obtained from the present statistical model is very similar to assumed interfacial failure curves used by previous researchers [8,9], and thus, it provides a physical interpretation of the previous work.

Rather than employing the relations of Eq (3) and (5) in the micromechanics model, the progressive failure of the interface depicted in Figure 2 is approximated by a straight line. Therefore, interfacial failure is modeled by decreasing interfacial stress with increasing displacement until complete interfacial failure occurs at the point where the failure curve intersects the displacement axis. If unloading occurs at any time, the stiffness of the interface remains constant while it is below the failure curve. The constants used to characterize the failure are the points at which it crosses the stress and displacement axes denoted by τ_{If} and v_{If}. In addition, the shear failure mechanisms are the same for the negative shear stress case, so a mirror image of the interfacial failure model is also used for negative interfacial shear stresses. This leads to the general equation for axial shear failure of the interface as follows:

$$
v_{I_{AVG}} = \begin{cases} S_I^v \tau_{I_{AVG}} & \text{for } v_{I_{AVG}} \le \left| v_I^* \right| \\[2em] v_{I_f} \left(\dfrac{\tau_{I_{AVG}}}{\left| \tau_{I_{AVG}} \right|} - \dfrac{\tau_{I_{AVG}}}{\tau_{I_f}} \right) & \text{for } v_{I_{AVG}} > \left| v_I^* \right| \end{cases}
\tag{6}
$$

where

$$
S_I^v = \frac{v_I^*}{\tau_{I_f}\left(1 - \dfrac{v_I^*}{v_{I_f}}\right)}
\qquad \text{and} \qquad
v_I^* = \max_{0 \le t \le t_0} \left| v_{I_{AVG}}(t) \right|
\tag{7}
$$

and the average interfacial displacement is represented as a function of time, t, through the present time, t_0.

The above model for shear failure at the interface along with the previous formulation for normal failure of the interface are now combined with a nonlinear micromechanics technique [7]. Also, since it is reasonable to assume that a given interface which has failed in the tangential direction (i.e. along the fiber) also possesses no strength in the normal direction (i.e. perpendicular to the fiber) and vice versa, the two mechanisms are coupled such that the same percentage of failure exists in each. Thus, although the individual failure criteria for the normal and tangential directions are considered separately, only a single value corresponding to the percent of failed interfaces is maintained throughout the calculations. A more realistic approach would seek to couple the two failure criteria. However, until a reasonable method of combining the two based on the actual physical mechanisms of interfacial failure is determined (*i.e.* similar to the statistical approach presented in this study), the two will be combined only through maintaining the same percentage of failed interfaces in each.

MICROMECHANICS TO LAMINATE ANALYSIS

The micromechanics formulation is a modified method of cells approach [5] where the representative volume element for a single ply is modeled by a unit cell consisting of six regions each of uniform stress [6]. A single fiber region, three matrix regions, and two infinitely thin interface regions, as shown in Figure 3, are employed where the fiber and matrix are assumed to possess equivalent normal strain in the fiber direction. Also, the average strain along the external faces of the analysis cell are set equal to the ply strain. Equilibrium is accomplished by ensuring that the average stress through any cross section of the unit cell is in equilibrium with the ply stress and by forcing the stresses normal to the faces between adjacent regions to be equal. Slip between adjacent regions is allowed only in the 2-3 plane where continuity of displacements is maintained only along the external faces of the unit cell. The resulting equations for the unit cell are essentially the same as the method of cells except for the axial shear response.

The laminate analysis utilizes the classical laminated plate theory assumption which states that any plane perpendicular to the midplane before deformation remains both plane and perpendicular to the midplane after deformation. The micromechanics equations for each ply are assembled according to this assumption which results in the following general form for the laminate analysis:

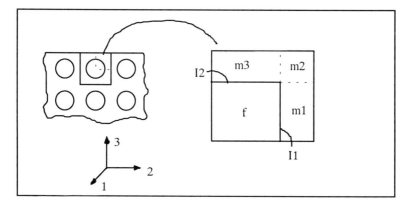

Figure 3. Schematic of the Micromechanics Model for a Single Ply

$$
\begin{bmatrix} & P & \end{bmatrix} \begin{Bmatrix} {}_o\varepsilon \\ \kappa \\ \sigma_{reg} \\ \vdots \end{Bmatrix} = \begin{Bmatrix} \vdots \\ f_{\Delta T} \\ \vdots \end{Bmatrix} + \begin{bmatrix} & P_N & \end{bmatrix} \begin{Bmatrix} N \\ M \end{Bmatrix} - \begin{bmatrix} & P_p & \end{bmatrix} \begin{Bmatrix} \vdots \\ \varepsilon^p_{reg} \\ \vdots \end{Bmatrix} \tag{8}
$$

where ${}_o\varepsilon$ and κ are the midplane strain and curvature, respectively, σ_{reg} represents the region stresses for each ply, $f_{\Delta T}$ is the thermal component, N and M are the applied forces and moments on the laminate, ε^p_{reg} represents the region plastic strains for each ply, and the matrices P, P_N, and P_p simply relate the various quantities according to the given assumptions [7].

The inelastic material behavior of the matrix is modeled by the unified viscoplastic theory of Bodner and Partom with directional hardening [10]. Also, the elastic and viscoplastic properties of the constituents were assumed to be temperature dependent. The constituent properties employed in the present study are listed in a previous paper [11]. Therefore, the nonlinearities accounted for in the present study consist of temperature dependent properties, matrix viscoplasticity, and interfacial failure. Results of this analysis for various layups of SCS6/Ti-15-3 composite laminates are now presented.

RESULTS

In order to gain insight into the effects on the calculated composite response of an interfacial model that allows for progressive failure, the results of four sets of calculations are presented in Figure 4 for a unidirectional SCS6/Ti-15-3 composite system undergoing a monotonically increasing in-plane shear load at room temperature. Identical loading rates of 10 MPa/s are applied to the four cases, but each case represents a separate model of the interface. Results of a strong (perfectly bonded), weak (unbonded), and a constant intermediate interfacial stiffness are compared to the results of the progressive interfacial failure model described above. As expected, the strong and weak bond cases differ significantly from one another with both the initial composite stiffness and yield point of the weak bond results being less than one-half that of the strong bond case. The only nonlinearity occurring in the strong bond calculation was the viscoplastic material model [11] because the calculation was at constant temperature and no interfacial failure occurred. The weak bond calculation, however, allowed for longitudinal slip between the matrix and fiber in addition to the viscoplastic effects. In addition, the other two curves also exhibit unique characteristics. For instance, the results of the case involving an interface possessing a constant stiffness intermediately chosen so that the behavior was neither strong nor weak displays only a mild nonlinearity due to the viscoplastic deformation of the matrix. On the other hand, the case with the progressive failure model exhibits significant initial nonlinearity as the effective interfacial stiffness is decreased throughout loading according to the statistical approximation of additional interfacial failures. This behavior is somewhat expected in experiments as shown in the following figures. Thus, if the constants which

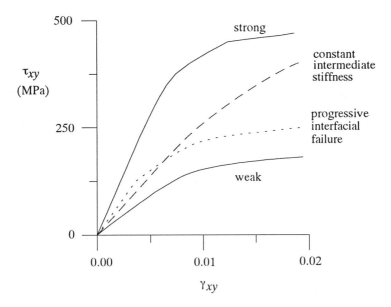

Figure 4. Effect of Various Interfacial Models on the In-Plane Shear Response of
Unidirectional SCS6/Ti-15-3

characterize the interfacial failure curve, τ_{If} and v_{If}, are identified, then the composite behavior with interfacial failure in a variety of loading environments can be approximated.

Therefore, comparisons of analytic results with experimental data were sought to
characterize the interface properties, but since data on the in-plane shear response of the
unidirectional composite is not available, the results from angle-ply layups were employed.
Lerch and Saltsman [12] have published the room and elevated temperature monotonic
response of the SCS6/Ti-15-3 composite system for various angle-ply layups. The response of the $[\pm45]_{2S}$ layup was used to determine the interfacial shear failure constants at
both room and elevated temperature. Table 1 presents both the shear and normal interfacial failure constants employed for this study. A study characterizing the normal failure
was previously completed [4], but the constants characterizing the normal failure have
been slightly modified by the authors since the completion of the previous study, and

Table 1. Interfacial Failure Constants for the SCS6/Ti-15-3 Composite System
(1% initial failure assumed before applied load; r = fiber diameter)

Temp (C)	τ_{If} (MPa)	v_{If}/r	σ_{If} (MPa)	u_{If}/r
25.	130.	0.1	95.	0.025
427.	75.	0.05	80.	0.025

therefore, are also listed here. In the previous study, no initial level of interfacial failure was assumed. On the other hand, in the present study it was assumed that the composite possessed an initial level of interfacial failure of one percent before any load was applied. Therefore, the failure constants between the present and previous study are slightly different. This one percent initial failure is somewhat arbitrary, but it is consistent with actual physical conditions of currently available MMCs. These MMCs, in the as-received condition, do contain interfacial failure and small cracks especially in regions where fibers are spaced very close to each other [13]. This occurs accidentally in manufacturing.

The results of the [±45]$_{2S}$ calculations as compared to the experimental data of Lerch and Saltsman are presented in Figure 5, and the additional results of various angle-ply layups at room temperature are given in Figure 6. A strong correlation with the experimental response is observed in all cases. The nonlinearities in the stress-strain response of the various layups are difficult to categorize as to whether they are a result of interfacial damage or matrix viscoplastic deformation. Therefore, Figure 7 presents the predictions for the through-the-thickness versus the longitudinal strain of the various layups at room temperature. With this plot the occurrence of interfacial damage and matrix viscoplastic deformation can be separated out. For instance, a decrease in the slope corresponds to interfacial damage while an increase in the slope signifies plasticity [1]. Each curve in Figure 7 displays an initial nonlinearity at a strain of approximately 0.002 which is characterized as a decrease in the slope, and therefore, it represents interfacial failure. Also, the interfacial failure in the [90] and [±60]$_S$ laminates was primarily the

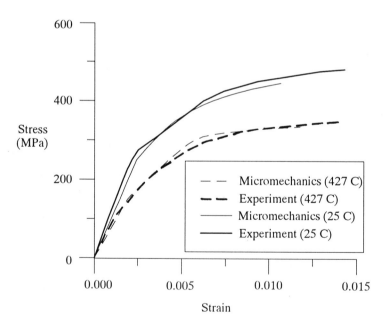

Figure 5. Monotonic Response of [±45]$_S$ SCS6/Ti-15-3 at 25 °C and 427 °C [12]

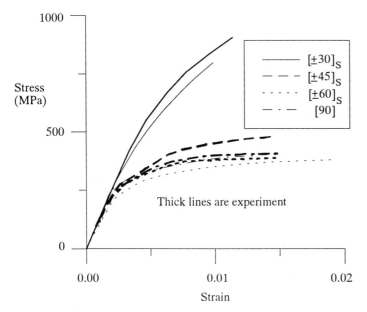

Figure 6. Room Temperature Response of Various Layups of SCS6/Ti-15-3 [12]

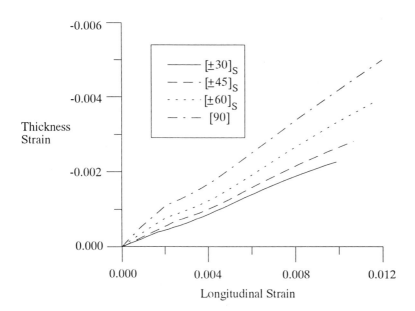

Figure 7. Micromechanics Calculations of Through-the-Thickness Versus Longitudinal Strain for Various Layups

result of failure normal to the interface while the [±45]ₛ and [±30]ₛ layups were controlled by shear failure along the interface. This distinction is important because separation of the fiber/matrix interface (failure normal to the fiber) has a greater impact on the through-the thickness strain than tangential slip (failure along the fiber) as is observed in Figure 7 where a greater decrease in slope occurs for the [90] and [±60]ₛ cases. In addition, upon further loading, plasticity becomes prevalent in all cases as the slope of each curve in Figure 7 is observed to increase at a strain of approximately 0.004. Also, it should be noted that the calculations presented in Figures 5 through 7 were performed at a strain rate of 10^{-4}/sec with a stress free state of 900 °C for each layup.

A quasi-static load-unload sequence at elevated temperature (427 °C) is presented in Figure 8 for the [±45]ₛ layup [14]. It should be noted that the slope of the stress-strain curve during unloading is controlled by the amount of interfacial damage that has occurred, and therefore, does not match the initial elastic modulus as would have occurred if matrix viscoplastic deformation was the only nonlinearity present [1]. A further comparison with experiment for the [±45]ₛ layup is presented in Figure 9 [15]. The results of an isothermal cyclic load to a maximum applied stress of 300 MPa is presented for the first 100 cycles. Each cycle had a period of 48 s with a minimum stress of 30 MPa and a constant temperature of 427 °C. The analysis demonstrates good agreement with the experiment in the range presented. Increasing the number of cycles further was considered inappropriate since the viscoplastic constants of the matrix material are valid up to a maximum strain of about 1% [11].

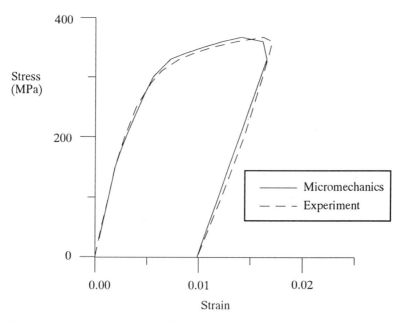

Figure 8. [±45]ₛ SCS6/Ti-15-3 Response to Load-Unload Sequence at 427 °C [14]

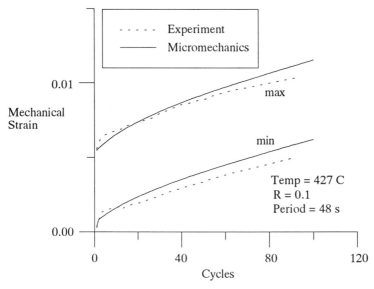

Figure 9. Max/Min Strain History for a [±45]$_S$ Layup of SCS6/Ti-15-3 Under Isothermal
Cyclic Load (Max Stress = 300 MPa) [15]

In addition to the angle-ply results, calculations of a quasi-isotropic layup,
[0/±45/90]$_S$, were also completed and are summarized in Figures 10 and 11. Three thermomechanical load sequences are presented which consist of isothermal, in-phase, and
out-of-phase cyclic fatigue loading. Each case assumed a stress free processing state of
900°C which was subsequently cooled to room temperature (25°C) in 800 s. The isothermal fatigue case was then heated to 427°C in 400 s followed by an applied mechanical
load that was cycled between 47.5 MPa and 475 MPa with a cycle period of 48 s while the
temperature was held constant. The in-phase case was heated to 149°C in 100 s following
cooldown, and then subjected to a combined thermal (149°C to 427°C) and mechanical
(47.5 MPa to 475 MPa) cyclic load where the temperature and mechanical load were increased and decreased simultaneously with a period of 48 s. Similarly, the out-of-phase
case was cycled between the same temperature and load only the temperature and mechanical load cycles were directly out-of-phase, so the temperature was decreased as the
mechanical load was increased. Also, the thermal preload for the out-of-phase case was
the same as for the isothermal fatigue. In addition, waveforms for cycling was triangular.
The interfacial failure constants employed were those listed in Table 1.
      The maximum and minimum mechanical strain as calculated for each cycle is pre-
sented in Figure 10 along with experimental data [16] for the first 10 cycles. These results
were obtained on a SUN SPARC 2 station with an average run time of approximately 3
minutes for cooldown and 10 TMF cycles. The isothermal and out-of-phase calculations
demonstrate strong correlation with their experimental counterparts. The in-phase results
for the minimum mechanical strain for each cycle also strongly correlates with experiment,

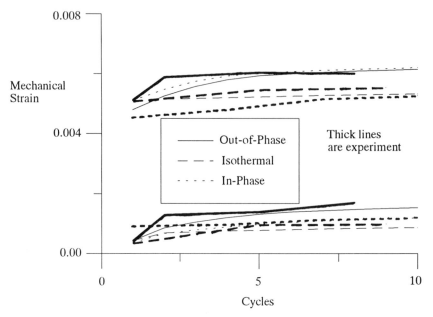

Figure 10. Max/Min Strain History for [0/±45/90]s Layup of SCS6/Ti-15-3 Subjected to Thermomechanical Fatigue (TMF) Loading [16]

but the maximum mechanical strain is approximately 15% greater than its experimental counterpart. Further, with more cycling, experiments showed that other damage mechanisms such as matrix and fiber cracks develop [16] which are not modeled in the present analysis. Also contributing to the difference between the analysis and experiment is the fact that the experiments did not acheive the exact TMF sequence for several cycles due to feedback control and software limitations [16].

Plots of mechanical strain in the loading direction as depicted in Figure 10 do not distinguish between the types of nonlinearities present. Therefore, Figure 11 presents a plot of the average transverse Poisson's ratio at maximum load for each cycle as calculated by the micromechanics model. The value plotted on the vertical axis is the negative of the width strain (perpendicular to the load in the plane of the laminate) at maximum load divided by the longitudinal strain at maximum load. An increase in the average Poisson's ratio with subsequent cycles would correspond to increasing plasticity while a decrease in the Poisson's ratio would correspond to greater interfacial damage. The curves in Figure 11 essentially stabilize within the first 10 cycles, and demonstrate that the most dominant mode of failure for the out-of-phase sequence is interfacial. This is in qualitative agreement with experimental observations [15]. Any interfacial damage in the isothermal fatigue was counteracted by viscoplastic deformation which prevented a significant drop in the Poisson's ratio. Therefore, the thermal portion of a thermomechanical fatigue load plays an important factor in the type of damage that may result which in turn will influence

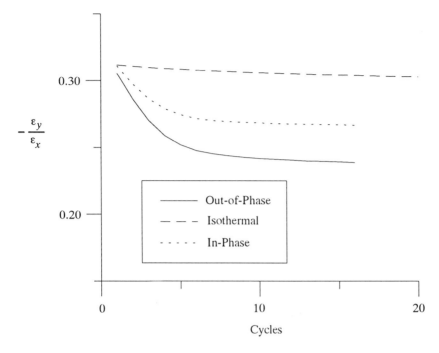

Figure 11. Model Predictions of Average Poisson's Ratio at Maximum Load for
[0/±45/90]$_S$ Layup of SCS6/Ti-15-3 Subjected to TMF Loading

the fatigue life. However, there are other damage mechanisms present during fatigue, and the final result is the net effect of all these. The present analysis is a step in this direction.

SUMMARY

This study examined the effects of fiber/matrix interfacial failure, both normal and tangential to the fiber, on a metal matrix composite system through an analytical technique. The criterion which controls the progression of interfacial failure is based on a statistical representation of the interfacial stresses and a constant interfacial failure strength. This failure criterion for the interface is coupled with a modified method of cells [5] micromechanics technique for laminated composites [7]. The fiber was assumed to be thermoelastic while a thermoelastic/viscoplastic material model for the matrix was assumed based on the unified theory of Bodner and Partom. The material properties for both constituents were temperature dependent [11]. Results from the present formulation for various layups of the SCS6/Ti-15-3 composite system were then compared to their experimental counterparts from previous studies [12,14-16].

The interfacial failure constants used to characterize the interface in shear were obtained from experimental comparisons with a [±45]$_S$ layup under monotonic loading at constant temperature. Experimental comparisons with a [90] layup were used to characterize the failure of the interface normal to the fiber. These same constants were then employed for the analysis of various layups and thermomechanical loading environments. Strong correlation with the experimental counterparts of previous studies indicate that the present formulation offers much promise as a means of characterizing the progressive failure of the interface during thermomechanical loading of a metal matrix composite. The model captured the progression of interfacial damage with subsequent cycles as the average in-plane Poisson's ratio for the laminate was observed to decrease with each cycle. These damage effects were found to significantly depend on the TMF load sequence where interfacial damage was most dominant when the mechanical and thermal loads were cycled out-of-phase. The present approach was also found to require minimal computational resources. Therefore, the approach is sufficiently tractable for use in the laboratory.

REFERENCES

[1]   Majumdar, B.S., and Newaz, G. M., "Inelastic Deformation of Metal Matrix Composites: Plasticity and Damage Mechanisms," *Philosophical Magazine A*, Vol. 66, No. 2, 1992, pp. 187-212.

[2]   Gayda, J., and Gabb, T.P., "Isothermal Fatigue Behavior of a [90]$_8$ SiC/Ti-15-3 Composite at 426°C," *International Journal of Fatigue*, Jan 1992, pp. 14-20

[3]   Santhosh, U., Ahmad, J., and Nagar, A., "Non-Linear Micromechanics Analysis Prediction of the Behavior of Titanium-Alloy Matrix Composites," *Fracture and Damage*, ASME, Vol 27, 1992.

[4]   Robertson, D. D., and Mall, S., "Micromechanical Analysis of Metal Matrix Composite Laminates With Fiber-Matrix Interfacial Damage," accepted for publication in *Composites Engineering*, (1995).

[5]   Aboudi, J., "Micromechanical Analysis of Composites by the Method of Cells," *Applied Mechanics Review*, Vol 42, No 7, July 1989, pp193-221.

[6]   Robertson, D., and Mall, S., "Micromechanical Analysis for Thermoviscoplastic Behavior of Unidirectional Fibrous Composites," *Composites Science and Technology*, 50 (1994), pp 483-496.

[7]   Robertson, D. D., and Mall, S., "A Nonlinear Micromechanics Based Analysis of Metal Matrix Composite Laminates," *Composites Science and Technology*, 52 (1994), pp 319-331.

[8] Needleman, A., "A Continuum Model for Void Nucleation by Inclusion Debonding," *Journal of Applied Mechanics*, Vol 54, 1987, pp 525-531.

[9] Tvergaard, V., "Effect of Fibre Debonding in a Whisker-Reinforced Metal," *Materials Science and Engineering*, Vol A125, 1990, pp 203-213.

[10] Stouffer, D.C., and Bodner, S.R., "A Constitutive Model for the Deformation Induced Anisotropic Plastic Flow of Metals," *International Journal of Engineering Science*, Vol 17, 1979, pp 757-764.

[11] Robertson, D. D., and Mall, S., "A Micromechanical Approach for Predicting Time-Dependent Nonlinear Material Behavior of a Metal Matrix Composite," AD-Vol 34/AMD-Vol 173, Thermomechanical Behavior of Advanced Structural Materials, ASME 1993, pp 85-96.

[12] Lerch, B.A, and Saltsman, J.F., "Tensile Deformation of SiC/Ti-15-3 Laminates," *Composite Materials: Fatigue and Fracture, Fourth Volume, ASTM STP 1156*, Stinchcomb and Ashbaugh, Eds., 1993, pp. 161-175.

[13] Mall, S., and Ermer, P.G., "Thermal Fatigue Behavior of a Unidirectional SCS6/Ti-15-3 Metal Matrix Composite," *Journal of Composite Materials*, Vol 25 - December 1991, pp. 1668-1687.

[14] Vaught, Wade H., "Thermomechanical Fatigue Characterization of an Angle Ply Metal Matrix Composite," M.S. Thesis, Air Force Institute of Technology, AFIT/GAE/ENY/91D-21, Dec 1991.

[15] Roush, J.T., Mall, S., Vaught, W.H., "Thermo-Mechanical Fatigue Behavior of an Angle-Ply SCS6/Ti-15-3 Metal Matrix Composite," *Composites Science and Technology*, 52, (1994), pp. 47-59.

[16] Hart, K.A., and Mall, S., "Thermomechanical Fatigue Behavior of a Quasi-Isotropic SCS6/Ti-15-3 Metal Matrix Composite," accepted for publication in ASME *Journal of Engineering Materials and Technology*, (1994).

W. S. Johnson, [1] Massoud Mirdamadi,[2] and John G. Bakuckas, Jr. [3]

DAMAGE ACCUMULATION IN TITANIUM MATRIX COMPOSITES UNDER GENERIC HYPERSONIC VEHICLE FLIGHT SIMULATION AND SUSTAINED LOADS

REFERENCE: Johnson, W. S., Mirdamadi, M., and Bakuckas, J., **"Damage Accumulation in Titanium Matrix Composites Under Generic Hypersonic Vehicle Flight Simulation and Sustained Loads,"** Thermomechanical Fatigue Behavior of Materials: Second Volume, ASTM STP 1263, M. Verrilli and M. Castelli, Eds., American Society for Testing and Materials, 1996.

ABSTRACT: The damage accumulation behavior of a $[0/90]_{2s}$ laminate made of Ti-15V-3Cr-3Al-3Sn (Ti-15-3) reinforced with continuous silicon-carbide fibers (SCS-6) subjected to a simulated generic hypersonic flight profile, portions of the flight profile, and sustained loads was evaluated experimentally. Portions of the flight profile were used separately to isolate combinations of load and time at temperature that influenced the fatigue behavior of the composite. Sustained load tests were also conducted and the results were compared with the fatigue results under the flight profile and its portions. The test results indicated that the fatigue strength of this materials system is considerably reduced by a combination of load and time at temperature.

KEY WORDS: fatigue, creep, thermal stresses

Titanium matrix composites (TMC), such as Ti-15V-3Cr-3Al-3Sn (Ti-15-3) reinforced with continuous silicon-carbide fibers (SCS-6), are being evaluated for use in hypersonic vehicles and advanced gas turbine engines where high strength-to-weight and high stiffness-to-weight ratios at elevated temperatures are critical. Such applications expose the composite to mechanical fatigue loading as well as thermally induced cycles. A large difference exists between the coefficient of thermal expansion of the fiber and the matrix, causing thermal residual stresses to develop upon consolidation of the composite. Unconstrained thermal cycling causes these residual stresses to cycle. Furthermore, mechanical fatigue can produce fatigue damage in the fibers, matrix, or fiber-matrix interfaces [1]. Potential TMC components will be subjected to a complex flight profile consisting of fatigue loading cycles, creep-fatigue loading cycles, and thermomechanical

---

[1]Professor, School of Materials Science and Engineering, Georgia Tech, Atlanta, GA.

[2]General Electric Co., Power Generation System, Schenectady, NY. Formerly, Senior Research Engineer, Analytical Services and Materials, Inc. Hampton, VA.

[3]Senior Research Engineer, Galaxy Scientific Corp., Pleasantville, NJ.

fatigue (TMF) loading at various elevated temperatures. Understanding and ranking the damage mechanisms induced by these different types of loading are very complex problems. There are other papers in this Special Technical Publication [2] and in [3] that contain the work of many other investigators concerned with the mechanism of damage in TMC's subjected to TMF loadings.

The objective of this study was to evaluate the fatigue behavior of this material system when subjected to a simulated flight profile. Since these flight profiles are representative of a generic hypersonic space vehicle, the design life is very short compared to a conventional aircraft, therefore the number of cycles and flights to failure are lower. Additionally, portions of the flight profile were evaluated experimentally to isolate critical combinations of load and time at temperature. In particular, the effect of hold time under constant load and elevated temperature was evaluated.

MATERIAL AND TEST SPECIMENS

A $[0/90]_{2s}$ SCS-6/Ti-15-3 laminate with a fiber volume fraction of 0.385 and a thickness of 1.68-mm was used in the present study. The SCS-6 fibers are continuous and have a 0.140-mm diameter. The composite laminates were made by hot-pressing Ti-15-3 foil between tapes of unidirectional SCS-6 silicon-carbide fibers held in place with molybdenum wires. As in [5], the composite in the present study was heat treated at $650^{o}C$ for one hour in air followed by an air quench to stabilize the matrix material. The test specimens used were straight-sided rectangular specimens, 152-mm x 12.7-mm x 1.68-mm, cut using a diamond wheel saw. Brass tabs (10-mm x 30-mm x 1-mm) were placed between the end of the specimen and the grips to avoid specimen failure in the serrated grips. The brass tabs were not bonded to the specimens but were held in place by the grips.

EXPERIMENTAL PROCEDURE

The temperatures and the load spectra used in this study are shown in Figure 1. Mechanical load cycles were applied at 1 Hz and thermal loading rates during heating and cooling were 2.8 and $1.4^{o}C$/sec, respectively.

The TMF test setup consisted of a 100-kN servo-hydraulic test frame with water-cooled grips, a load profiler, a 5-kW induction generator controlled by a temperature profiler, and a nitrogen supply tank. The load and temperature spectra were independently programmed into the load and temperature profiler. The temperature profiler was modified to accept a command signal from the load profiler to initiate the temperature spectrum at any desired point in the load profile. More details on the TMF test setup are given in [4].

ANALYTICAL METHODS

A micromechanics based model, VISCOPLY, was used to predict the strain response of the composite. Details of the formulation and numerical implementation of the VISCOPLY program are given in [5]. Briefly, the VISCOPLY program uses the vanishing fiber diameter (VFD) model to calculate the orthotropic properties of a ply.

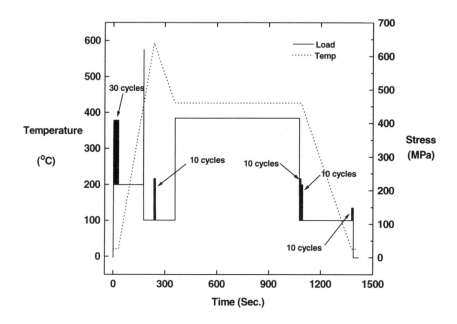

Figure 1. Generic Hypersonic Flight Profile

The ply properties are then used in a laminated plate theory to predict the overall laminate response. Both the fiber and the matrix can be described as a thermo-viscoplastic material. The viscoplastic theory implemented in the VISCOPLY program is valid for high temperature, nonisothermal applications. This viscoplastic theory assumes the existence of an equilibrium stress-strain response which corresponds to the theoretical lower bound of the dynamic response. In addition, this theory further assumes that the elastic response is rate independent and that inelastic rate-dependent deformation takes place if the current stress state is greater than the equilibrium stress.

Complex combinations of thermal and mechanical loads can be readily modeled using the VISCOPLY program. Sequential jobs can be run for varying order and rate of load and temperature. Fiber and matrix average stresses and strains and the overall composite response under thermomechanical loading conditions can be calculated. Due to the assumptions made in the VFD model, local micromechanical effects such as the lateral constraint of the matrix due to the presence of the fibers and stress concentrations in the fibers and matrix are not accounted for in the VISCOPLY program. Nevertheless, the VISCOPLY program has been successfully used to predict the global thermomechanical response within 10% of the experimental results in this TMC . A more complete report on the thermo-viscoplastic stress-strain response of these laminates subjected to the generic hypersonic flight profiles can be found in ref. [5].

RESULTS AND DISCUSSIONS

The TMF data for the entire flight profile, Profile A, and portions of the flight profile, Profiles B through E, are shown in Table 1. The results for the sustained load tests, Profile F, are also given in Table 1. The TMF results will be discussed first, followed by the sustained load tests results, and lastly, the variation of strain with time results will be discussed.

TABLE 1-- Thermomechanical Fatigue Data.

Profile ----- Temp Load	Maximum Stress S_{max} (MPa)	Maximum Temp. T_{max} ($^\circ$C)	Profile/ Cycle Time (Sec)	Number of flights/cycle to failure	Time to Failure (Sec)
A	414	593	1388	300*	416400
	620	593	1388	68	94384
	620	593	1388	74	102712
B	620	427	1	2220	2220
C	620	593	1348	300*	404400*
	620	593	1348	300*	404400*
D	414	427	721	92	66332
	414	427	721	92	71379
E	414	593	1348	300*	404400
	414	593	1348	118	159064
	414	593	1348	127	171196
F	414	427	NA	NA	39600

* Test specimen did not fail.

Fatigue Results

The entire flight profile, Profile A in Table 1, was used with 100% applied stress equal to 414 MPa for one test and the load equal to 620 MPa for two tests. (All the stresses in the flight profile were ratioed the same amount.) The 414 MPa test did not fail after 300 flights and was stopped. For the 620 MPa tests, one test specimen failed at 68 flights, one at 74 flights. The remainder of the tests were directed towards determining the combination of load and time at temperature that caused the fatigue failure for the flight profile where 100% load equaled 620 MPa.

An isothermal fatigue test with continuous cycling, Profile B in Table 1, was conducted at 427°C with a maximum applied stress of 620 MPa at a frequency of 1 Hz and a stress-ratio of 0.2. The fatigue life under Profile B was 2220 cycles. The fatigue life under Profile A averaged 71 flights where the stress was cycled to a maximum of 620 MPa (at 427°C) only once during each flight. Therefore, under Profile A, approximately 71 cycles to 620 MPa occurred during the life of the test specimen. Since 2220 cycles occurred under Profile B, the fatigue at the maximum applied stress of 620 MPa was not the main contributor to the fatigue failure of the composite. The combination of the maximum stress with the temperature profile was examined next. A fatigue test, Profile C in Table 1, was conducted at the maximum applied stress of 620 MPa with the same temperature profile as in Profile A. When the temperature reached 427°C, the load was cycled with a maximum stress of 620 MPa and a stress ratio of 0.2, then was held at the minimum load for the rest of the temperature cycle. The sustained load level was 112 MPa. Two tests were conducted under this profile and the test specimens survived 300 cycles without failure. These results indicated that the fatigue cycles at 620 MPa combined with the temperature profile were also not the main contributor to the fatigue of the composite.

An isothermal fatigue test with a 720 second hold-time imposed at a maximum applied stress of 414 MPa, Profile D in Table 1, was conducted at 427°C to investigate the effect of the load at time and temperature. The fatigue lives under this loading condition were 92 and 99 cycles. These results indicate that a hold-time at the maximum load with a constant temperature had a significant effect on the fatigue life of the composite. To examine the effect of the temperature profile at a maximum applied stress of 414 MPa, Profile E (Table 1), was used. The lower-level sustained loads were at a stress level of 112 MPa. The time periods are consistent with those load level in the full flight profile shown in Figure 1 where the temperature was at 427°C or higher. The first test specimen had not failed after 300 cycles. Subsequent tests had fatigue lives of 118 cycles and 127 cycles. The authors have no explanation for the scatter in the data for profile E. Unexpectedly, Profile E appeared to be less damaging than Profile D and this result warranted further investigation.

A comparison of all fatigue results shown in Table 1 indicates that the change in temperature occurring during the flight profile did not have a significant detrimental effect on the fatigue life. However, holding a significant load at a constant temperature did have a significant detrimental effect on the fatigue life. Nicholas, et al. [6] also showed that time at temperature while under load was a major contributor to fatigue.

<u>Sustained Load Tests Results</u>--The strain-time responses from the three sustained load tests conducted were continuously recorded and the results are shown in Figure 2. One test was taken to specimen failure (separation) while the other two tests were stopped prior to specimen failure to assess the damage states. The total accumulated strain of 0.009, 0.0097 and 0.0103 (laminate failure) were measured from each of the three tests. As shown in Figure 2, thermal strains were induced due to unconstrained thermal expansion during the initial thermal loading to 427°C. Additional mechanical strain was accumulated due to the subsequent mechanical loading to 414 MPa. During this initial mechanical loading, a knee in the stress-strain response occurred at an applied stress of 94 MPa as noted in the insert in Figure 2. It was first noted in ref. [7] that the knee in the stress-strain response in cross-plied TMC's corresponded to fiber-matrix debonding in the 90°plies. Thus, it was assumed in this study that fiber-matrix debonding occurred during the initial mechanical loading and contributed to the strain measured immediately after the maximum stress of 414 MPa was reached. In all three specimens, the strain increased thereafter with time at temperature and load. The sudden increase in the strain measured at near the end of the third test is associated with fiber breakage that occurred during catastrophic fracture of the specimen.

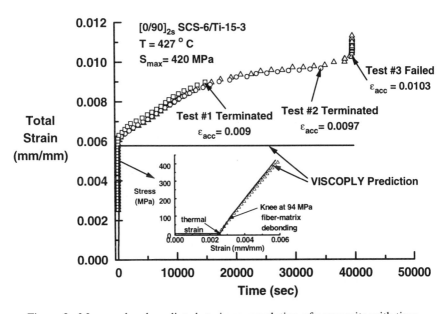

Figure 2. Measured and predicted strain accumulation of composite with time.

Also shown in this figure is the strain-time response predicted using the VISCOPLY program. In the prediction, fiber-matrix debonding was simulated by reducing the transverse modulus in the 90° fibers as described in detail in [4]. Fiber-matrix debonding was assumed to occur at 94 MPa during the initial mechanical loading at 427°C; the stress at which the knee in the stress-strain curve was observed as shown in Figure 2. The strain-time response predictions revealed no time-dependent behavior. The viscoplastic constants for the matrix material were determined at these temperatures on neat specimens of Ti-15-3 heat treated exactly like the composite. The basis matrix creep curves also showed no time dependent response at the stress levels calculated in the matrix at $427^{\circ}C$ and 414 MPa. This would suggest that global viscoplastic deformation was not occurring in the composite. However, localized viscoplastic deformation could be operating at these loading conditions which is not accounted for in the VISCOPLY program. Nevertheless, this localized effect is unlikely to cause such a large difference between the measured and predicted global strains. Thus, damage mechanisms other than fiber-matrix debonding and localized viscoplastic deformation are contributing to the measured global strain as discussed in the subsequent section.

In order to determine if fiber breakage occurred in the 0°plies, the matrix material was removed from the heated test section of the specimens using an acid etching technique as described previously. As shown in the photomicrograph in Figure 3a, the 0°fibers of the outer ply were intact and, thus, there was no evidence of fiber breaks at a strain of 0.009. The second test which accumulated a total strain of 0.0097 was sectioned and a 15-mm length heated test section of the specimen was acid etched and the 90° fibers were removed. As shown in the photomicrograph in Figure 3b, all 0°fibers were in place indicating no fiber breakage in any of the four 0°plies. The fibers are free standing in the photograph in Figure 3b. All matrix is etched away. If there were broken fibers, they would have fell out! Thus, fiber breakage was not a possible mechanism for strain accumulation up to a strain level of 0.0097.

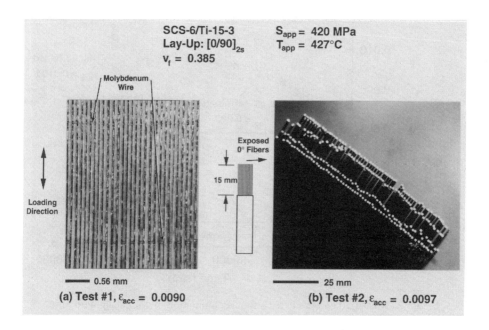

Figure 3. Photomicrographs showing exposed fibers after testing in two specimens. No fiber breakage was observed.

In order to determine if matrix cracking occurred, cross-section cuts of 45° with respect to the loading direction were made from the three tests which accumulated strains of 0.0090 and 0.0097 and 0.0103 (laminate failure strain) and are shown in Figures 4a, 4b and 4c, respectively. As shown in these figures, matrix cracks are evident which appear to have initiated from the debonded surfaces of the fibers in the inner 90° plies, Figure 4a. The presence of these cracks was somewhat surprising to the authors since they grew under sustained load in a rather ductile matrix material. One explanation for their initiation and subsequent growth is that they started in a highly stressed region next to the debonded fiber in the brittle reaction zone. This brittle reaction zone is shown and discussed in [7]. As the accumulated strain increased with time at temperature and load, these matrix cracks progressed transversely to the 0° fibers and linked-up, Figures 4b and 4c. In addition, as the accumulated strain increased, the matrix crack opening displacements increased. These section cuts also revealed (not shown here) no matrix cracks on the surface. Thus, matrix cracking occurred primarily in the interior of the specimens. The layer of 0 degree fibers in the surface ply essentially blocks the cracks from growing to the surface.

Photomicrographs of the fracture surface morphology of the third test specimen which accumulated a total strain of 0.0103 (laminate failure strain) are shown in Figure 5. A global view of the fracture surface is shown in the SEM micrograph in Figure 5a. This micrograph indicates a step type fracture surface. Fiber pull-out is also evident in the figure. The surface of all pulled-out fibers were clean of matrix material indicating fiber-matrix debonding. In general, the fracture surface of the matrix material surrounding the outer 0° fibers had a ductile dimpled

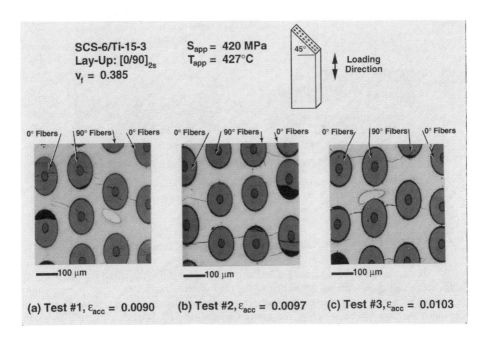

Figure 4. Photomicrographs showing matrix cracks in the three specimens after testing.

texture while in the interior regions, the surface had a cleavage-type morphology. This observation further substantiates previous observations that matrix cracks initiated and grew from the debonded surfaces in the 90° plies however did not progress past the outer 0° fibers to the surface. A detailed view of the fracture surface, marked "A" in Figure 5a, is shown in the photomicrograph in Figure 5b. As shown in this figure, a gap exists in the fiber-matrix interface region indicating debonding of fibers in the 90° plies. A photomicrograph of the fracture surface of the area marked "B" in Figure 5a is shown in Figure 5c. As seen in this figure, river like patters are shown emanating from the 90° fiber suggesting that matrix cracks initiated and grew from the fiber-matrix interface in the 90^oplies. Also shown in this figure, the matrix material located at the step face regions between the 0° fibers had a shear mode dimpled morphology. This indicates that fiber failure preceded matrix failure at that location. This is, of course, expected since the static strain to failure of the fiber is considerably less than that of the matrix.

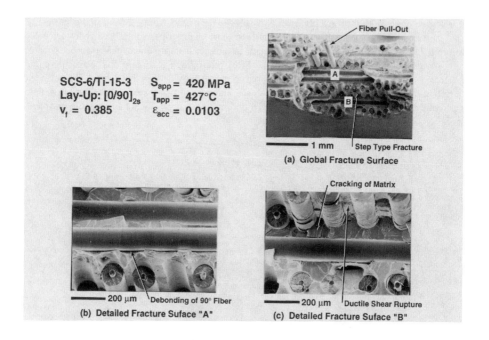

Figure 5. Photomicrographs of the fracture surface morphology

The observations made from Figure 2 through 5 suggest that the initial mechanism of the strain accumulation was due to debonding of the fiber-matrix interface in the 90^oplies, resulting in the formation of a gap in the debonded region. Then, matrix cracks initiated from the debonded surfaces in the 90^oplies and progressed transversely toward 0^ofibers. The strain accumulation increased with time at temperature and load as a result of damage growth in the form of fiber-matrix debonding and matrix cracking. This is summarized in Figure 6.

<u>Variation of Strain with Time</u>--The strain response (minimum and maximum strains) of the composite subjected to Profiles A through E (Table 1) is shown in Figure 7 as a function of time. The indicated failure strain is the strain recorded just before catastrophic failure (the onset of

rapid 0^0 fiber failure) of the test specimen. As seen in Figure 2, the time to failure was dependent on the profile used. However, the magnitude of the failure strain decreased with increased time at temperature for all the profiles tested (except Profile C which was a run-out test shown by a horizontal arrow in the figure). This implies that the strength of the SCS-6 fiber is decreased by the exposure to temperature while under load. This was also suggested in ref. [8]. This may be attributed to further reaction between the silicon carbide in the fiber and the titanium matrix. This further reaction results in brittle interphase region that can easily crack and cause significant stress concentrations in the fiber thus reducing the apparent fiber strength.

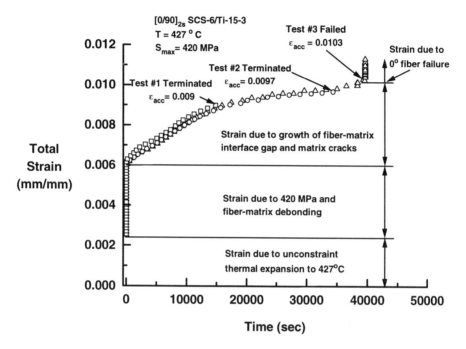

Figure 6. Mechanisms of strain accumulation associated with strain-time response.

Of interest is also the level of the stabilized minimum strain. The thermal residual stresses that are introduced into the fibers during the initial cool down from the stabilization aging is approximately on average -710 MPa [9]. During the fatigue loading, the fiber-matrix interfaces of the 90^0 fibers fail and cracks develop in the matrix. Because of this damage the matrix can no longer support the tensile thermal residual stresses. Since the matrix can not support the residual stresses the residual stresses in the fiber are relieved. Thus this stabilized minimum strain value of just under 0.002 represents the tensile elongation of the 0^0 fibers due to the release of their compressive residual stresses.

The variations in maximum strains with time at 427^0C for Profiles A, D, E and F (Table 1) are shown in Figure 7. Each of these profiles contained a hold time at 427^0C at an applied stress of 414 MPa. For Profiles A and E the elapsed time does not include the time to reach the temperature of 427^0C. As seen in the figure, the total strain at the failure under sustained loading condition (Profile F) was also about 1%. If the time at temperature at the maximum applied stress of 414 MPa was the only mechanism controlling the fatigue behavior of this material system, one would expect a single strain history curve for Profiles A, D, E, F. However, Figure 7 shows that all strain data did not form a single curve and, therefore, other factors besides time at temperature must influence the fatigue life. The fact that the sustained test failed earlier than the other loading profiles that contained cyclic loads and cyclic temperatures in addition to the time spent under load at 427°C is quite mysterious and needs to be examined further.

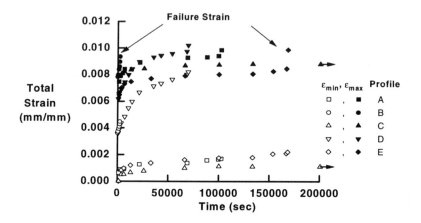

Figure 7.   Variation of total strain as a function of time.

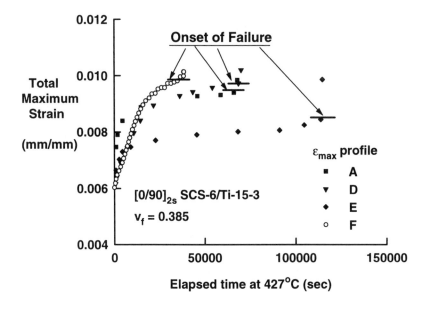

Figure 8.   Variation of strain with elapsed time at 427°C.

SUMMARY

Fatigue behavior of a [0/90]$_{2s}$ SCS-6/Ti-15-3 laminate subjected to a generic hypersonic flight simulation profile was evaluated experimentally. Portions of the flight profile were used to identify critical combinations of load and time at temperature occurring during the flight profile. The test results indicated that holding a significant load at temperature was detrimental to the fatigue life of this composite. However, the combination of a temperature profile and mechanical load was unexpectedly less damaging than an isothermal condition with a sustained mechanical load and warrants further investigation. The failure strain for all the profiles used was shown to decrease with time at temperature under load. This implied a decrease in fiber strength with time at temperature.

The mechanisms contributing to the observed strain accumulation under sustained load were identified using microscopic evaluations of sectioned surfaces, exposed fibers after matrix etching and fracture surfaces. Results are summarized in Figure 6. The initial strain was associated with the unconstrained thermal expansion and applied mechanical loads which was accurately predicted using the VISCOPLY program. During the initial mechanical load phase, fiber-matrix debonding on the 90°plies was assumed to have occurred and contributed to the global strain. Fiber-matrix debonding was simulated in the predictions. The measured strain continued to increase with time at temperature and load, however, predictions made using the VISCOPLY program indicated no global time-dependent response. This would suggest that global viscoplastic deformation was not occurring and, thus, damage mechanisms other than fiber-matrix debonding and global viscoplastic deformation are contributing to the measured strain. Microscopic examinations revealed matrix cracks which initiated, grew, and coalesced from the debonding surfaces of the 90° plies and, thus, contributed to the increase in the measure strain with time at temperature and load. Due to this damage growth process, the stress in the 0° fibers increased with time at temperature and load. Eventually, the stresses transferred to the fibers became sufficiently high to cause fiber breakage resulting in a substantial increase in the strain associated with sudden catastrophic fracture of the specimen.

The strain-time response of the TMC tested exhibited classical characteristics of creep behavior. However, results from this study indicated that the strain accumulation was due primarily to internal damage in the form of fiber-matrix debonding and matrix cracking and not deformation associated with conventional creep. It is therefore important that the mechanistic process is carefully evaluated and understood. Otherwise, the material behavior could be empirically correlated using incorrect modeling assumptions.

ACKNOWLEDGMENT

The authors would like to acknowledge that all of this work was conducted in the Mechanics of Materials Branch, Materials Division, at NASA Langley Research Center. The first author was employed as a Senior Research Engineer and the second and third authors were both NRC Resident Research Associates. Financial support was furnished by the NASP Program.

REFERENCES

1. Johnson, W. S., " Damage Development in Titanium Metal Matrix Composites Subjected to Cyclic Loading," Composites, Vol. 24, No. 3, 1993, pp. 187-196. (Keynote presentation Fatigue and Fracture of Inorganic Composites, Cambridge, England, April, 1992).

2. *Thermo-Mechanical Fatigue Behavior of Materials: 2nd Volume*, ASTM STP 1263, M.J. Verrilli and M.G. Castelli, Eds., American Society for Testing and Materials, Philadelphia, 1995.

3. *Life Prediction Methodology for Titanium Matrix Composites*, ASTM STP 1253, W. S. Johnson, J. M. Larsen, and B. N. Cox, Eds., American Society for Testing and Materials, Philadelphia, 1995

4. Mirdamadi, M. and Johnson, W. S., "Experimental Techniques for Hypersonic Flight Simulation Testing of Titanium Matrix Composites," Proceeding of the 1994 SEM Spring Conference and Exhibits, June 6-8, 1994, Baltimore, MD, pp. 679-684.

5. Johnson, W. S., Mirdamadi, M. and Bahei-El-Din, Y. A., "Stress-Strain Analysis of a $[0/90]_{2s}$ Titanium Matrix Laminate Subjected to a Generic Hypersonic Flight Profile," *Journal of Composites Technology and Research*, Vol. 15, No. 4, Winter 1993, pp. 297-303.

6. Nicholas, T., and Johnson, D. A., Time and Cycle-Dependent Aspects of Thermal and Mechanical Fatigue in a Titanium Matrix Composite, *Thermo-Mechanical Fatigue Behavior of Materials: 2nd Volume*, ASTM STP 1263, M.J. Verrilli and M.G. Castelli, Eds., American Society for Testing and Materials, Philadelphia, 1995.

7. Johnson, W. S., Lubowinski, S. J., and Highsmith, A. L., "Mechanical Characterization of SCS-6/Ti-15-3 Metal Matrix Composites at Room Temperature," *Thermal and Mechanical Behavior of Ceramic and Metal Matrix Composits*, ASTM STP 1080, Kennedy, Moeller, and Johnson, Eds. Phila., PA, 1990, pp. 193-218.

8. Jeng, S. M., Yang, C. J., Alassoeur, P., and Yang, J.-M., *Composite Design and Manufacture, and Application*, Paper 25-C, ICCM/VIII, S. W. Tsai and G. S. Springer, Eds, 1991.[7] Bigelow, C. A., "Thermal Residual Stresses in a Silicon-Carbide/Titanium [0/90] Laminate," *Journal of Composites Technology and Research*, Vol. 15, No. 4, Winter 1993, pp. 304-310.

9. Bigelow, C. A., "Thermal Residual Stresses in a Silicon-Carbide/Titanium [0/90] Laminate," *Journal of Composites Technology and Research*, Vol. 15, No. 4, Winter 1993, pp. 304-310.

Sait Z. Aksoy[1], John Gayda[2] and Timothy P. Gabb[2]

FATIGUE BEHAVIOR OF [0]₈ SCS-6/Ti-6Al-4V COMPOSITE SUBJECTED
TO HIGH TEMPERATURE TURBOSHAFT DESIGN CYCLES

REFERENCE: Aksoy, S. Z., Gayda, J., and Gabb, T. P., **"Fatigue Behavior of [O]₈ SCS-6/Ti-6Al-4V Composite Subjected to High Temperature Turboshaft Design Cycles,"** Thermomechanical Fatigue Behavior of Materials: Second Volume, ASTM STP 1263, Michael J. Verrilli and Michael G. Castelli, Eds., American Society for Testing and Materials, 1996.

ABSTRACT: Mission cycle fatigue testing of a [0]₈ SCS-6/Ti-6Al-4V composite was performed. The mission cycle simulates the stress-temperature-time profile in a ring-reinforced impeller of a turboshaft engine. The cycle has a fourteen minute period and attains a peak stress and temperature of 1100 MPa and 427°C respectively. A fatigue life of 9,528 cycles was achieved. While this life was less than the design goal, 15,000 cycles, a moderate increase in fiber content coupled with a small decrease in peak stress or temperature would probably achieve the stated design goal.

KEYWORDS: fatigue, titanium matrix composite, silicon carbide fiber

---

Performance and thrust-to-weight goals of advanced gas turbine engines require significant advances in material capabilities. Titanium matrix composites (TMCs) offer significant opportunities in compressor components to achieve performance improvements by reducing weight and increasing speed. The application of TMCs to compressor components gives rise to a number of significant design challenges. The structural challenge arising from a performance-driven solution is to obtain design objectives without sacrificing durability. In addition to anisotropic behavior of TMCs, advanced centrifugal compressors are subjected to severe thermal-structural loading conditions during the engine mission cycle. The most damaging stresses and strains are those induced by steep thermal gradients, which occur during the startup and shutdown transients. While conventional isothermal fatigue lives of

---

[1]Principle Engineer, Life Structures and Dynamics, Allied Signal, Stratford, CT 06497.
[2]Research Engineer, Materials Division, NASA Lewis Research Center, Cleveland, Ohio 44135.

current TMCs appear encouraging, conventional thermomechanical fatigue (TMF) lives are a source of concern [1]. However, none of these fatigue tests are fully representative of realistic service conditions. Therefore, design based on isothermal or thermomechanical fatigue data should be verified by component rig testing or component specific fatigue testing before committing to full scale component development. This is more crucial for components using emerging materials with little or no field experience. One approach to coping with this problem is to perform component specific fatigue testing using test waveforms which simulate the stress temperature history at life limiting locations of a component. Laboratory specimen testing is a very economical and efficient manner to provide a designer with relavent data during the development stage for a given component.

The objective of this study was to assess the utility of component specific thermomechanical fatigue testing for design verification. For this purpose, two types of test cycles were used: a) a full mission cycle simulating the operating stress and temperature state of a life limiting loacation in a TMC reinforced impeller, b) a simplified version of the mission cycle employing the same peak stress and temperature as the full mission cycle. The maximum principal stress of the TMC ring which reinforces the impeller in an advanced turboshaft engine was used to define the loading conditions for the mission tests. Flat specimens were then used to assess component fatigue life under realistic service conditions.

TMC REINFORCED CENTRIFUGAL COMPRESSOR DESIGN

Structural integrity of any engine component must be assessed according to design guidelines which define criteria for burst and fatigue failure. The most important failure mode for the impeller is low cycle fatigue. For this application, the mission cycle shown in Fig. 1 was established to simulate

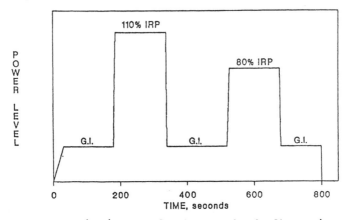

FIG. 1--Mission cycle for turboshaft engine.

operating conditions which a turboshaft engine would be subjected to during field operation. The speeds and metal temperatures were generated using this information. Finite element heat transfer was used to approximate metal temperatures. This information was then input into a finite element stress program which was used to determine the cyclic stress components of the TMC reinforced impeller during operation. Once stresses were calculated, the life for each

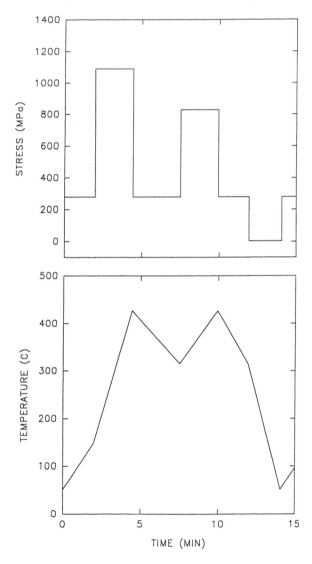

FIG. 2--Stress-temperature profile of the LCF mission cycle.

major and minor cycle was determined. Since the life prediction procedure for this emerging material system has not been fully calibrated by field or component spin tests, mission testing on flat TMC specimens was conducted to provide some kind of verification program before fabricating any component hardware.

The stress versus time and temperature versus time waveforms for the full mission cycle are presented in Fig. 2. Note the

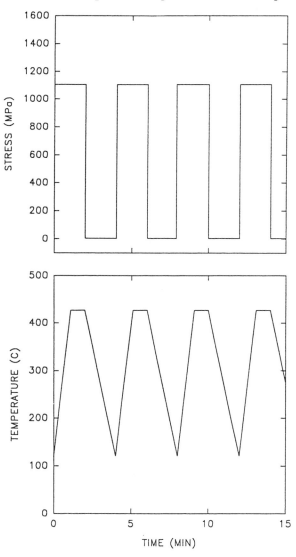

FIG. 3--Waveforms for the simplified mission cycles.

Flat, dogbone test specimens were cut from the composite panel by means of electro-discharge machining followed by diamond grinding of the machined edges. The gage section of the dogbone specimens was about 8 mm wide and 25 mm long with the fiber direction parallel to the specimen length. The specimen thickness was essentially that of the composite panel, about 1.7 mm. All fatigue tests in this study were conducted using a uniaxial, load-controlled servohydraulic test system equipped with hydraulic wedge grips and a high temperature, axial extensometer. Direct induction heating of the test specimens was employed to obtain the desired heating and cooling rates in the mission cycle.

During mission testing stress-strain hysteresis loops were periodically recorded with an X-Y plotter and a continuous record of strain versus time was obtained with a strip chart recorder. After testing, SEM examination of the fracture surfaces were done to help identify the operative failure mechanism.

RESULTS AND DISCUSSION

Tensile testing of the SCS-6/Ti-6Al-4V composite produced a failure strain of 0.95% and an ultimate strength of 1365 MPa at 25°C. These values are consistent with good quality TMC with a 35% fiber content [2].

Isothermal fatigue tests were also run to assess TMC quality and used as a comparative measure to assess the severity of the mission cycle. The isothermal fatigue tests were run under load-control with an R-ratio of 0.05 and a moderately rapid frequency, 0.2 Hz. A test temperature of 371°C was chosen to represent a high end, average temperature for the LCF mission cycle. The results of these tests are plotted in Fig. 5 and

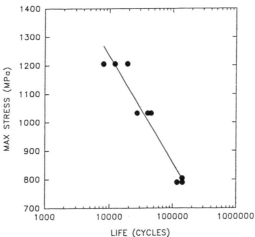

FIG. 5--371°C isothermal fatigue life line.

show the TMC to have a typical isothermal fatigue life of about 22,000 cycles at 1100 MPa, the peak stress target in mission testing. From the standpoint of isothermal fatigue, the TMC life is slightly greater than that desired for man-rated turbomachinery, 15,000 cycles, and therefore the 1100 MPa stress target would not appear to be an unreasonable goal for the LCF mission cycle. Further, the isothermal fatigue lives for this material appear to be equivalent to that for other TMCs tested under similar conditions [3].

As is the case for elevated temperature isothermal fatigue testing, the initial cycle of the LCF mission test exhibited a measurable amount of permanent deformation, approximately 0.03% strain. From that point on a single stress-strain loop showed no measurable permanent deformation, although over many cycles an increase in peak strain was noted. A typical stress-strain loop, after shakedown, is depicted in Fig. 6,

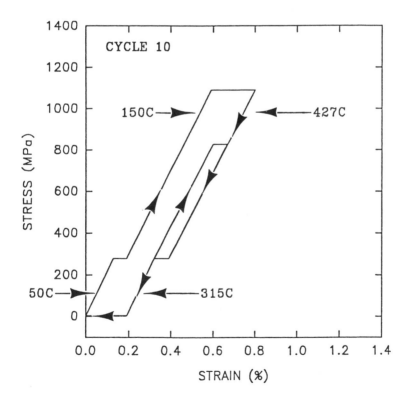

FIG. 6--Typical stress-strain loop for the LCF mission cycle.

and as one might expect, the response is much more complex than that for an isothermal test. Nevertheless, the stress-strain response for the LCF mission cycle is logical, but does require some degree of explanation. Starting at the zero stress, zero strain point in Fig. 6, the specimen load is increased under isothermal conditions to produce the first inclined segment. Next, the temperature is increased while the stress is held constant producing the first horizontal segment. Additional stress and temperature changes are sequentially imposed producing the stress-strain response indicated by the arrows. All horizontal segments are for the most part thermal strains generated by temperature changes at constant load, while those strains associated with the inclined segments are mechanical strains resulting from changes in applied load at constant temperature. The temperature for each inclined segment is noted for clarity, while the heating and cooling rates for each horizontal segment can be obtained from the temperature versus time waveform, Fig. 2. Loading rates were approximately 400 MPa/sec which corresponds to that used in isothermal fatigue tests.

As previously stated, peak strain did increase in the LCF mission test with continued cycling. This is most evident by plotting peak strain versus cycle count ,Fig. 7. Some fraction

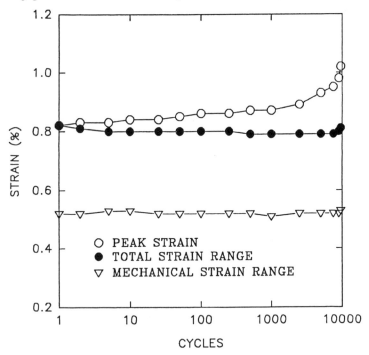

FIG. 7--Evolution of peak strain and strain ranges during the LCF mission cycle.

of this shift, especially toward the onset of testing, could be associated with strain ratchetting. This phenomenon has been attributed to matrix relaxation at elevated temperatures and stresses which produces a subsequent increase in fiber stress/strain as well as composite strain [4,5]. However, toward the end of the test the rapid increase in peak strain could also be associated with the growth of large cracks.

Cracking should affect specimen compliance and therefore should produce an increase in strain range. Plotting total strain range or mechanical strain range versus cycle does show a modest increase at the very end of the test ,Fig. 7. Mechanical strain range was defined as the summation of strains produced on loading at 50°C and 150°C in Fig. 6. The lack of a pronounced increase in either total or mechanical strain range over the majority of the test suggests that much of the increase in peak strain is associated with ratchetting.

Deformation response in the simplified mission test was similar in many respects to the LCF mission test. The initial cycle exhibited a permanent deformation of about 0.02% strain, but thereafter no measurable permanent deformation was observed for any given cycle, as seen in Fig. 8. Peak strain

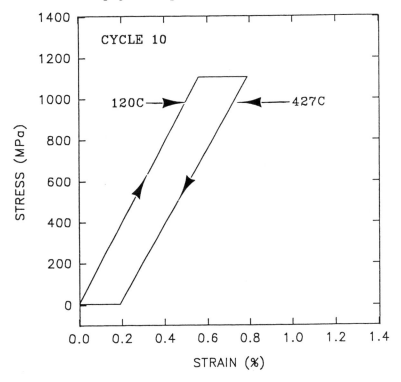

FIG. 8--Stress-strain loop for the simplified mission cycle.

did, however, increase with continued cycling as seen in Fig. 9, and as with the LCF mission test, very little change in total strain range or mechanical strain range was observed until the very end of the test. While many characteristics of these two mission cycles were similar, the stress-strain response of the simplified mission cycle was, by design, much less complex. As seen in Fig. 8, the stress-strain loop of the simplified mission cycle is produced by loading at minimum temperature, followed by heating to maximum temperature at peak stress, and then holding peak stress and temperature for 72 seconds, after which the specimen is unloaded, and then cooled to minimum temperature at zero load. While the time per cycle is shorter for the simplified mission cycle, under five minutes versus fourteen minutes for the LCF mission cycle, the time at peak stress and temperature, 72 seconds, is greater for the simplified mission cycle as the LCF mission cycle

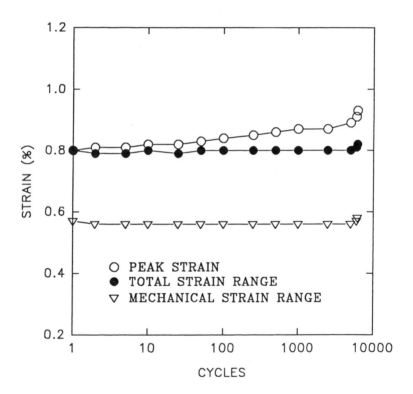

FIG. 9--Evolution of peak strain and strain ranges during the simplified mission cycle.

attains peak stress and temperature and immediately ramps stress and temperature down, Fig. 2. The simultaneous application of peak stress and temperature for a sustained period of time may accentuate time-dependent damage mechanisms and therefore make the simplified mission cycle a more severe test.

The cyclic lives of the isothermal fatigue test, simplified mission test, and the LCF mission test are compared in Fig. 10. In all cases life is defined as cycles to complete separation/fracture of the test specimen. The results of these tests show the simplified mission cycle produced a fatigue life of 6,241 cycles while the LCF mission cycle lasted 9,528 cycles. In comparison, the typical isothermal fatigue life at comparable peak stress levels was about 22,000 cycles. As the

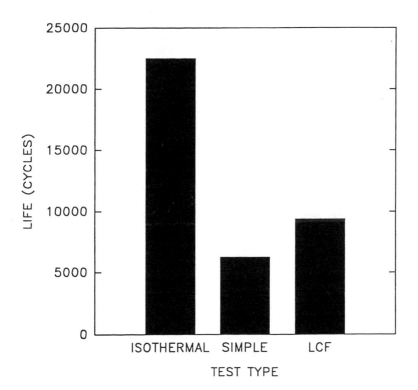

FIG. 10--Fatigue life comparison of the isothermal, simplified mission cycle and the LCF mission cycle.

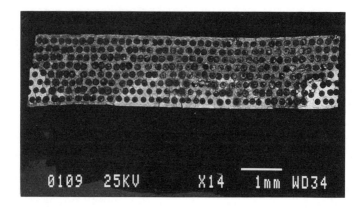

FIG. 11--Fracture surface of the LCF mission test. Note
initiation sites at machined edges and surface.

time per cycle at elevated temperatures and stresses is longer
for both mission cycles in comparison to the isothermal
fatigue test, the shorter cyclic life of either mission cycle
should not be totally unexpected. Further, the life deficit of
the simplified mission test compared with the LCF mission test
could reflect additional damage associated with the sustained
dwell at peak temperature and stress. However, the difference
in fatigue life between the mission tests, less than twofold,
is also well within the normal range of scatter for fatigue.
More testing would be necessary to confirm or refute this
hypothesis concerning the severity of the dwell. Adopting a
conservative design philosophy, it would appear that the
simplified mission cycle is a viable test to assess TMC life
under realistic service conditions without grossly
underestimating service life, while significantly reducing
test time in comparison to the LCF mission cycle.

The fracture mode of the LCF mission cycle, Fig. 11, and the
simplified mission cycle were similar to that of isothermal
tests [6], with initiation sites at cut edge fibers and other
surface anomolies, such as spot welds used to attach
thermocouple leads, followed by large regions of flat fatigue
crack propagation which transition to a more ductile, tensile
overload region.

While conventional TMF testing is often employed to evaluate
generic material behavior, for specific applications such as
the ring-reinforced impeller, mission testing may be
preferable for several reasons. First, in-phase and out-of-
phase TMF tests can produce differing failure modes in TMCs.
At high stresses, in-phase TMF fracture is predominantly
associated with tensile overload, while out-of-phase TMF

stress waveform reflects that along the fiber direction of the TMC core in a ring reinforced impeller. This LCF mission cycle is a complex nonisothermal fatigue test with a fourteen minute cycle period and a peak temperature and stress of 427°C and 1100 MPa respectively. Note that the temperature change lags the stress change throughout the cycle. This is logical as changes in engine RPM produce an instantaneous stress response while component temperature changes more gradually. A simplified version of the LCF mission cycle shown in Fig. 3 was also run employing the same peak temperature and stress to capture the most pertinent features of the full LCF mission cycle and yet reduce total test time. Note the simplified mission cycle ramps stress up at minimum temperature and ramps stress down at maximum temperature. This philosophy was adopted as engine components tend to power up cold and shutdown hot. In addition a 72 second dwell at peak temperature and stress was included in the simplified mission cycle to accentuate time-dependent damage. A 72 second dwell was selected as it yields 300 hot hours at peak temperature and stress after 15,000 cycles, the design criteria used for this turboshaft engine application.

MATERIAL AND TEST PROCEDURE

A current generation TMC, SCS-6/Ti-6Al-4V, was employed in this study. This composite was fabricated by Textron Specialty Materials Division using a conventional foil-fiber-foil approach. The composite had an eight ply, unidirectional layup with a fiber content of 35% by volume. The α-ß microstructure of the matrix is presented in Fig. 4. Details of the reaction

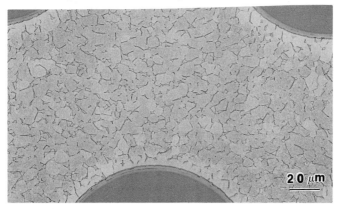

FIG. 4--Microstructure of SCS-6/Ti-6Al-4V composite.

zone at the fiber-matrix interface are also shown. Overall the integrity of the composite was judged to be satisfactory based on ultrasonic inspection.

fracture is characterized by an environmentally-assisted surface cracking mechanism [1,7]. Second, realistic operating conditions in turbomachinery are rarely all in-phase or out-of-phase in nature. These observations suggest that the utility of conventional TMF tests is questionable for assessing fatigue durability of TMCs under actual service conditions when compared with more realistic mission testing. However, conventional TMF tests are undoubtedly valuable tools for assessing fundamental material behavior and development of robust fatigue life models.

CONCLUSION

In summary, a LCF mission test, which attempts to reproduce the stress-temperature-time profile in the TMC core of a ring-reinforced impeller, was run on a $[0]_8$ SCS-6/Ti-6Al-4V composite. The resulting life, 9,528 cycles, was encouraging, although it did not achieve a 15,000 cycle life, a common life goal employed for man-rated turbine engines. However, a moderate increase in fiber content coupled with a small decrease in peak stress or temperature may result in a fatigue life which meets or surpasses that goal.

REFERENCES

[1] Gabb, T.P., Gayda, J., Bartolotta, P.A. and Castelli, M.G., "A Review of Thermomechanical Fatigue Damage Mechanisms in Two Titanium and Titanium Aluminide Matrix Composites," International Journal of Fatigue, Vol. 15, No. 5, Sept. 1993, pp 413-422.

[2] Lerch, B.A. and Saltsman, J., "Tensile Deformation Damage in SiC Reinforced Ti-15V-3Cr-3Al-3Sn," NASA TM 103620, Washington, DC, 1991.

[3] Castelli, M.G. and Gayda, J., "An Overview of Elevated Temperature Damage Mechanisms and Fatigue Behavior of a Unidirectional SCS-6/Ti-15-3 Composite," Reliability, Stress Analysis and Failure Prevention, ASME, New York, 1993.

[4] Gayda, J., Gabb, T.P. and Freed, A.D., "The Isothermal Fatigue Behavior of a Unidirectional SiC/Ti Composite and the Ti Alloy Matrix," Fundamental Relationships Between Microstructure and Mechanical Properties of Metal Matrix Composites, TMS-AIME, Warrendale, PA, 1989.

[5] Aksoy, S., "Thermomechanical Damage Development in SiC(SCS-6)/Ti-6Al-4V Metal Matrix Composite," International Gas Turbine and Aeroengine Congress and Exposition, ASME, New York, 1992.

[6] Jeng, M.S., Alassoeur, P., Yang, J.M. and Aksoy, S.,

"Fracture Mechanisms of Fiber-Reinforced Titanium Alloy Matrix Composites, Part IV: Low Cycle Fatigue", <u>Material Science and Engineering</u>, Vol. A148, 1991, pp 67-77.

[7] Jeng, M.S., Yang, J.M. and Aksoy, S., "Damage Mechanisms of SCS-6/Ti-6Al-4V Composites Under Thermomechanical Fatigue", <u>Materials Science and Engineering</u>, Vol. A156, 1992, pp 117-124.

Richard W. Neu[1]

THERMOMECHANICAL FATIGUE DAMAGE MECHANISM MAPS FOR METAL MATRIX COMPOSITES

REFERENCE: Neu, R. W., **"Thermomechanical Fatigue Damage Mechanism Maps for Metal Matrix Composites,"** *Thermomechanical Fatigue Behavior of Materials: Second Volume, ASTM STP 1263*, M. J. Verrilli and M. G. Castelli, Eds., American Society for Testing and Materials, 1996.

ABSTRACT: Thermomechanical fatigue (TMF) damage mechanism and life maps are generated for unidirectional SCS-6/Timetal 21S undergoing repeated uniaxial loadings oriented along the fiber direction. The maps show the constant life contours as well as the dominant damage mechanisms as a function of maximum cyclic temperature and maximum applied cyclic stress. The maps are constructed using models that describe the individual damage mechanisms. Presently, the damage models include time, temperature, and stress-temperature phasing dependencies but are limited to repeated uniaxial loadings with constant amplitudes and positive stress ratios. Both in-phase and out-of-phase TMF maps are constructed with maximum cyclic temperatures ranging from 300° to 900°C and maximum applied cyclic stresses ranging from 0 to 2000 MPa. The temperature range, stress ratio, and cycle period (500°C, 0.1, 3 min., respectively) are held constant. In addition, maps are generated for isothermal fatigue using the same stress ratio and cycle period. The maps are useful for giving the overall picture of the TMF behavior over a wide range of stress and temperature combinations. The connection between TMF and isothermal fatigue is illustrated.

KEY WORDS: metal matrix composites, titanium matrix, silicon carbide fibers, thermomechanical, fatigue, elevated temperature, micromechanics

INTRODUCTION

Thermomechanical fatigue (TMF) life prediction can be quite challenging, because the life depends on a number of factors. These include the type of cycle, the maximum temperature, temperature range, maximum stress (strain), stress (strain) ratio, environment, frequency, and the phasing of the thermal and mechanical strain. Further adding to the complexity, different locations in a component may experience different combinations of multiaxial stress and temperature. The combined effects of all these factors dictate how damage will accumulate. A number of damage mechanisms or modes

[1]Formerly, NRC associate, Wright Laboratory Materials Directorate, Wright-Patterson AFB, OH 45433-7817; presently, assistant professor, Woodruff School of Mechanical Engineering, Georgia Institute of Technology, Atlanta, GA 30332-0405.

describing the synergistic effects of some of these factors have been identified from experiments, but determining which mechanism will occur for a particular combination of stress and temperature is not always clear.

One way to visualize the effect of stress and temperature on the life and damage mechanisms is through the use of maps that allow visualization of two independent variables at once. Maps have been found to be useful for illustrating the effects of stress and temperature on the deformation mechanisms in many monolithic materials [1] as well as the effects of stress and notch length on the bridging and crack growth mechanisms in titanium matrix composites [2].

This paper describes the procedure for constructing the life and damage mechanism maps to illustrate the TMF behavior of a titanium matrix composite undergoing a uniaxial loading oriented along the fiber direction. The procedure involves identifying the possible fatigue damage mechanisms based on experimental observations. The present experiments include both isothermal fatigue and TMF. Then prediction models that describe each of the identified mechanisms are adopted or derived. The present models are limited to uniaxial constant amplitude cycling with positive stress ratios. The maps are constructed using these prediction models. Some recommendations for possible extensions of the mechanism maps are given.

MATERIAL

The maps shown in this paper illustrate the behavior of a unidirectional SCS-6/ Timetal 21S composite. SCS-6 denotes a 142 μm-diameter silicon carbide fiber produced by Textron Specialty Materials. Timetal 21S is a beta-titanium alloy with the composition of Ti-15Mo-3Nb-3Al-0.2Si (wt %). The composite panels are assembled by alternating layers of rolled titanium foils with fiber mats of SCS-6 woven with a Ti-Nb ribbon, and then consolidated by hot isostatic pressing. The thickness of the resulting 4-ply composite panels is about 0.9 mm and the fiber volume fraction (V_f) is about 0.35. More details on the material and test procedure are given in Refs. [3, 4]

FATIGUE DAMAGE MECHANISMS

As a first step, the damage mechanisms that operate in titanium matrix composites are identified. Three possible mechanisms are [5]: (1) fatigue of the matrix, (2) surface-initiated fatigue-environment damage in the matrix, and (3) fiber-dominated damage. Each mechanism may include synergistic interactions of fatigue, creep, and the environment.

Fatigue of the Matrix

This mechanism describes fatigue of the matrix in absence of environmental effects. The matrix response is controlled by both the applied mechanical load as well as the thermal residual stresses and the mechanical strain induced from the mismatch in the coefficient of thermal expansion (CTE) between the fiber and matrix. Time-dependent effects may still appear because the matrix is viscoplastic. The fatigue mechanism of the matrix is akin to low cycle fatigue of the monolithic matrix material [5-7]. Typical of this type of fatigue, crack initiation occurs at surface asperities or flaws [8]. The composite also contains a number of built-in flaws. These include sites at the machined

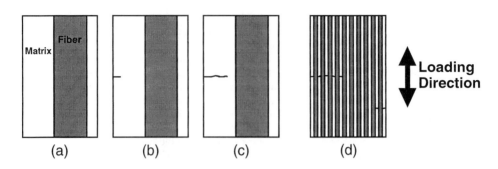

FIG. 1 - *Fatigue of matrix mechanism.*

fibers on the specimen edges and at the fiber/matrix interface [8-11]. There are generally relatively few cracks that propagate during cycling [8, 12]. A schematic illustrating the damage progression for this mechanism is shown in Fig. 1. Final fracture occurs when these matrix fatigue cracks have propagated to such a length that the remaining matrix and fibers can no longer carry the load. Under high applied stress, the fibers may break in the wake of the matrix crack [13]. Under low applied stress, the crack may propagate completely through the matrix with the fibers carrying all the load by bridging the crack as shown in Fig. 1(d). In this situation fatigue can continue by interfacial fatigue until fiber strength decreases and failure occurs [14]. The final fracture surface generally contains only one or two initiation sites and relatively large areas of fatigue cracks in the matrix [12]. It should be noted that even though fibers must eventually fail, the rate-controlling damage accumulation mechanism during most of the life is fatigue of the matrix.

Surface-initiated Fatigue-Environment Damage

This damage mechanism describes the synergistic influences of the environmental attack (in this case, oxygen attack) and matrix fatigue. The characteristic test for this mechanism is out-of-phase (OP) TMF when the maximum temperature of the cycle is sufficiently high for the kinetics of environmental attack to operate. Oxygen diffuses into the surface of the matrix during cycling. The rate of the diffusion is dependent on the temperature. The surface layers become embrittled [15] or in some cases a discernible oxide layer forms, typically consisting of TiO_2 [16-19] (Fig. 2(a)). Whether TiO_2 forms or oxidation embrittlement occurs in the outer layer does not seem to significantly affect the characteristics of the mechanism. The matrix undergoes strain cycling similar to the fatigue of matrix mechanism. As the adherent brittle layer grows thicker, the maximum strain that it is capable of supporting becomes smaller. At a critical thickness the adherent layer fractures. This generally occurs at a cycle less than one-tenth of the total cycles to failure (N_f) [19]. Because the brittle layer is adherent, crack spacing is regular (Fig. 2(b)). Each crack creates a path for the environment, so the crack tip rapidly becomes embrittled (Fig. 2(c)). Once this embrittled zone reaches a critical depth, the crack advances (Fig. 2(d)). As this process continues, these cracks can reach the outer ply fibers (Fig. 2(e)). At low applied stress, crack bridging is generally prevalent [3, 19-22]. However, the fiber strength decreases rapidly when they are directly exposed to the

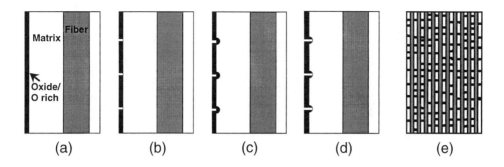

FIG. 2 - *Surface-initiated fatigue-environment damage mechanism.*

environment [17, 23]. Even so, no fiber breaks could be found at 0.9 N_f [19], and very few fiber fracture events using acoustic emission had been recorded by 0.8 N_f [22]. Gabb et al. [12] showed that this is indeed a synergistic mechanism requiring both environmental penetration and mechanical cycling operating nearly simultaneously to produce the highest damage accumulation rate. They showed that running a number of thermal cycles followed by a number of low temperature fatigue cycles results in a much lower damage accumulation rate. Eventually at a location of one of the many matrix cracks, failure occurs. Multiple initiation sites are typically found on the fracture surface [24]. Scanning electron microscope photographs of the damage produced by this mechanism are shown in a number of papers [3, 8, 19, 21, 24-28].

Under stress-control cycling, the maximum strain typically increases while the minimum stays constant resulting in an overall stiffness degradation with cycling [3, 19, 24, 26, 28]. A simple model accounting for progressive cracking of the matrix during OP TMF was able to match the experimental strain accumulation [3]. The effective composite CTE remains constant to about 0.5 N_f and decreases as matrix crack grows through the thickness [19]. The composite CTE approaches the fiber CTE as the matrix cracks grow.

Fiber-dominated Damage

This mechanism is described as a progression of fiber fractures. A characteristic test is an in-phase (IP) TMF test conducted at sufficiently high stress such that the fibers are loaded to a stress approaching the distribution of the individual fiber strengths. On the initial loading some of the weakest fibers may fail [22]. During the very early cycling ($< 0.1 N_f$), matrix stress ratchets down, with fiber stress increasing due to the viscoplastic behavior of the matrix [29, 30]. Fiber failures occur periodically on subsequent cycles, promoted by the increasing fiber stress (Fig. 3). The rate of the increasing fiber stress is dependent on both temperature range and maximum temperature through the deformation response. As early as 0.1 N_f fractured fibers have been observed [19]. The exact fiber degradation mechanism is not completely clear, but may be either frictional rubbing at the fiber/matrix reaction layer or carbon-rich coating [14] or an environmental attack to the components of the reaction and carbon layers [31] or possibly a combination of these mechanisms [23]. As a function of percent life, the fiber fracture events occur at a near constant rate [22]. The environment appears to have very little influence on this

 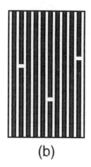 

(a)   (b)   (c)

FIG. 3 - *Fiber-dominated damage mechanism.*

mechanism [18]. Failure occurs when a sufficient number of fiber breaks have occurred and the matrix can no longer withstand the load that is not carried by the fibers. The damage observed at 0.5 N_f and 0.9 N_f through microscopic examination is nearly indistinguishable [19]. The matrix regions of the fracture surface show no fatigue cracks and the morphology is typical of ductile rupture [3, 24-26].

Under stress-control cycling, this damage mechanism is characterized by both the minimum and maximum strain increasing rapidly when <0.1 N_f and very slowly above 0.1 N_f [3, 19, 24-26]. A small stiffness degradation and CTE reduction are also observed during the early cycling [19]. Very little stiffness reduction occurs after 0.1 N_f. A simple model accounting for both creep of the matrix and progressive fiber breakage matched strain accumulation in the characteristic IP TMF experiments [3]. The stiffness remains nearly constant [3, 19].

PREDICTION MODEL

Two analyses are conducted to predict the life. First, the response of the fiber and matrix is determined using micromechanics. Then the number of cycles to failure is determined using models that are dependent on the constituent response as well as the environmental conditions. The models in the second analysis describe the individual damage mechanisms.

The response of the fiber and matrix is computed using the uniaxial stress model coded in FIDEP [32]. The matrix is considered viscoplastic and is modeled using the Bodner and Partom formulation with directional hardening [33, 34]. The fiber is thermoelastic. The material parameters and details of the model are given in Ref. [35]. The difference in average axial response between the simpler uniaxial stress model and a multiple concentric cylinder model when modeling TMF behavior has been shown to be minimal as long as an accurate viscoplastic model is used to describe the matrix response [35]. The simpler model is used to reduce computation time. The fiber volume fraction is included in the analysis. Also, the processing history is included to obtain the initial state of the material. The analysis for each condition is conducted for ten thermal and mechanical loading cycles. The response of the constituents at cycle ten is used as the approximate stabilized behavior. Some justifications for using cycle 10 are given in Refs. [36, 37].

The interactions among the three identified damage mechanisms described in the previous section are difficult to isolate. The simplest approach is to assume that the damage mechanisms do not interact with one another. Then the damage mechanism that gives the maximum damage per cycle controls the life, and the damage per cycle, D, is given by

$$D = \max(D^{Mfat}, D^{Menv}, D^F) \qquad (1)$$

where D^{Mfat} is the damage per cycle for the fatigue of matrix mechanism, D^{Menv} is the damage per cycle for surface-initiated fatigue-environment mechanism, and D^F is the damage per cycle for the fiber-dominated mechanism. Writing D in this form does not imply that there is not synergism between micromechanisms within a damage mode. For example, synergism between fatigue and environment is contained in D^{Menv}. Since D represents the damage per cycle, the number of cycles to failure, N_f, is

$$N_f = 1/D \qquad (2)$$

The approach of predicting life based on the different possible damage mechanisms rather than based on a single correlating parameter or single empirical relationship was found to be critical for TMF life prediction modeling [36].

Some justification for the form of eqn. (1) can be seen from the fatigue life data of experiments designed to give both IP and OP TMF components, and therefore two distinct damage mechanisms, within one cycle [38]. In particular, the IP cycle induced a fiber-dominated damage mechanism, while the OP cycle induced a surface-initiated fatigue-environment damage mechanism. The experiments indicated that the life is controlled by the mechanism that gives the highest damage rate. Furthermore, the experiments indicated that these two mechanisms do not significantly interact since the damage rate of the dominant damage mechanism was not significantly influenced by the damage rate of the other mechanism, which is consistent with eqn. (1).

One of the most challenging tasks is to derive or choose the functions for D^{Mfat}, D^{Menv}, and D^F. Some simplifications are made to facilitate the derivations or choosing of these terms. First, each term describes the average damage per cycle. Therefore, the damage rate is independent of the amount of damage. The amount of damage is most likely a nonlinear function of $D*N$, where N is the number of cycles. Using this simplification, the focus of the modeling is then on what controls the damage progression during most of the life. The events during the initial cycles and final cycles are indirectly incorporated in the empirical constants.

Fatigue of the Matrix

Since D^{Mfat} is controlled during most of the life by crack initiation, a low cycle fatigue relationship applied to the matrix response is chosen. Since mean stresses are significant due to the tensile thermal residual stresses in the matrix, a relationship including mean stress effect must be used. One relationship is the Coffin-Manson with mean stress correction (Morrow relationship) with $1/D^{Mfat}$ replacing N_f [39],

$$\frac{\Delta\varepsilon_m^{(m)}}{2} = \frac{\sigma_f' - \sigma_m^{(m)}}{E^{(m)}}\left(\frac{2}{D^{Mfat}}\right)^b + \varepsilon_f'\left(\frac{2}{D^{Mfat}}\right)^c \qquad (3)$$

The damage is dependent on the matrix mechanical strain range, $\Delta \varepsilon_m{}^{(m)}$, and the mean matrix stress, $\sigma_m{}^{(m)}$. The empirical constants, σ_f', b, ε_f', c, are determined from low cycle fatigue tests on the matrix material conducted at room temperature, and $E^{(m)}$ is the modulus of the matrix at room temperature. The parameters for Timetal 21S are as follows: $\sigma_f' = 3500$ MPa, b = -0.21, $\varepsilon_f' = 0.5$, c = -0.75, and $E^{(m)} = 100000$ MPa. Since the number of possible crack nucleation sites is greater in the composite compared to smooth fatigue specimens due to the brittle phases at the fiber/matrix interface and due to the nonuniform local stress fields, the fatigue constants using the matrix material data are most likely an upper bound.

Surface-initiated Fatigue-Environment Damage

The surface-initiated fatigue-environment damage term is based on the fatigue-environment interaction of cracks initiating at the surface and growing inward. Therefore, the D^{Menv} term is a bit more complicated because it is not only dependent on the matrix mechanical strain range, $\Delta \varepsilon_m{}^{(m)}$, describing the fatigue process, but is also dependent on time, temperature, and kinetics of oxidation to describe the severity of the environmental attack. Synergistic effects due to the stress-temperature phasing are also incorporated through the ratio of the thermal and mechanical strain rates. Phasing has been shown to be an important parameter [12, 18]. A possible form for this term was derived in Ref. [5] and will be used here. The final form of the equation is

$$D^{Menv} = \left[\frac{C_{crit}}{\Phi^{Menv} D_{eff}} \right]^{-1/\beta} 2^{a/\beta} \, t_c{}^{(1-a/\beta)} \left(\Delta \varepsilon_m{}^{(m)} \right)^{(2+a)/\beta} \qquad (4)$$

where t_c is the period of a cycle, Φ^{Menv} is the effective phasing of the cycle and is a function of the ratio of the thermal strain rate, $\dot{\varepsilon}_{th}$, and mechanical strain rate, $\dot{\varepsilon}_m$,

$$\Phi^{Menv} = \frac{1}{t_c} \int_0^{t_c} \exp\left[-\frac{1}{2} \left(\frac{(\dot{\varepsilon}_{th}{}^{(m)}/\dot{\varepsilon}_m{}^{(m)}) - M^{Menv}}{\xi^{Menv}} \right)^2 \right] dt, \qquad (5)$$

where M^{Menv} represents the $\dot{\varepsilon}_{th}{}^{(i)}/\dot{\varepsilon}_m{}^{(i)}$ phasing that is the most damaging for this mechanism and ξ^{Menv} is a measure of the severity of the phasing effect between OP TMF and isothermal fatigue. D_{eff} is the effective oxidation constant and is described by an Arhennius-type expression averaged over a cycle,

$$D_{eff} = \frac{1}{t_c} \int_0^{t_c} D_o \exp\left(-\frac{Q^{Menv}}{RT} \right) dt . \qquad (6)$$

where D_o is the diffusion coefficient, Q^{Menv} is the apparent activation energy for the diffusion, and R is the gas constant. The remaining symbols (C_{crit}, β, a) are empirical parameters. The D^{Menv} parameters for SCS-6/Timetal 21S are determined from key OP TMF and isothermal fatigue tests and are taken from Ref. [36].

Fiber-dominated Damage

The fiber-dominated damage mechanism is primarily dependent on the fiber stress, σ^F. Matrix stress relaxation and the attendant transfer of load to the fibers is accounted for in the micromechanics modeling. The fiber-dominated damage term also includes the effect of fiber strength degradation [22]. Any growth of the reaction zone and environmental attack of the carbon-rich layers [31] can reduce fiber strength and are dependent on time at temperature. The phasing of stress and temperature can also influence the rate of damage accumulation. Strength reduction due to frictional rubbing and degradation of the interface could be partially influenced by all of these factors, though the precise mechanism is not known. Including all these factors, an expression to describe the damage per cycle is given by [5]

$$DF = \int_0^{t_c} A^F \ \Phi^F \ \exp\left(-\frac{Q^F}{RT}\right) \left[\frac{\sigma^{(f)}}{\sigma_T}\right]^m \ dt \tag{7}$$

where σ_T is the average fiber strength and tends to be a fitting parameter since fiber strength distributions vary, Q^F is the apparent activation energy for temperature-dependent degradation of the fiber, and m and A^F are empirical parameters. The phasing factor, Φ^F, is influenced by both the phasing of the matrix and fiber through the fraction η and is given by [5]

$$\Phi^F = \eta \ \Phi^{F(m)} + (1 - \eta) \ \Phi^{F(f)} \tag{8}$$

where

$$\Phi^{F(i)} = \frac{1}{t_c} \int_0^{t_c} \exp\left[-\frac{1}{2}\left(\frac{(\dot{\varepsilon}_{th}^{(i)}/\dot{\varepsilon}_m^{(i)}) - M^F}{\xi^F}\right)^2\right] \ dt, \quad (i) = (m) \text{ or } (f) \tag{9}$$

M^F represents the $\dot{\varepsilon}_{th}^{(i)}/\dot{\varepsilon}_m^{(i)}$ phasing that is the most damaging for the fiber-dominated mechanism and ξ^F is a measure of the phasing effect between IP TMF and isothermal fatigue. The D^F parameters for SCS-6/Timetal 21S are primarily determined from key IP TMF tests and are taken from Ref. [36]. More details on the determination of the parameters are given in Refs. [5, 40].

Correlations and Predictions

A comparison of the model to experimental IP and OP TMF lives is shown in Fig. 4. For the IP TMF conditions shown in this figure, the lives are controlled by the fiber-dominated damage mechanism. The lives are very sensitive to the maximum applied stress. The OP TMF lives are controlled by the surface-initiated fatigue-environment damage mechanism. OP TMF is life-limiting at lower maximum applied stresses. The baseline 150°-650°C data was used to determine some of the empirical constants, so these are marked as correlations. Since the effect of temperature was determined from isothermal fatigue experiments [36, 40], the OP curve for the maximum temperature of 815°C is a prediction. Additional correlations and predictions for isothermal fatigue and TMF tests that show the effects of maximum temperature, temperature range, and frequency are given in Ref. [36].

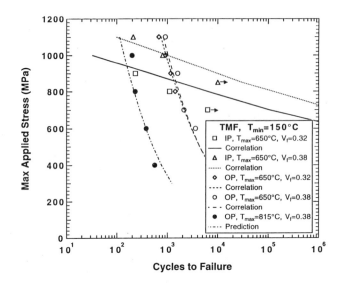

FIG. 4 - *Correlations and predictions for IP and OP TMF lives.*

LIFE AND MECHANISM MAPS

Creating the Maps

A unique way to examine the effect of stress and temperature on the TMF behavior is through life and damage mechanism maps. The model described in the previous section can be applied to any stress-temperature-time combination, though it is presently limited constant amplitude triangular cycling. To create a map, a grid is overlaid on a plot of maximum temperature and maximum applied stress (Fig. 5). At each grid intersection below the ultimate tensile strength (UTS), a life analysis is conducted to obtain the cycles to failure as well as the dominant damage mechanism. The UTS on the map is the UTS of the material at the temperature when the maximum stress is applied. For each map, the temperature range (ΔT), stress ratio (R), frequency (f), and atmosphere (environment) are kept constant. Maximum temperature (T_{max}) is chosen as an independent variable instead of ΔT, because life has be shown to be more dependent on T_{max} [8, 24, 40].

Out-of-Phase TMF Map

Contours of life, the UTS, and shaded regions indicating the dominant damage mechanism are shown in Fig. 6 for OP TMF. The analyses are for TMF cycles conducted in air atmosphere with $\Delta T = 500°C$, $R = 0.1$, and $f = 5.56 \times 10^{-3}$ Hz (3 min. cycle period). Each contour curve represents a decade difference in life from 10^2 to 10^7 cycles. The change in slope in the life contours near 400°C is indicative of a change in the damage

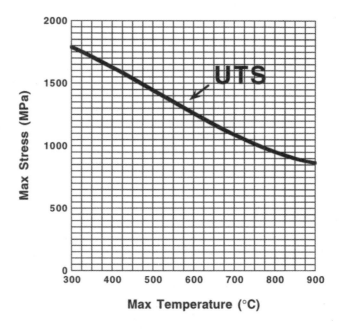

FIG. 5 - *Analysis grid for the TMF life and damage mechanism maps.*

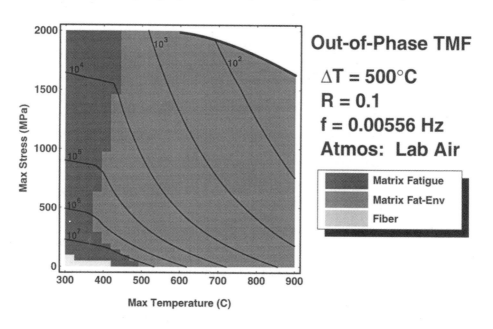

FIG. 6 - *Out-of-phase TMF life and damage mechanism map.*

mechanism. At 400°C the mechanism changes from fatigue of the matrix (noted as Matrix Fatigue in the figure) to surface-initiated fatigue-environment damage (noted as Matrix Fat-Env) as the maximum temperature increases.

Failure is typically defined in the laboratory as the cycle when the specimen separates into two pieces. However, when the applied stress is small, complete separation may not occur. In this case failure is defined as a critical level of damage. This is consistent with the modeling that bases the damage rate on the mechanism occurring during most of the life. The rate-controlling process for OP TMF is the growth of the matrix cracks. At high applied stress, fiber degradation and individual fiber failures occur as a last step before complete specimen separation. At low stress, the final transient resulting in the individual fiber failures may not occur, yet the distribution of matrix cracks is indicative of failure.

In-Phase TMF Map

The IP TMF life and damage mechanism map is shown in Fig. 7. For IP TMF, the region where life is greater than 10^7 cycles is much larger. The UTS curve is lower than OP TMF, because fracture under IP TMF occurs at the higher temperature of the cycle when the strength of the matrix is lower. All three damage mechanisms appear on this map. The fiber-dominated mechanism (noted as Fiber in the figure) occurs when the maximum stress is relatively high. Most IP tests that have been conducted in the laboratory fall in this region. The life lines are close together indicating that the life is especially sensitive to the maximum stress. Life is less sensitive to the maximum stress when the fatigue of matrix mechanism operates at lower T_{max}. The border between these mechanisms is dependent on both T_{max} and S_{max}. The map also indicates that the

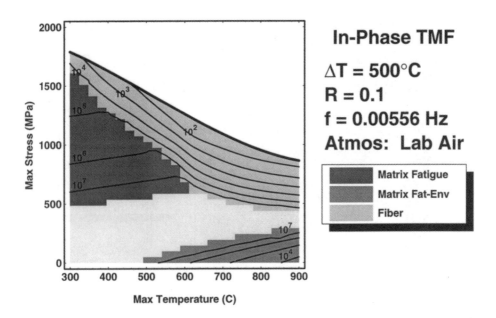

FIG. 7 - *In-phase TMF life and damage mechanism map.*

surface-initiated fatigue-environment damage mechanism occurs at high temperatures when the maximum stress is very low. This may seem contradictory, since the life decreases with decreasing stress. This is easily explained. Even though the overall composite phasing is IP, the matrix phasing in this region is OP. The difference arises from the mismatch in the fiber and matrix CTEs during thermal cycling. The initial axial residual stress in the matrix is tensile. As the temperature increases, the matrix mechanical strain decreases, producing a local OP condition in the matrix. At very high T_{max}, the life is longest when $S_{max} \approx 350$ MPa. At this stress, the matrix strain that is induced by the applied mechanical cycling cancels the matrix strain that is induced by thermal cycling, so there is no driving force for matrix fatigue. Since the fiber stress is also small, the life is long. It should also be noted that the OP and IP maps must coincide at an applied stress of zero when only thermal cycling is occurring. The IP TMF map reveals that it is really incorrect to say that IP TMF is fiber-dominated. IP TMF is fiber-dominated only in that region on the map where most of the IP experiments are typically conducted.

Isothermal Fatigue Map

Finally, a map for isothermal fatigue is shown in Fig. 8. Isothermal fatigue is essentially a special case of TMF. It is the case when $\Delta T = 0$. All three damage mechanisms also appear on this map. The elevated temperature fatigue mechanism transitions from fiber-dominated at high stress to surface-initiated fatigue-environment damage at lower stresses. At temperatures below 300°C, the dominant damage mechanism is fatigue of the matrix. At higher applied stress levels, the isothermal fatigue map looks similar to the IP TMF map. As the maximum applied stress is decreased, the

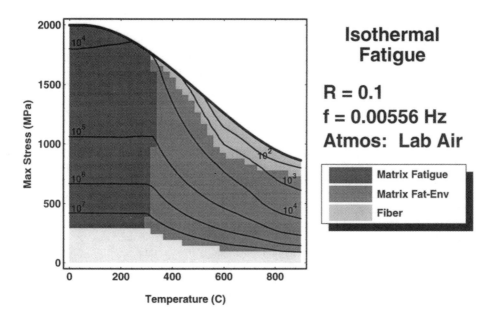

FIG. 8 - *Isothermal fatigue life and damage mechanism map.*

isothermal fatigue map begins to look more like an OP TMF map. However, the lives for isothermal fatigue conditions are generally higher than those for OP TMF when the surface-initiated fatigue-environment damage operates. This illustrates the detrimental effect that the combined thermal and mechanical cycling has on the damage accumulation rate.

DISCUSSION

To develop a map showing a large number of stress and temperature conditions, of course, requires interpolation and extrapolation of the expected behavior. Since the damage models are based on specific mechanisms, the interpolations and extrapolations are somewhat reliable. Nevertheless, additional tests can be easily identified to verify different regions on the maps.

The maps are based on observations in the 4-ply SCS-6/Timetal 21S composite system and the damage mechanisms are consistent with studies conducted on this system [3, 12, 19, 22, 28, 40]. The mechanism maps are also consistent with the observed damage mechanisms found in experimental studies on other titanium-based composites reinforced with silicon carbide fibers conducted under tension-tension loading. Even though different composite fabrication techniques were used in many cases, and different frequencies and ΔTs were imposed, the damage mechanisms observed in these investigations are consistent with those given in the maps. However, the life contours and mechanism borders may be shifted somewhat. A number of experimental observations are discussed in light of the damage mechanism maps.

First, a number of investigations have compared IP and OP TMF behavior. Except as noted, a fiber-dominated damage mechanism was reported for IP TMF tests and a surface-initiated fatigue-environment damage mechanism was reported for OP TMF tests. One study compared 8-ply SCS-6/Ti-15-3 (Ti-15V-3Cr-3Al-3Sn) composites and SCS-6/Ti-14Al-21Nb composites (also known by its at %, SCS-6/Ti-24Al-11Nb), which were cycled under typical simultaneous TMF as well as sequential nonisothermal fatigue [24]. The SCS-6/Ti-15-3 composite was tested at temperatures of 150°-550°C, while SCS-6/Ti-14Al-21Nb composite was tested at 150°-815°C. For these temperature cycles, the TMF damage mechanisms reported for these titanium- (i.e., Ti-15-3) and titanium-aluminide-based (i.e., Ti-14Al-21Nb) systems were shown to be similar and corresponded to those given in the maps (Figs. 6 and 7) [24].

In another study on an 8-ply SCS-6/Ti-14Al-21Nb composite, both IP and OP TMF tests were cycled from 425° to 815°C [18] with S_{max} = 480 MPa under the IP loading and S_{max} = 550 MPa under OP. The dominant damage mechanism for IP was reported to be fiber-dominated and no effect of environment (air versus argon) was noted. In the OP tests, matrix cracking at the surface and at the interface in the outer row of fibers was noted. In an inert atmosphere, the OP life was greater, especially as test time increased. These observations are consistent with the OP mechanism map (Fig. 6). The map suggests the environment should affect the life for these OP conditions.

When T_{max} of an OP TMF cycle is sufficiently low, a transition from a surface-initiated fatigue-environment mechanism to fatigue of the matrix mechanism is expected. This behavior was observed in OP TMF tests conducted on an 8-ply SCS-6/Ti-6Al-4V composite [41]. The maximum applied stress was 1090 MPa and T_{min} = 120°C. When $T_{max} \leq 450°C$, the failure mechanism was matrix fatigue. When $T_{max} \geq 550°C$, failure was controlled by a surface-initiated fatigue-environment mechanism. This indicates that the transition occurs when temperature is between 450°C and 550°C and is consistent with the map (Fig. 6).

In another investigation on an 8-ply SCS-6/Ti-6Al-4V composite [27], both OP TMF and IF tests were conducted. The conditions for all the tests were in the following ranges: T_{max} = 370 to 650°C with T_{min} = 120°C and S_{max} = 828 to 1180 MPa. Observations indicated that a surface-initiated fatigue-environment mechanism occurred in all these tests. The mechanism maps for OP TMF and IF (Figs. 6 and 8) suggest that this type of mechanism is expected for these test conditions.

On a 9-ply SCS-6/Ti-15-3 composite, TMF tests were conducted at a temperature cycle of 93°-538°C and isothermal fatigue (IF) tests were conducted at 427°C [26]. S_{max} ranged from 827 MPa to 1241 MPa. The mechanisms observed in TMF tests were consistent with the maps for these conditions. The IF tests showed both extensive matrix cracking and fiber damage, which is expected since the tests were conducted in a region on the map (Fig. 8) near the transition from fiber-dominated to surface-initiated fatigue-environment damage. Similar observations were noted for an 8-ply SCS-6/Ti-14Al-21Nb composite [25].

IF tests were conducted on an 8-ply SCS-6/Ti-15-3 composite at a couple of temperatures [8]. When the peak temperature was at or above 300°C, fatigue life was dramatically reduced compared to the life at 150°C. As noted in the isothermal fatigue mechanism map (Fig. 8), 300°C is roughly the temperature when the transition from matrix fatigue to a surface-initiated fatigue-environment damage mechanism is expected and a sharp drop in life occurs. It was noted that cracks initiate at the fiber-matrix interface from environmental attack by oxygen diffusion through the matrix to the interface [18]. This may indicate that near the matrix fatigue and fatigue-environment boundary on the map, there may be a synergistic mechanism of crack initiation at the fiber/matrix interface due to a fatigue-environment interaction. Cracks initiate at the fiber/matrix interface in the matrix fatigue mechanism and apparently the environmental attack, especially of the outer plies, increases the damage rate in this range of temperatures

In the same study, lives under nonisothermal OP conditions were lower than isothermal fatigue tests conducted at the maximum temperature of the nonisothermal cycle. The nonisothermal mechanism was surface-initiated fatigue-environment damage. Comparing the lives of OP TMF and isothermal fatigue for a given temperature on the maps (Figs. 6 and 8), the OP TMF cycle has a lower life by at least an order of magnitude. This is consistent with the difference in reported experimental lives.

IF tests were conducted at temperatures ranging from 425°C to 815°C on 3-ply and 8-ply SCS-6/Ti-14Al-21Nb composites [21, 42], and tests were conducted at 650°C on an 8-ply SCS-6/Ti-15-3 composite [20]. At the highest stresses both intermittent fiber cracking and surface-initiated fatigue-environment matrix damage were reported. This corresponds to the fiber-dominated damage region on the IF map (Fig. 8). In this region, life is very sensitive to the maximum applied stress level. Surface-initiated fatigue-environment damage was observed when the maximum applied stress was at intermediate levels. This corresponds to the appropriate region on the IF map. Even when runout was reached at lower stress, surface-initiated matrix cracking damage was found in the specimen [21]. The map suggests that this type of damage should exist. A fatigue test conducted in vacuum at 425°C in the intermediate stress range was a runout [21]. This indicated that environmental effects limit the fatigue life at temperatures as low as 425°C, which is consistent with the IF map. This temperature falls within the region where environment influences life.

The OP and IP TMF maps coincide when the applied stress is zero. Surface-initiated fatigue-environment damage has been shown to occur during thermal cycling of SCS-6/Ti-14Al-21Nb composites [16-18] and SCS-6/Timetal 21S composites [15], which is consistent with the predicted damage mechanism (Figs. 6 and 7). The maps indicate that thermal cycling is more damaging than isothermal exposure with no applied load (compare Figs. 6 and 7 to Fig. 8). This is consistent with a recent study that indicated the residual strength decreases more rapidly under thermal cycling compared to isothermal exposures [15].

Room temperature fatigue studies have been conducted on a number of the titanium-based composite systems: 8-ply SCS-6/Ti-15-3 [11, 43], 6-ply SCS-6/Ti-15-3 [10], 8-ply SCS-6/Ti-6Al-4V [10, 11], and 8-ply SCS-6/Ti-25-10 (Ti-25Al-10Nb-3V-1Mo) [10]. The basic mechanism is fatigue of the matrix. The location of crack initiation in the matrix is dependent on the composite system and processing [11, 43]. When the maximum applied stress is near the UTS, the mechanism changes to a cyclic accumulation of fiber failures [10]. This emphasizes that the borders between mechanisms and near the UTS are somewhat fuzzy. When operating near the UTS, a progressive fiber breakage mechanism, which is controlled by the fiber strength distribution, is expected [10]. This room temperature fiber-dominated mechanism is not modeled at this time.

RECOMMENDATIONS FOR FUTURE WORK

Further extensions of the maps include identifying the amount of interaction between mechanisms at or near the borders between mechanisms. The extent of interaction and the size of the interaction region at the borders is dependent on the particular damage mechanisms. It was noted earlier that experiments designed to cause both fiber-dominated damage and surface-initiated fatigue-environment damage mechanisms to operate within one cycle suggest that the life is controlled solely by the mechanism that gives the highest damage rate, which is consistent with eqn. (1) [38]. However, the interactions between other mechanisms still need to be explored and the size of the interaction region should be identified.

Other possible extensions for the maps include developing multiaxial damage laws including the anisotropy of the damage, developing nonlinear damage laws that can account for a series of damage accumulation events, rather than just the dominant one, developing maps for more general cycle types including stress reversals, varying amplitude service/mission type cycles, and developing damage mechanism relationships for newly identified damage mechanisms in other materials or fiber layups and improving the damage expressions for known damage mechanisms. If necessary, eqn. (1) could be cast as a more complex function to account for interactions between certain damage mechanisms if interaction regions between mechanisms are wide and to account for nonlinear damage accumulation. The damage expressions given in eqns. (3), (4), and (7) are not unique. Other damage expressions may be adopted. It should be kept in mind that the damage expressions must be valid over a range of temperatures, so mechanistic-based models that are dependent on time, temperature, and stress-temperature phasing are generally required.

SUMMARY AND CONCLUSIONS

The procedure for constructing life and damage mechanism maps has been described. First, it involves identifying the dominant damage mechanisms of the material. Then empirical or semi-empirical expressions are adopted or derived to describe how the life (or damage rate) is dependent on the key factors of each identified mechanism. Generally, in metal matrix composites, the life is dependent on the stabilized response of the constituents as well as the environment and phasing of the thermal and mechanical strain. Analyses are conducted for each combination of maximum applied cyclic stress and maximum cyclic temperature. Each analysis involves computing a life for each possible mechanism. The dominant damage mechanism is the one that gives the lowest life. Life contours and the controlling damage mechanisms are shown on the maps.

In this study, maps were generated for the unidirectional SCS-6/Timetal 21S composite undergoing a uniaxial constant amplitude repeated loading oriented along the fiber direction. Three mechanisms were identified: (1) fatigue of the matrix, (2) surface-initiated fatigue-environment damage, and (3) fiber-dominated damage. Maps were constructed for three cycle types: (1) OP TMF, (2) IP TMF, and (3) isothermal fatigue.

The maps provide an overall picture of the TMF behavior of the material over a wide range of stress and temperature conditions that may be missed in studies that focus on too narrow of a range. The maps clearly show the sensitivity of different combinations of stress and temperature. The connection between isothermal fatigue and TMF can also be better understood through the maps. The mechanisms and life trends given in the maps are consistent with those observed in other titanium- and titanium-aluminide-based matrix composites with silicon carbide fibers. The maps can be useful for identifying the severity of different cycle types and stress-temperature combinations as well as identifying the expected damage mechanisms.

ACKNOWLEDGMENTS

This work was conducted at Wright Laboratory Materials Directorate, Wright-Patterson AFB, Ohio. The support extended by the National Research Council, Washington, DC, through their Associateship Program is gratefully acknowledged. Dr. T. Nicholas is thanked for his insightful discussions.

REFERENCES

[1]   Frost, H. J. and Ashby, M. F., *Deformation-Mechanism Maps*, Pergamon Press, Oxford, 1982.

[2]   Zok, F. W., Connell, S. J., and Du, Z. Z., "Fatigue Maps for Titanium Matrix Composites," *Life Prediction Methodology for Titanium Matrix Composites, ASTM STP 1253*, W. S. Johnson, J. M. Larsen, and B. N. Cox, Eds., American Society for Testing and Materials, Philadelphia, 1995.

[3]   Neu, R. W. and Nicholas, T., "Effect of Laminate Orientation on the Thermomechanical Fatigue Behavior of a Titanium Matrix Composite," *Journal of Composites Technology & Research*, Vol. 16, No. 3, July 1994, pp. 214-224.

[4]   Nicholas, T., Russ, S. M., Neu, R. W., and Schehl, N., "Life Prediction of a [0/90] Metal Matrix Composite Under Isothermal and Thermomechanical Fatigue," *Life*

Prediction Methodology for Titanium Matrix Composites, ASTM STP 1253, W. S. Johnson, J. M. Larsen, and B. N. Cox, Eds., American Society for Testing and Materials, Philadelphia, 1995.

[5]   Neu, R. W., "A Mechanistic-based Thermomechanical Fatigue Life Prediction Model for Metal Matrix Composites," *Fatigue and Fracture of Engineering Materials and Structures*, Vol. 16, No. 8, 1993, pp. 811-828.

[6]   Herrmann, D. J., Ward, G. T., and Hillberry, B. M., "Prediction of Matrix Fatigue Crack Initiation from Notches in Titanium Matrix Composites," *Life Prediction Methodology for Titanium Matrix Composites, ASTM STP 1253*, W. S. Johnson, J. M. Larsen, and B. N. Cox, Eds., American Society for Testing and Materials, Philadelphia, 1995.

[7]   Hillberry, B. M. and Johnson, W. S., "Prediction of Matrix Fatigue Crack Initiation in Notched SCS-6/Ti-15-3 Metal Matrix Composites," *Journal of Composites Technology & Research*, Winter 1992, pp. 221-224.

[8]   Gayda, J., Gabb, T. P., and Lerch, B. A., "Fatigue-Environment Interactions in a SiC/Ti-15-3 Composite," *International Journal of Fatigue*, Vol. 15, No. 1, 1993, pp. 41-45.

[9]   Gabb, T. P., Gayda, J., and MacKay, R. A., "Isothermal and Nonisothermal Fatigue Behavior of a Metal Matrix Composite," *Journal of Composite Materials*, Vol. 24, June 1990, pp. 667-686.

[10]   Jeng, S. M., Alassoeur, P., Yang, J.-M., and Aksoy, S., "Fracture Mechanisms of Fiber-Reinforced Titanium Alloy Matrix Composites, Part IV: Low Cycle Fatigue," *Materials Science and Engineering*, Vol. A148, 1991, pp. 67-77.

[11]   Greaves, I., Yates, J. R., and Atkinson, H. V., "The Role of the Interface in the Initiation of Fatigue Cracks in SCS-6/Titanium MMCs," *Composites*, Vol. 25, No. 7, 1994, pp. 692-697.

[12]   Gabb, T. P. and Gayda, J., "Matrix Fatigue Cracking Mechanisms of a2 TMC for Hypersonic Applications," NASA Technical Memorandum 106506, 1994; also, *Life Prediction Methodology for Titanium Matrix Composites, ASTM STP 1253*, W. S. Johnson, J. M. Larsen, and B. N. Cox, Eds., American Society for Testing and Materials, Philadelphia, 1995.

[13]   Larsen, J. M., Jira, J. R., John, R., and Ashbaugh, N. E., "Crack Bridging Effects in Notch Fatigue of SCS-6/Timetal 21S Composite Laminates," *Life Prediction Methodology for Titanium Matrix Composites, ASTM STP 1253*, W. S. Johnson, J. M. Larsen, and B. N. Cox, Eds., American Society for Testing and Materials, Philadelphia, 1995.

[14]   Walls, D. P. and Zok, F. W., "Interfacial Fatigue in a Fiber Reinforced Metal Matrix Composite," *Acta Metallurgica et Materialia*, Vol. 42, No. 8, 1994, pp. 2675-2681.

[15]   Revelos, W. C., Jones, J. W., and Dolley, E. J., "Thermal Fatigue of a SiC/Ti-15Mo-2.7Nb-3Al-0.2Si Composite," *Metallurgical and Materials Transactions*, 1995 (in press).

[16]   Russ, S. M., "Thermal Fatigue of Ti-24Al-11Nb/SCS-6," *Metallurgical Transactions A*, Vol. 21A, June 1990, pp. 1595-1602.

[17]   Revelos, W. C. and Smith, P. R., "Effect of Environment on the Thermal Fatigue Response of an SCS-6/Ti-24Al-11Nb Composite," *Metallurgical Transactions A*, Vol. 23A, Feb. 1992, pp. 587-595.

[18]   Bartolotta, P. A., Kantzos, P., Verrilli, M. J., and Dickerson, R. M., "Environmental Degradation of an Intermetallic Composite During

Thermomechanical Fatigue," *Fatigue 93*, Vol. 2, Eds. J. P. Bailon and J. I. Dickson, 1993, pp. 1001-1006.

[19]   Castelli, M. G., "Characterization of Damage Progression in SCS-6/Timetal 21S [0]$_4$ Under Thermomechanical Fatigue Loading," *Life Prediction Methodology for Titanium Matrix Composites, ASTM STP 1253*, W. S. Johnson, J. M. Larsen, and B. N. Cox, Eds., American Society for Testing and Materials, Philadelphia, 1995.

[20]   Pollock, W. D. and Johnson, W. S., "Characterization of Unnotched SCS-6/Ti-15-3 Metal Matrix Composites at 650°C," *Composite Materials: Testing and Design (Tenth Volume), ASTM STP 1120*, Glenn C. Grimes, Ed., American Society for Testing and Materials, Philadelphia, 1992, pp. 175-191.

[21]   Brindley, P. K. and Draper, S. L., "Failure Mechanisms of 0° and 90° SiC/Ti-24Al-11Nb Composites Under Various Loading Conditions," *Structural Intermetallics*, Ed. by R. Darolia, J. J. Lewandowski, C. T. Liu, P. L. Martin, D. B. Miracle, and M. V. Nathal, The Minerals, Metals & Materials Society, 1993, pp. 727-737.

[22]   Neu, R. W. and Roman, I., "Acoustic Emission Monitoring of Damage in Metal Matrix Composites Subjected to Thermomechanical Fatigue," *Composites Science and Technology*, Vol. 52, 1994, pp. 1-8.

[23]   Blatt, D., Karpur, P., Stubbs, D. A., and Matikas, T. E., "Observations of Interfacial Damage in the Fiber Bridged Zone of a Titanium Matrix Composite," *Scripta Metallurgica et Materialia*, Vol. 29, 1993, pp. 851-856.

[24]   Gabb, T. P., Gayda, J., Bartolotta, P. A., and Castelli, M. G., "A Review of Thermomechanical Fatigue Damage Mechanisms in Two Titanium and Titanium Aluminide Matrix Composites," *International Journal of Fatigue*, September 1993, pp. 413-422.

[25]   Russ, S. M., Nicholas, T., Bates, M., and Mall, S., "Thermomechanical Fatigue of SCS-6/Ti-24Al-11Nb Metal Matrix Composites," *Failure Mechanisms in High Temperature Composite Materials*, AD-Vol. 22/AMD-Vol. 122, Ed. by G.K. Haritos, G. Newaz, and S. Mall, ASME, New York, 1991, pp. 37-43.

[26]   Castelli, M. G., Bartolotta, P., and Ellis, J. R., "Thermomechanical Testing of High-Temperature Composites: Thermomechanical Fatigue (TMF) Behavior of SiC(SCS-6)/Ti-15-3," *Composites Materials: Testing and Design (Tenth Volume), ASTM STP 1120*, Glenn C. Grimes, Ed., American Society for Testing and Materials, Philadelphia, 1992, pp. 70-86.

[27]   Jeng, S. M., Yang, J.-M., and Aksoy, S., "Damage Mechanisms of SCS-6/Ti-6Al-4V Composites Under Thermal-Mechanical Fatigue," *Materials Science and Engineering*, Vol. A156, 1992, pp. 117-124.

[28]   Castelli, M. G., "Thermomechanical Fatigue Deformation, Damage and Life of SCS-6/Timetal 21S [0]$_4$," *International Journal of Fatigue*, 1995. (submitted).

[29]   Kroupa, J. L. and Neu, R. W., "The Nonisothermal Viscoplastic Behavior of a Titanium-Matrix Composite," *Composites Engineering*, Vol. 4, No. 9, 1994, pp. 965-977.

[30]   Mirdamadi, M., Johnson, W. S., Bahei-El-Din, Y. A., and Castelli, M. G., "Analysis of Thermomechanical Fatigue of Unidirectional Titanium Metal Matrix Composites," *Composite Materials: Fatigue and Fracture, Fourth Volume, ASTM STP 1156*, W. W. Stinchcomb and N. E. Ashbaugh, Eds., American Society for Testing and Materials, Philadelphia, 1993, pp. 591-607.

[31]   Gambone, M. L. and Wawner, F. E., "The Effect of Elevated Temperature Exposure of Composites on the Strength Distribution of the Reinforcing Fibers," *Intermetallic Matrix Composites III*, Volume 350, Edited by J. A. Graves, R. R. Bowman, and J. J. Lewandowski, MRS, Pittsburgh, PA, 1994, pp. 111-118.

[32]   Coker, D., "FIDEP2 User's Manual," WL-TR-95-xxxx, Wright-Patterson AFB, Ohio, 1995.

[33]   Chan, K. S., Bodner, S. R., and Lindholm, U. S., "Phenomenological Modeling of Hardening and Thermal Recovery in Metals," *Journal of Engineering Materials and Technology*, Vol. 110, Jan. 1988, pp. 1-8.

[34]   Neu, R. W., "Nonisothermal Material Parameters for the Bodner-Partom Model," *Material Parameter Estimation for Modern Constitutive Equations*, L. A. Bertram, S. B. Brown, and A. D. Freed, Eds., MD-Vol. 43/AMD-Vol. 168, ASME, 1993, pp. 211-226.

[35]   Neu, R. W., Coker, D., and Nicholas, T., "Cyclic Behavior of Unidirectional and Cross-ply Titanium Matrix Composites," *International Journal of Plasticity*, 1994. (submitted)

[36]   Neu, R. W. and Nicholas, T., "Methodologies for Predicting the Thermomechanical Fatigue Life of Unidirectional Metal Matrix Composites," *Advances in Fatigue Lifetime Predictive Techniques (3rd Symposium), ASTM STP 1292*, M. R. Mitchell and R. W. Landgraf, Eds., American Society for Testing and Materials, Philadelphia, 1996.

[37]   Coker, D., Neu, R. W., and Nicholas, T., "Analysis of the Thermoviscoplastic Behavior of [0/90] SCS-6/Timetal 21S Composites," *Thermo-Mechanical Fatigue Behavior of Materials: 2nd Volume, ASTM STP 1263*, M. J. Verrilli and M. G. Castelli, Eds., American Society for Testing and Materials, Philadelphia, 1995.

[38]   Stubbs, D. A. and Russ, S. M., "Examination of the Correlation Between NDE-Detected Manufacturing Abnormalities and Thermomechanical Fatigue Life" Proceedings of the Structural Testing Technology at High Temperature - II, the Society for Experimental Mechanics, Publ., Bethel, CT, Nov. 1993, pp. 165-174.

[39]   Morrow, J., "Cyclic Plastic Strain Energy and Fatigue of Metals," *Internal Friction, Damping, and Cyclic Plasticity, ASTM STP 378*, 1965, pp. 45-87.

[40]   Neu, R. W. and Nicholas, T., "Thermomechanical Fatigue of SCS-6/TIMETAL®21S Under Out-of-Phase Loading," *Thermomechanical Behavior of Advanced Structural Materials*, W. F. Jones, Ed., AD-Vol. 34/AMD-Vol. 173, ASME, 1993, pp. 97-111.

[41]   Aksoy, S., "Stiffness Degradation in Metal Matrix Composites Caused by Thermomechanical Fatigue Loading," *Journal of Engineering for Gas Turbines and Power*, Vol. 116, No. 3, July 1994.

[42]   Bartolotta, P. A. and Brindley, P. K., "High-Temperature Fatigue Behavior of a SiC/Ti-24Al-11Nb Composite," *Composite Materials: Testing and Design (Tenth Volume), ASTM STP 1120*, G. C. Grimes, Ed., American Society for Testing and Materials, Philadelphia, 1992, pp. 192-203.

[43]   Johnson, W. S., Lubowinski, S. J., and Highsmith, A. L., "Mechanical Characterization of Unnotched SCS-6/Ti-15-3 Metal Matrix Composites at Room Temperature," *Thermal and Mechanical Behavior of Metal Matrix and Ceramic Matrix Composites, ASTM STP 1080*, J. M. Kennedy, H. H. Moeller, and W. S. Johnson, Eds., American Society for Testing and Materials, Philadelphia, 1990, pp. 193-218.

Dale L. Ball[1]

An Experimental and Analytical Investigation of Titanium Matrix Composite Thermomechanical Fatigue

REFERENCE: Ball, D. L., "An Experimental and Analytical Investigation of Titanium Matrix Composite Thermomechanical Fatigue," *Thermomechanical Fatigue Behavior of Materials: Second Volume, ASTM STP 1263*, Michael J. Verrilli and Michael G. Castelli, Eds., American Society for Testing and Materials, 1996.

ABSTRACT: The results of complementary experimental and analytical investigations of thermomechanical fatigue of both unidirectional and crossply titanium matrix composite laminates are presented . Experimental results are given for both isothermal and thermomechanical fatigue tests which were based on simple, constant amplitude mechanical and thermal loading profiles. The discussion of analytical methods includes the development of titanium matrix composite laminate constitutive relationships, the development of damage models and the integration of both into a thermomechanical fatigue analysis algorithm. The technical approach begins with a micro-mechanical formulation of lamina response. Material behavior at the ply level is based on a mechanics of materials approach using thermo-elastic fibers and an elasto-thermo-viscoplastic matrix. The effects of several types of distributed damage are included in the material constitutive relationships at the ply level in the manner of continuum damage mechanics. The modified ply constitutive relationships are then used in an otherwise unmodified classical lamination theory treatment of laminate response. Finally, simple models for damage progression are utilized in an analytical framework which recalculates response and increments damage sizes at every load point in an applied thermal/mechanical load history. The model is used for the prediction of isothermal fatigue and thermomechanical fatigue life of unnotched, crossply $[0°/90°]_s$ titanium matrix composite laminates. The results of corresponding isothermal and thermomechanical fatigue tests are presented in detail and the correlation between experimental and analytical results is established in certain cases.

KEYWORDS: damage mechanics, metal matrix composites, microcrack mechanics, titanium matrix composites, thermomechanical fatigue

---

[1] Engineering Specialist, Lockheed Fort Worth Company, Fort Worth TX 76101

NOMENCLATURE:

A	crack area
b	scale parameter of fiber Weibull strength distribution
B	coefficient of secondary creep power law
c	crack length,
	modulus of fiber Weibull strength distribution
C_c	coefficient of creep crack growth power law
C_f	coefficient of fatigue crack growth power law
C_{ijkl}	stiffness tensor
D^f_i	fiber damage index
D^i_i	fiber/matrix interface damage index
D^m_{ij}	matrix damage index
e	total complementary energy density
E_L	longitudinal modulus of elasticity (E_{11})
E_T	transverse modulus of elasticity (E_{22})
G	strain energy release rate
G_{LT}	shear modulus
J_2	second invariant of the stress deviator tensor
K_I	mode I (opening mode) stress intensity factor
K_{II}	mode II (sliding mode) stress intensity factor
M_i	applied resultant moments
M_{ij}	compliance matrix (contracted notation)
M_{ijkl}	compliance tensor
n	exponent of secondary creep power law
N_i	applied resultant forces
p	exponent of fatigue crack growth power law
q	exponent of creep crack growth power law
q_i	ply total strain
Q_{ij}	reduced stiffness matrix for plane stress orthotropic lamina (material coordinates)
R	stress ratio (cycle minimum/cycle maximum)
S_{ij}	stress deviator tensor
t	time, laminate thickness
T	temperature
u	elastic strain energy density
V^f	fiber volume fraction
w	work of inelastic response per unit volume
w^{ce}	work of microcrack extension per unit volume
α_{ij}	coefficient of thermal expansion
β	stress intensity boundary correction factor
ε_i	strain (contracted notation)
ε_{ij}	strain tensor
ν	frequency
ν_{ij}	Poisson's ratio
σ_i	stress (contracted notation)

σ_{ij} stress tensor

The use of advanced metal matrix composite (MMC) materials in hypersonic aircraft structure will require the development of the various analytical tools required to support the design process. A key element of this design process is the evaluation of durability for all structural components and the evaluation of damage tolerance for the selected number of components which are deemed to be safety critical. In both cases the ability to predict the long term response of critical components to the mechanical and thermal load spectra typical of hypersonic flight is required. It is because of this and related requirements that a great deal of both experimental and theoretical research has been taking place in the area of thermomechanical fatigue (TMF) and thermal cycling of continuous fiber reinforced metal matrix composites in general and of titanium matrix composites (TMC) in particular [1-3].

The primary objectives of this study were the development of unnotched thermomechanical fatigue analysis methods for TMC laminates and components and the development of a computer code capable of predicting the thermomechanical fatigue behavior of TMC laminates in a hypersonic vehicle environment. The program included both experimental and analytical investigations of titanium matrix composite thermomechanical fatigue. The schematic shown in Fig. 1 illustrates the major features of these related efforts.

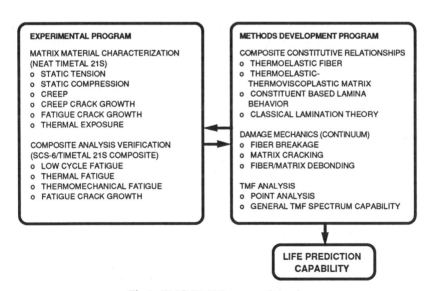

Fig 1. TMC TMF Program Overview.

EXPERIMENTAL PROGRAM:

There were two objectives in the experimental portion of this program. The first was to characterize the matrix material in neat laminate form so as to provide inputs to the micromechanics-based model. The second was to characterize the unnotched isothermal and thermomechanical fatigue behavior of unidirectional and crossply composite materials in simple cyclic tests suitable for verification of the analysis methods. Only the results from the latter activity are discussed herein.

Materials and Procedures

The composite material system studied was a silicon fiber reinforced beta titanium: SCS-6/Timetal 21S. The reinforcing fiber, SCS-6[2], is a 140μm diameter silicon carbide (SiC) filament with a carbon core and a carbon rich coating. The matrix material, Timetal[3], is a metastable beta titanium alloy with approximate chemistry, Ti-13.9Mo-2.8Nb-3.5Al (weight percent). The composite laminates were manufactured by TEXTRON Specialty Materials Division by interleaving four fiber mats in between five titanium foils and then consolidating the assembly in a hot isostatic press. The fiber mat had 129 ends per inch and was cross woven in an over-one under-one (box) weave with 0.002 inch diameter Molybdenum wire at five picks per inch in the fill direction. The nominal thickness of the finished unidirectional $[0°]_4$ panels was 0.914 mm while that of the crossply $[0°/90°]_s$ panels was 0.965 mm. The calculated fiber volume fractions were 35 percent and 33 percent for the unidirectional and crossply panels respectively. Test specimens were cut from the panels using wire EDM. The specimen configuration was a 12.7mm by 152mm dogbone with a 7.62mm by 29.2mm uniform (K_t=1) gage section. A 76.2mm radius was provided between the grip and gage sections and all edges in the gage and radius regions were diamond ground. This configuration was developed and used successfully in an earlier program [4]. All of the specimens (excluding 12 room temperature LCF test coupons) were vacuum heat treated at 621°C for 8 hours. None of the test specimens were coated.

Constant amplitude isothermal low cycle fatigue (LCF) tests were conducted for both longitudinal $[0°]_4$ and crossply $[0°/90°]_s$ composites. The programmed mechanical loading for these tests had a triangular waveform (loading rate = unloading rate = constant) with R=0.1. All of the tests were conducted on closed loop servohydraulic machines under load control at a frequency of 0.33 Hz (20 cpm). Strains were measured continuously using a standard LVDT type extensometer with 12.7 mm gage length. The LCF tests were conducted at 25°C, 480°C and 650°C. The specimens were radiant heated by induction heated susceptors. The temperature control signal was taken using Chromel Alumel thermocouples that were spring loaded to the specimen. The room temperature tests were conducted in a laboratory air environment while the elevated temperature tests were conducted in an inert (argon) environment. The environmental

---

[2] The SCS-6 fiber is produced by TEXTRON Specialty Materials Division, Lowell MA, USA.
[3] Timetal is a registered trademark of Titanium Metals Corporation (TIMET).

chamber required for the inert environment tests utilized a pumping and purging technique in which a fixed volume of argon gas was repeatedly circulated through a gettering system and then sampled for oxygen content prior to its re-entry into the retort. Nominal operation of this system yielded a residual oxygen content of 0.1 part per million.

A series of constant amplitude thermomechanical fatigue (TMF) tests were conducted using both $[0°]_4$ and $[0°/90°]_s$ specimens. The mechanical loading and strain feedback were accomplished in the same manner as for the LCF tests with the exception that the loading frequency was reduced to match that of the thermal cycling. The temperature cycling was constant amplitude with T_{min}=316°C and T_{max}=650°C and a triangular waveform. Heating and cooling rates were typically in the range of 6°C/sec to 11°C/sec. The specimens were radiant heated by induction heated susceptors and convection cooled by forced air. The TMF tests included load-temperature programs in which the thermal and mechanical load cycles were in-phase (IP), 90° out-of-phase (90°OP) and 180° out-of-phase (180°OP). In each case, before starting the test, the specimen was subjected to a complete thermal cycle with zero mechanical load and the resulting thermal strains were recorded. This response history was digitized and became the thermal null signal. This technique was also employed in [4]. As with the LCF tests, the temperature control signal was taken using Chromel Alumel thermocouples that were spring loaded to the specimen. All of the tests were conducted in a laboratory air environment with the exception of two constant load tests which were run in argon.

Numerous researchers have shown that a number of damage mechanisms act simultaneously during thermal and/or mechanical cycling of TMCs [5-8]. As a result it is difficult to isolate and study the effects of any one of them. This is particularly true for unnotched configurations. The addition of damage mode interaction and environmental effects further complicates an already difficult situation. Because the focus of this study was on generally applicable bulk or continuum damage models, only those macroscopic response characteristics which were considered to be indicators of the initiation and evolution of bulk damage were measured during the experimental program. This was accomplished primarily by recording stress-strain hysteresis loops at periodic intervals during each test. These data indicated changes in laminate strain (at the cycle minimum and cycle maximum) with number of applied cycles and changes in laminate modulus with number of applied cycles. In addition, for selected tests, probes were mounted above and below the gage section and an electric current was passed through the specimen continuously during the test. In this technique the current passing through the specimen was held constant and the voltage drop between the probes was measured. As the test progressed the development of various forms of damage (cracking) caused a decrease in the bulk electrical conductivity which in turn, caused an increase in the voltage measured across the probes. Furthermore, changes in potential drop were observable even when little or no change in laminate modulus had taken place. No attempt was made during the experimental program to characterize the physical nature of the various forms of damage or their interactions at the microscopic level. The model that these experiments were intended to support is best characterized as an engineering model; that is to say, the model is phenomenological at the macroscopic level.

Isothermal Low Cycle Fatigue

Longitudinal [0°]₄:

In the longitudinal unidirectional configuration the load is applied in the direction of the fibers. The results of tests conducted at 25°C, 480°C and 650°C are shown in Fig 2. Under these conditions the fatigue behavior is dominated by the fibers and the composite fatigue strengths are quite high (as compared with monolithic Timetal 21S and crossply SCS-6/Timetal 21S). Note that the 480°C and 650°C tests were run in argon, as discussed above. The agreement between the 650°C data and the results of similar tests conducted in air [9] (also shown in Fig 2) implies that oxidation is not a life limiting factor under these conditions. (The Timetal 21S chemistry was formulated specifically for oxidation resistance.) The initiation and growth of cracks in the matrix at the Mo crossweave was the predominant failure mechanism in the current longitudinal LCF tests; there were no surface initiated matrix cracks. Crack initiation occurs at the Mo crossweave sites for two reasons: first, for loading in the direction of the fiber, the crossweave introduces a stress concentration and second, the diffusion of Mo into the surrounding matrix during consolidation causes embrittlement of the titanium. In all cases significant fiber pullout was observed on the fracture surfaces. This is typical of fiber dominated behavior in which extensive fiber cracking leads to fiber failure followed by tensile shear of the matrix.

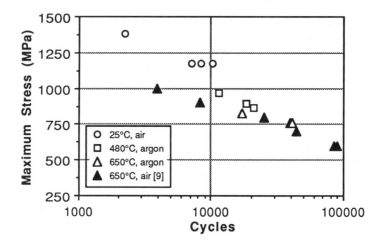

Fig 2. SCS-6/Timetal 21S, [0°]₄,
Isothermal Low Cycle Fatigue, R=0.1.

Crossply [0°/90°]$_s$:

Fatigue results for unidirectional configurations are only useful in so far as they are representative of lamina response and therefore may be used either as direct input for macro-mechanics based fatigue models or as verification for micro-mechanics based models. In many applications, TMC laminates are subjected to loading in more than one direction and therefore require more than one direction of reinforcement.

The results of 25°C, 480°C and 650°C isothermal LCF tests results are shown in Fig 3. It is not conclusive from the limited data shown, but comparison with other room temperature data [10] shows that there is a reduction in fatigue life with increasing temperature. This is thought to result primarily from the low fatigue resistance of the 90° plies. Earlier studies have shown that the fibers are very weakly bonded to the matrix in typical SCS-6/Timetal 21S systems [6, 11]. Apparently there is little or no interphase material at the fiber/matrix interface and what material is there fails at low stress on the first application of load. In addition to the chemical bond at the interface, there is a mechanical bond which arises due to the difference between the coefficients of thermal expansion (CTE) of the fiber and the matrix. At room temperature the matrix exerts both radial and longitudinal compressive forces on the fiber. These forces relax however as the temperature is increased. As a result of these phenomena virtually no load is carried across the fibers in the 90° plies in tension. At low temperatures applied loads must first overcome the residual stresses at the fiber/matrix interface, after which point the fiber serves only to prop the hole in the matrix. At elevated temperatures, the transverse load is immediately carried by the matrix alone (which is permeated with holes).

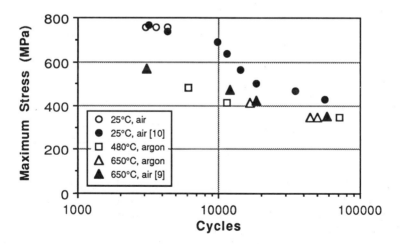

Fig 3. SCS-6/Timetal 21S, [0°/90°]$_s$,
Isothermal Low Cycle Fatigue, R=0.1.

The resulting high stress concentrations coupled with the matrix material's low resistance to creep are the primary causes for the reduction in fatigue life with elevated temperature observed in Fig 3. In all cases failure of these specimens was caused by matrix cracks which initiated at the fiber/matrix interface of the 90° fibers (at the point of peak strain concentration, e.g. 90° from the load direction). This behavior was also reported by Castelli [6]. As was the case with the longitudinal unidirectional material, comparison of the 650°C/argon results with 650°C/air results from other sources [10] seems to indicate that environmental attack plays a minor role under these conditions. See Fig 3.

Thermomechanical Fatigue

Longitudinal [0 °]₄:

Nine thermomechanical fatigue tests with a temperature cycle from 316°C to 650°C were performed using unidirectionally reinforced material. In all of the tests, matrix cracks were found to initiate and grow at the Mo crossweave. The in-phase (IP) tests were characterized by gross strain ratcheting and relatively little change in laminate compliance with repeated cycling. These results are consistent with those of Castelli [7] who showed that the dominant failure mechanism under these conditions was fiber breakage. (Note that the crossweave for the material used in [7] was TiNb. As a result the composite did not suffer the embrittling effects of Mo diffusion and it apparently did not exhibit preferential crack initiation at crossweave sites.) The current IP TMF tests were similar to the isothermal LCF tests in that no matrix surface cracking was observed. In contrast to the IP tests, the 180° OP tests were characterized by little or no strain ratcheting but significant changes in compliance which increased with accumulated cycles. In this case, significant surface cracking was observed, particularly at the higher stress levels. The implication here is that cracking at the cross-weave is a localized (restricted to one plane) damage mechanism that is not readily manifested by changes in macroscopic compliance, whereas surface cracking is a bulk phenomenon that is indicated by changes in laminate compliance. The 90° OP tests fell in between these two extremes with moderate strain accumulation occurring in all three tests but compliance changes occurring only during high stress test. Even though only a small number of tests were performed, the data indicate the presence of a typical IP-OP crossover at approximately 2500 cycles and 700 MPa. See Fig 4. Castelli reported this crossover at 800 cycles and 1000 Mpa [7]. The lower crossover stress found in the current study is attributed primarily to the Mo crossweave.

Two additional tests were performed in which unnotched [0°]₄ specimens were held at constant load while the temperature was cycled between 316° and 650°C. The tests were run in argon. Due to limitations on the cooling rates obtainable in the inert environment, the cycle time for these tests was 11.6 minutes. The first test specimen failed after four thermal cycles at a constant load of 758 MPa. The second ran for 2734 thermal cycles with a mechanically applied stress of 690 Mpa. In both cases the failure was characterized by extensive fiber pullout accompanied by tensile shear of the matrix. No matrix surface cracking was observed in either case.

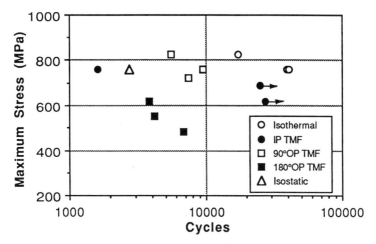

Fig 4. SCS-6/Timetal 21S [0°]4 Thermomechanical Fatigue.
Temperature Cycle=316°C to 650°C, frequency=1 cpm.
Stress Ratio: R=0.1, frequency=1 cpm.
Environment: laboratory air.

Crossply [0°/90°]ₛ:

Six TMF tests with the same 316°C to 650°C temperature cycle were performed using [0° /90°]$_s$ material. As expected with the crossply configuration, the presence of the 90° plies introduced large scale matrix cracking which initiated at the 90° fiber/matrix interface. This was in addition to the cracking at the Mo crossweave in the 0° plies. Certain features of the crossply results corresponded with those obtained for the longitudinal uniaxial material. Specifically no surface initiated matrix cracks were observed in the IP tests and the 180°OP tests exhibited little or no strain accumulation and moderate compliance changes. Unfortunately, insufficient data were gathered to support further conclusions regarding [0°/90°]$_s$ TMF behavior. A much more comprehensive study of the failure mechanisms in SCS-6/Timetal 21S crossply TMF was performed by Castelli [8]. In addition, Castelli has rigorously defined TMF testing procedures and specified the manner in which composite property changes may be properly measured during TMF testing [7, 8, 12]. The results from the current program are shown in Fig 5.

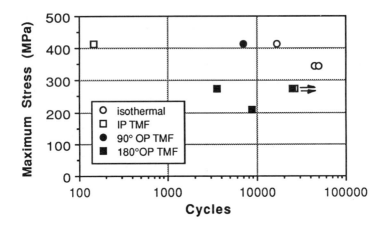

Fig 5. SCS-6/Timetal 21S [0°/90°]s Thermomechanical Fatigue.
Temperature Cycle=316°C to 650°C.
Stress Ratio: R=0.1.
Environment: laboratory air.

ANALYTICAL PROGRAM:

The technical approach taken here for the analysis of titanium matrix composite laminates relies on both micro- and macro-mechanical modeling techniques for the formulation of composite response. The fibers are assumed to be thermo-elastic while the matrix is taken to be elasto-thermo-viscoplastic. The effects of several types of distributed damage are included in the material constitutive relationships at the ply level. Loading may consist of both thermal and mechanical loads and the response is, in general, not linear. The introduction of irreversible strains causes the response to become path dependent and as a result the analysis must be performed incrementally.

In the following discussion, the superscripts f and m indicate the fiber and matrix phases respectively while subscripts generally indicate the components of vectors or tensors. The summation convention is used throughout unless otherwise noted.

Constituent Behavior

Fiber

Fiber response is calculated assuming stain control in the 1-direction and stress control in all others. For strain control the fiber incremental constitutive relationships are [13, 14].

$$d\sigma_{ij}^f = C_{ijkl}^f \left(d\varepsilon_{kl}^f - \alpha_{kl}^f dT\right) + d\sigma_{ij}^{f,T} \tag{1}$$

where the differential thermal stress, which arises due to the temperature dependence of the elastic constants and CTEs, is given by

$$d\sigma_{ij}^{f,T} = \left(\frac{\partial C_{ijkl}^f}{\partial T}\left(\varepsilon_{kl}^f - \alpha_{kl}^f(T - T_o)\right) - C_{ijkl}^f \frac{\partial \alpha_{kl}^f}{\partial T}(T - T_o)\right) dT \tag{2}$$

Under stress control the response strain increments are

$$d\varepsilon_{ij}^f = M_{ijkl}^f d\sigma_{kl}^f + \alpha_{ij}^f dT + d\varepsilon_{ij}^{f,T} \tag{3}$$

where the differential thermal strain is

$$d\varepsilon_{ij}^{f,T} = \left(\frac{\partial M_{ijkl}^f}{\partial T}\sigma_{kl}^f + \frac{\partial \alpha_{ij}^f}{\partial T}(T - T_o)\right) dT \tag{4}$$

The fiber is assumed homogeneous and in general, transversely isotropic. However transverse properties are very difficult to obtain and are not generally available. As a result, isotropy is normally assumed. Fiber response stress and strain increments are found directly by integrating over the time interval from t_{n-1} to t_n.

Matrix

As with the fiber, the matrix response is taken to be strain controlled in the 1-direction. In incremental form the constitutive equation used for the 1-direction is

$$d\sigma_{ij}^m = C_{ijkl}^m \left(d\varepsilon_{kl}^m - \alpha_{kl}^m dT - d\varepsilon_{kl}^{m,p} - d\varepsilon_{kl}^{m,c}\right) + d\sigma_{ij}^{m,T} \tag{5}$$

In this case the differential thermal stress is given by

$$d\sigma_{ij}^{m,T} = \left(\frac{\partial C_{ijkl}^m}{\partial T}\left(\varepsilon_{kl}^m - \alpha_{kl}^m(T - T_o) - \varepsilon_{kl}^{m,p} - \varepsilon_{kl}^{m,c}\right) - C_{ijkl}^m \frac{\partial \alpha_{kl}^m}{\partial T}(T - T_o)\right) dT \tag{6}$$

Note that $\varepsilon^{m,p}{}_{kl}$ and $\varepsilon^{m,c}{}_{kl}$ are the total accumulated plastic and creep strains up to but not including the current increment of loading. For all other directions, the incremental stress-strain relations are written as

$$d\varepsilon_{ij}^m = M_{ijkl}^m d\sigma_{kl}^m + \alpha_{ij}^m dT + d\varepsilon_{ij}^{m,p} + d\varepsilon_{ij}^{m,c} + d\varepsilon_{ij}^{m,T} \tag{7}$$

with the differential thermal strain given by

$$d\varepsilon_{ij}^{m,T} = \left(\frac{\partial M_{ijkl}^m}{\partial T} \sigma_{kl}^m + \frac{\partial \alpha_{ij}^m}{\partial T}(T - T_o) \right) dT \tag{8}$$

The plastic and creep strain increments are found using the method of successive elastic solutions [15, 16]. The plastic (time-independent) strain increments are found assuming a J_2 yield surface with kinematic hardening

$$f = \left(S_{ij}^m - \beta_{ij}^m \right)\left(S_{ij}^m - \beta_{ij}^m \right) - \frac{2}{3}\left(\sigma_o^m \right)^2 \tag{9}$$

where $S^m{}_{ij}$ is the deviatoric stress, $\beta^m{}_{ij}$ are the coordinates of the yield surface center and $\sigma^m{}_o = \sigma^m{}_o(T)$ is the yield strength. The flow rule is given by

$$d\varepsilon_{ij}^{m,p} = \frac{3 d\varepsilon^{m,p}}{2\left(\sigma_e^m \right)_{n-1}} \left(S_{ij}^m - \beta_{ij}^m \right) \tag{10}$$

The creep (time dependent) strain increments are found in a similar manner with the exception that all time dependent strains are treated as being irreversible so there is no loading function required. Creep strain rates are estimated using a power law for secondary, unidirectional creep

$$\dot{\varepsilon}_{ij}^{m,c} = \frac{3B\left(\sigma_e^m \right)^n}{2\left(\sigma_e^m \right)} S_{ij}^m \tag{11}$$

and increments are found by integrating over the time step

$$\Delta\varepsilon_{ij}^{m,c} = \int_{t_{n-1}}^{t_n} \dot{\varepsilon}_{ij}^{m,c} dt \tag{12}$$

The total matrix strain at the end of the interval is simply

$$\left(\varepsilon_{ij}^m \right)_n = \Delta\varepsilon_{ij}^m + \left(\varepsilon_{ij}^m \right)_{n-1} \tag{13}$$

Matrix response stresses are found in a similar manner:

$$\left(\sigma_{ij}^m\right)_n = \Delta\sigma_{ij}^m + \left(\sigma_{ij}^m\right)_{n-1} \tag{14}$$

Lamina Behavior

Material behavior at the ply level is approximated using simple mechanics of materials formulations.

Undamaged Ply

The constitutive relationships for an undamaged lamina are dependent on the mechanical properties and volume fractions of the constituent phases.   Assuming plane stress orthotropy the five independent elastic constants are found as [17]

$$E_L^{ply} = V^f E_L^f + V^m E_L^m \tag{15}$$

$$E_T^{ply} = \left[\frac{V^f}{E_T^f}\left(1 - v_{LT}^f v_{TL}^f\right) + \frac{V^m}{E_T^m}\left(1 - v_{LT}^m v_{TL}^m\right) + \frac{\left(V^f v_{LT}^f + V^m v_{LT}^m\right)}{\left(V^f E_L^f + V^m E_L^m\right)}\right]^{-1} \tag{16}$$

$$G_{LT}^{ply} = \left[\frac{V^f}{G_{LT}^f} + \frac{V^m}{G_{LT}^m}\right]^{-1} \tag{17}$$

$$G_T^{ply} = \left[\frac{V^f}{G_T^f} + \frac{V^m}{G_T^m}\right]^{-1} \tag{18}$$

$$v_{LT}^{ply} = V^f v_{LT}^f + V^m v_{LT}^m \tag{19}$$

Likewise, the two coefficients of thermal expansion are

$$\alpha_L^{ply} = \frac{\left(V^f E_L^f \alpha_L^f + V^m E_L^m \alpha_L^m\right)}{\left(V^f E_L^f + V^m E_L^m\right)} \tag{20}$$

$$\alpha_T^{ply} = V^f\left[\alpha_T^f - v_{LT}^f\left(\alpha_L^{ply} - \alpha_L^f\right)\right] + V^m\left[\alpha_T^m - v_{LT}^m\left(\alpha_L^{ply} - \alpha_L^m\right)\right] \tag{21}$$

The temperature dependence of these constants is affected through the use of temperature dependent constituent properties.   The construction of the compliance and stiffness matrices for a transversely isotropic lamina is straightforward, see reference [18] for example.

Damage Models

At least three mechanical damage mechanisms are known to operate at the ply level in TMC laminates; they are fiber breakage, fiber-matrix debonding and matrix cracking [5-8], [19]. The extent to which the damage sites remain distributed, or become localized is largely a function of geometric configuration and load application. Structural components or regions of structural components in which acting damage mechanisms remain widely and nearly homogeneously distributed, are best analyzed by incorporating the effects of damage into the bulk material properties.   A number of researchers have made considerable progress in developing this continuum approach by characterizing various damage mechanisms and the impact that they have on composite response.   Notable works in the area of dispersed microcracking are those of Allen and Harris [20, 21], Kachanov [22], Laws and Dvorak [23] and Talreja [24]. The use of these, or any other, models in a design environment, however, will require applicability to general structure subjected to sequences of random, thermal/mechanical loadings.

In the current analysis, damage mechanisms such as fiber breakage, fiber/matrix debonding and matrix microcracking are assumed to operate independently; interactions among the elements of a given distribution, and interactions between distributions are not addressed. The distribution of damage sites for each is assumed statistically homogeneous within the Representative Volume Element (RVE).

Fiber Breakage

The failure of fibers at sites which are widely dispersed and uniformly distributed within a given region is modeled using a temperature dependent Weibull fiber strength distribution [25] with the assumption that such fiber breakage is a Poisson process (time independent). A fiber damage index is defined such that $D^f_1=0$ indicates that the fibers are fully effective within the RVE, i.e. no fiber breaks, while $D^f_i=1$ indicates that so many fibers are broken that the fibers are no longer effective.   The fiber damage index in the 1-direction is assumed to be given by the fiber longitudinal strength distribution at a given fiber stress and temperature

$$D^f_1 = 1 - \exp\left[-\left(\frac{\sigma^f_1}{b}\right)^c \right] \tag{22}$$

The temperature dependence of $D^f_1=D^f_L$ is introduced via the temperature dependence of the Weibull modulus, $c=c(T)$, and scale parameter, $b=b(T)$. $D^f_2=D^f_3=D^f_T$ is the density of split fibers.   No data are currently available on this damage mode.   In view of the weakness of the chemical bond at the fiber/matrix interface, the fiber splitting damage mode was not considered in the current analysis.

If the fiber stress in a given ply is tensile, then the effective elastic constants and coefficients of thermal expansion for that ply are modified by replacing each occurrence of the term $V^f E^f_L$ in equations (15) through (21) with the term $V^f(1-D^f_L)E^f_L$. If the fiber is in compression, the ply properties are not altered.

Fiber/Matrix Debonding

Separation at the fiber/matrix interface is treated in a manner similar to that of fiber breakage. In general, one may assume that the chemical bond which exists at the interface has some distribution in strength and then that that distribution may be used to define a fiber/matrix interface damage index for each ply. In the current analysis the interphase strength distribution is taken to be a step function.

$$D^i_T=0 \text{ for } \sigma^i_2<\sigma^i_0$$

$$D^i_T=1 \text{ for } \sigma^i_2\geq\sigma^i_0 \tag{23}$$

Note that the radial stress at the fiber/matrix interface does not become tensile until the compressive radial stress due to the CTE mismatch has been overcome. As before, these damage indices may be used to define effective or reduced ply properties. If the applied stress in the matrix is greater than the residual stress due to the fiber/matrix CTE mismatch then

$$E^{ply}_T = \left(1-D^i_T\right)E^{ply,o}_T + \frac{D^i_T E^m_T}{V^f K^m_{t\varepsilon_2} + V^m} \tag{24}$$

otherwise, the ply transverse modulus is not altered. The superscript 'o' indicates the initial, undamaged value. This expression is nothing more than a weighted average of the undamaged transverse modulus and an estimate of what the transverse modulus would be with voids in the place of the fibers. In practice, there is no data available on the strength of this bond, for the calculations given below it is taken to be zero ($\sigma^i_0=0$).

Matrix Microcracking

Consider two populations of microcracks: 1-direction cracks which are normal to the fiber direction, and 2-direction cracks which are parallel to the fiber direction; see Fig 6. We assume that a ply RVE may be defined by selecting an area large enough such that the number of cracks within the region is representative of the ply. The incremental complementary energy density (per ply), Δe, is taken as the sum of elastic and inelastic (plastic and creep) increments plus an additional term, the incremental work of microcrack extension, which we seek to define.

$$\Delta e = \Delta u + \Delta w + \Delta w^{ce} \tag{25}$$

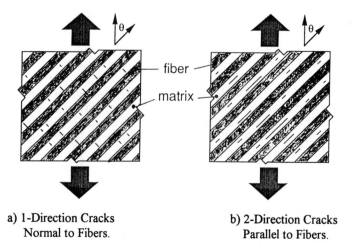

a) 1-Direction Cracks
Normal to Fibers.

b) 2-Direction Cracks
Parallel to Fibers.

Fig 6. Schematic of Lamina Microcracks.

The energy of microcrack extension is given by the integral of the crack driving force over the crack area increment.

$$\Delta w^{ce} = \frac{1}{V} \int G \, da \qquad (26)$$

where V=Wht (width · height · thickness) is the volume of the RVE. Note that under load controlled conditions (e.g. non-zero load point displacements) G is interpreted as crack driving force, i.e. the energy required to create new crack surface area. This energy may be calculated for each crack in the population of N_1 normal cracks and for each crack in the population of N_2 parallel cracks. If we sum over all cracks (both populations) we obtain the energy input required to go from crack system 'a' to crack system 'b'

$$\Delta w^{ce} = \frac{1}{V} \left[\sum_{i=1}^{N_1} \int_a^b G_{1i} \, dA_{1i} + \sum_{j=1}^{N_2} \int_a^b G_{2j} \, dA_{2j} \right] \qquad (27)$$

where the subscripts '1' and '2' denote normal crack and parallel crack terms respectively. Now returning to the expression for incremental complementary energy density, if we consider a path increment which contains no unloading, we may take

$$\Delta q_i = \frac{\partial}{\partial \sigma_i} \Delta e \qquad (28)$$

where Δq_i are the ply total strain increments. (For unloading increments we assume elastic unloading and no crack extension.) For a ply a distance z from the midplane of a laminate subjected to resultant forces and moments we have

$$\Delta q_i = \Delta \varepsilon_i + z \Delta \kappa_i \tag{29}$$

and, using equation (28)

$$\Delta q_i = \frac{\partial}{\partial \sigma_i} \Delta u + \frac{\partial}{\partial \sigma_i} \Delta w + \frac{\partial}{\partial \sigma_i} \Delta w^{ce} \tag{30}$$

The first term on the right hand side of (30) is, by definition, given by the product of the undamaged elastic compliance matrix and the stress increment vector.

$$\frac{\partial}{\partial \sigma_i} \Delta u = M_{ij}^o \Delta \sigma_j \tag{31}$$

where contracted (Voigt) notation has been used. The second term represents the inelastic strain increment, $\Delta \varepsilon_i^I$. And, assuming Δw^{ce} is proportional to σ^2, the third term may be interpreted as a strain that arises due to a change in compliance

$$\frac{\partial}{\partial \sigma_i} \Delta w^{ce} = \delta M_{ij} \Delta \sigma_j \tag{32}$$

By equating equations (29) and (30) and using the results (31) and (32) we get

$$\Delta \varepsilon_i + z \Delta \kappa_i - \Delta \varepsilon_i^I = M_{ij}^o \Delta \sigma_j + \delta M_{ij} \Delta \sigma_j \tag{33}$$

The compliance change is found by utilizing the relationship given in equation (32). Our ability to calculate this change is limited only by our ability to determine the crack driving forces (or strain energy release rates) in equation (27). For example, if we assume that all of the 1-direction and all of the 2-direction cracks are completely contained within the matrix phase and if we further assume that the matrix phase behaves elastically, at least for the current load increment, then we may calculate G as

$$G = \frac{1}{E^m} \left(K_I^2 + K_{II}^2 \right) \qquad \text{for plane stress} \tag{34}$$

where we have considered the fact that each of these cracks may be loaded in both Modes I and II. The assumption that the matrix is elastic and the use of the stress intensity factor, K, as a fracture parameter is reasonable at temperatures below 480°C. At higher temperatures Δw^{ce} should be written in terms of the J integral [26] or C* integral [27]. This adds a considerable degree of complexity to the formulation of Δw^{ce} and its derivatives. This extension to the methodology has not been performed. Currently the errors introduced by these approximations are mitigated by decreasing load increment size and revising ply properties at the conclusion of each increment.

Using equation (34) in equation (27) yields

$$\Delta w^{ce} = \frac{1}{V}\left[\sum_{i=1}^{N_1}\int_a^b \frac{1}{E^m}\left(K_I^2\right)_{1i}dA_{1i} + \sum_{i=1}^{N_1}\int_a^b \frac{1}{E^m}\left(K_{II}^2\right)_{1i}dA_{1i}\right. \tag{35}$$

$$\left. + \sum_{j=1}^{N_2}\int_a^b \frac{1}{E^m}\left(K_I^2\right)_{2j}dA_{2j} + \sum_{j=1}^{N_2}\int_a^b \frac{1}{E^m}\left(K_{II}^2\right)_{2j}dA_{2j}\right]$$

And then writing the stress intensity factor in terms of the matrix stresses σ^m_1, σ^m_2 and σ^m_6, we get

$$\Delta w^{ce} = \frac{1}{V}\left[\sum_{i=1}^{N_1}\int_a^b \frac{2\pi}{E^m}\left(\sigma_1^m\beta_{I_{1i}}\right)^2 tc_{1i}dc_{1i} + \sum_{i=1}^{N_1}\int_a^b \frac{2\pi}{E^m}\left(\sigma_6^m\beta_{II_{1i}}\right)^2 tc_{1i}dA_{1i}\right. \tag{36}$$

$$\left. + \sum_{j=1}^{N_2}\int_a^b \frac{2\pi}{E^m}\left(\sigma_2^m\beta_{I_{2j}}\right)^2 tc_{2j}dc_{2j} + \sum_{j=1}^{N_2}\int_a^b \frac{2\pi}{E^m}\left(\sigma_6^m\beta_{II_{2j}}\right)^2 tc_{2j}dc_{2j}\right]$$

where β_I and β_{II} are the geometric correction factors for modes I and II. As indicated in equation (32), this expression is now differentiated with respect to each of the components of stress to find the effective incremental strains due to the compliance change. For example, the first component is found as

$$\Delta\sigma_1\delta M_{11} = \frac{\partial\left(\Delta w^{ce}\right)}{\partial\sigma_1} = \frac{\partial\left(\Delta w^{ce}\right)}{\partial\sigma_1^m}\frac{\partial\sigma_1^m}{\partial\sigma_1} = \left[\frac{4\pi}{Wh}\sum_{i=1}^{N_1}\int_a^b \frac{\sigma_1^m}{E^m}\beta_{I_{1i}}^2 c_{1i}dc_{1i}\right]\left(\frac{M_{11}^{ply}}{M_{11}^m}\right) \tag{37}$$

Similar results are obtained for the 2- and 6- components. For plane stress orthotropy the 3-, 4- and 5-components are taken to be zero.

The two terms on the right hand side of (33) may be used to define a 'damaged' compliance matrix

$$\Delta\varepsilon_i + z\Delta\kappa_i - \Delta\varepsilon_i^I = M_{ij}^*\Delta\sigma_j \tag{38}$$

For an orthotropic ply in plane stress, we may also define an effective (damaged), reduced stiffness matrix, allowing the expression

$$\Delta\sigma_j = Q_{ij}^*\left[\Delta\varepsilon_j + z\Delta\kappa_j - \Delta\varepsilon_j^I\right] \tag{39}$$

It is convenient to define a matrix damage index such that $D^m_{ij}=0$ for an undamaged ply and $D_{ij}=1$ for a failed ply. When the ply is undamaged we have $M^*_{ij}=M^\circ_{ij}$ (or $Q^*_{ij}=Q^\circ_{ij}$) and when the ply has failed we have $M^*_{ij}=\infty$ (or $Q^*_{ij}=0$); that is

$$M^*_{ij} = \frac{M^\circ_{ij}}{1-D^m_{ij}} \qquad \text{(indices not summed)} \qquad (40)$$

or, for the effective, reduced stiffnesses

$$Q^*_{ij} = \left(1-D^m_{ij}\right)Q^\circ_{ij} \qquad \text{(indices not summed)} \qquad (41)$$

The matrix damage indices may be written in terms of the damaged and undamaged compliances as

$$D^m_{ij} = 1-\left[\frac{M^*_{ij}}{M^\circ_{ij}}\right] \qquad \text{(indices not summed)} \qquad (42)$$

With the required incremental solution technique we have (moving from crack state 'a' to crack state 'b')

$$M^b_{ij} = M^a_{ij} + \delta M_{ij} \qquad (43)$$

By dividing each term in (43) by M°_{ij} we get

$$\frac{1}{1-D^{m,b}_{ij}} = \frac{1}{1-D^{m,a}_{ij}} + \frac{\delta M_{ij}}{M^\circ_{ij}} \qquad \text{(indices not summed)} \qquad (44)$$

and solving for $D^{m,b}_{ij}$ we have the matrix damage index at crack state 'b' given the matrix damage index at crack state 'a', the original, undamaged compliance and the change in compliance in going from state a to state b.

$$D^{m,b}_{ij} = 1-\left[\frac{1}{1-D^{m,a}_{ij}} + \frac{dM_{ij}}{M^\circ_{ij}}\right]^{-1} \qquad \text{(indices not summed)} \qquad (45)$$

As an example, the damage index is calculated for a single crack in a homogeneous, isotropic, finite width strip the Appendix. The correspondence between this damage model and the traditional stress analysis for this configuration is demonstrated.

Laminate Behavior

Ply response is found by rotating the lamina effective compliances or effective, reduced stiffnesses to a common (body) coordinate system. Given the simplifying assumptions that the analysis is quasi-static, that thermal conductivity and mechanical deformation are uncoupled and that modulii and CTEs are constant at the interval mean value for each increment, ply stress increments are related to laminate mid-plane strain and curvature increments and ply thermal strain increments as

$$\Delta\sigma_i = \overline{Q}_{ij}^* \Delta\varepsilon_j^\circ + \overline{Q}_{ij}^* z \Delta\kappa_j - \overline{Q}_{ij}^* \alpha_j \Delta T - \overline{Q}_{ij}^* \Delta\varepsilon_j^I \tag{46}$$

where the $\overline{Q}_{ij}^*$ are the effective, transformed, reduced stiffnesses. The errors introduced via the above stated approximations are ameliorated by reducing mechanical-thermal load increment size and/or time interval size. Note that inelastic strains, including time independent, plastic strains and time dependent, viscoplastic strains are included in $\Delta\varepsilon_j^I$. As equation (46) implies, the irreversible terms are found through an iterative process of successive elastic solutions.

Laminate response is now calculated using classical lamination theory [28]. Laminate extensional, coupling and bending stiffness matrices, A_{ij}^*, B_{ij}^* and D_{ij}^* are defined in the usual manner. Note that these stiffness matrices now include the effects of damage. Resultant force and moment increments are found by integrating the stress increments over the laminate thickness.

$$\Delta N_i = \int_{-t_{lam}/2}^{t_{lam}/2} \Delta\sigma_i \, dz \tag{47}$$

$$\Delta M_i = \int_{-t_{lam}/2}^{t_{lam}/2} \Delta\sigma_i z \, dz \tag{48}$$

Each of the Individual strain increments in (46) is treated in this manner, thus defining 'effective' force and moment increments. The incremental laminate constitutive relations are

$$\{\Delta N\} = \left[A^*\right]\{\Delta e\} + \left[B^*\right]\{\Delta k\} + \left\{\Delta N^T\right\} + \left\{\Delta N^I\right\} \tag{49}$$

$$\{\Delta M\} = \left[B^*\right]\{\Delta e\} + \left[D^*\right]\{\Delta k\} + \left\{\Delta M^T\right\} + \left\{\Delta M^I\right\} \tag{50}$$

With the stresses and strains being history dependent, the analysis must be performed on a point-to-point basis, with the state at the previous load point serving as the starting point for the current load increment [28]. As a result, the total forces and moments at the current load point, n, are found by adding the calculated increments to the total resultant forces and moments from the previous load point, n-1.

$$\{N\}_n = \{\Delta N\} - \{\Delta N^T\} - \{\Delta N^I\} - \{N^I\}_{n-1} + \{N\}_{n-1} \qquad (51)$$

$$\{M\}_n = \{\Delta M\} - \{\Delta M^T\} - \{\Delta M^I\} - \{M^I\}_{n-1} + \{M\}_{n-1} \qquad (52)$$

where $\{\Delta N\}$ and $\{\Delta M\}$ are the total force and moment increments, $\{\Delta N^T\}$ and $\{\Delta M^T\}$ are the effective thermoelastic force and moment increments and $\{\Delta N^I\}$ and $\{\Delta M^I\}$ are the effective inelastic force and moment increments which occur during the time interval from t_{n-1} to t_n. (The brackets { } indicate vector quantities.) Finally, $\{N^I\}_{n-1}$ and $\{M^I\}_{n-1}$ are the effective forces and moments due to pre-existing inelastic strains.

Before the matrix microcracking model can be used for laminate analysis, it must be modified to account for the fact that cracks in the interior plies of a laminate are restrained from opening by the adjacent plies. A suitable crack opening displacement (COD) restraint factor with general dependence on the angle between fiber orientations of adjacent plies is not currently available. For the current analysis a simple factor suggested by Talreja [24]

$$f_c = t_{ply} / t_{laminate} \qquad (53)$$

is included in the stress intensity modification factor, β, for internal plies.

Life Prediction

Given each of the three damage mechanisms discussed above, rules are now defined for their evolution with respect to time. In the case of fiber breakage, each time ply response is calculated, the longitudinal fiber stress is checked against the Weibull cdf at the appropriate temperature. The density of broken fibers is estimated from the cdf and checked against the prior value. If it is larger, indicating that more fibers have broken, then the ply longitudinal modulus is recalculated and the increment is repeated in order to redistribute the load. This process may be repeated within a given load step or over the course of many load steps until the broken fiber density equals unity, at which time all of the fibers in the ply are assumed to be broken. When this occurs the ply generally fails and the remaining plies must pick up the additional load.

In a similar fashion applied stress in the matrix is continually checked against the residual stress in the matrix at the fiber/matrix interface. The first time this stress exceeds the sum of the interface bond strength and the residual stress, the bond is broken and the transverse modulus is modified as shown above. For each subsequent event, the applied stress has only to overcome the residual stress for the modulus to be modified.

The subcritical growth of matrix microcracks is modeled using an empirical relationship which takes the crack growth increment to be the linear sum of a cycle dependent (fatigue) component and a time dependent (creep) component [29].

$$\Delta c = \left(\frac{dc}{dN}\right)\Delta N + \int_{t_{n-1}}^{t_n}\left(\frac{dc}{dt}\right)dt \tag{54}$$

Standard power law relationships are used for both the fatigue and the creep crack growth rates.

$$\frac{dc}{dN} = C_f(\Delta K)^p \tag{55}$$

$$\frac{dc}{dt} = C_c\left(K_{max}\right)^q \tag{56}$$

The creep crack growth rate parameters, C_c and q, were determined by least squares fit of equation (56) to creep crack growth rate data which were generated during the materials characterization portion of this program, see Fig. 1. The fatigue crack growth rate parameters, C_f and p, were determined in a similar manner with the exception that a calculated creep component was subtracted from the measured isothermal fatigue crack growth rate data prior to the regression. Specifically, equation (54) was rewritten for the special case of constant amplitude cycling at constant temperature

$$\frac{dc}{dN}_{fatigue} \approx \frac{dc}{dN}_{total} - \int_0^{\frac{1}{v}}\left(\frac{dc}{dt}\right)dt \tag{57}$$

and the regression was performed on the $(dc/dN)_{fatigue}$ values.

Example Calculations

This methodology was evaluated by calculating TMC laminate response and time to failure for a number of thermal/mechanical loading scenarios. In all cases the fiber breakage, fiber/matrix debonding and matrix microcracking damage mechanisms were activated in a cyclic laminate analysis. The evaluation of the life prediction methodology was limited to an assessment of the degree to which calculated times to failure agreed with experimental results. No attempt was made to determine whether or not each of the individual damage models was representative of the actual damage mechanisms occurring in the composite. It is important to note particularly that neither the physical (chemical) or mechanical effects of the Mo crossweave on crack initiation or growth were addressed.

Because the model is micromechanics based, only the material properties of the constituents are required for an analysis. In general, these properties must be specified over the full range of temperatures being addressed. The following calculations were made using the assumption that the fiber is elastic and isotropic with the mechanical and physical properties given in Table 1. The matrix was taken to be linear strain hardening with the properties shown in Table 2. Each problem included cooldown from the consolidation cycle at a constant cooling rate of 0.133°C/sec (e.g. 2 hour linear cooling from 980°C to 25°C) to insure that the CTE mismatch residual stresses were accommodated. The physical bond at the fiber/matrix interface was assumed to have zero strength; the only bond at the interface was the mechanical bond due to the CTE mismatch. The composite was modeled as a 6-ply laminate with two 40.6μm thick, monolithic (V^f=0) face sheets and four 208μm thick, V^f=0.39, fiber reinforced plies.

Table 1. SCS-6 Silicon Carbide Fiber Mechanical and Physical Properties.

	T (°C)			
	25	480	650	1010
Modulus, E (GPa)	400.0	400.0	400.0	400.0
Shear modulus, G (GPa)	158.6	158.6	158.6	158.6
Poisson's ratio	0.25	0.25	0.25	0.25
CTE, α (10^{-6} m/m°C)	3.53	4.00	4.28	4.73
Tensile Strength (GPa)	3.861	3.764	3.621	3.230
Strength distribution Weibull scale parameter, b (GPa)	4.652	4.503	4.322	3.438
Strength distribution Weibull modulus, c	5.531	5.531	5.531	5.531

Table 2. Timetal 21S Mechanical and Physical Properties.

	T (°C)			
	25	480	650	1010
Modulus, E (Gpa)	75.8	78.6	70.3	34.5
Shear modulus, G (Gpa)	28.3	29.0	26.2	13.1
Poisson's ratio	0.35	0.35	0.35	0.35
CTE, α (10^{-6} m/m°C)	6.10	9.38	9.74	12.38
Plastic modulus, H (Gpa)	1.551	0.490	0.062	0.007
Elongation	0.045	0.056	0.210	0.5
Secondary creep coefficient[2], B	0.000E+00[1]	1.000E-30	4.783E-18	-
Secondary creep exponent[2], n	1.000[1]	4.320	5.571	-
Creep crack growth coefficient[3], C_c	0.000E+00[1]	0.000E+00[1]	2.173E-14	-
Creep crack growth exponent[3], q	1.000[1]	1.000[1]	7.744	-
Fatigue crack growth coefficient[4], C_f	1.082E-10	6.928E-10	6.537E-11	-
Fatigue crack growth coefficient[4], p	2.497	2.096	3.665	-

[1]estimated
[2]time in sec, strain in m/m, stress in MPa
[3]data in m/sec vs MPa√m units
[4]data in m/cycle vs MPa√m units

SCS-6/Timetal 21S [0°/90°]s Isothermal LCF

The isothermal low cycle fatigue of a [0°/90°]$_s$ laminate at 25°C was simulated in the following manner. Both 1-direction and 2-direction matrix microcracks were assumed to exist in all plies and their mean initial size was taken to be 14μm (one tenth of the fiber diameter). Microcrack area number densities ($n_1=N_1$/Wh, $n_2=N_2$/Wh) were initially varied between $n_1=n_2=10$ and $n_1=n_2=100$ until reasonable agreement with the test results was achieved.

The general progression of the analysis was as follows. 2-direction cracks in the 90° plies grew subcritically for approximately 95% of the predicted life. During this time little or no growth of the 1-direction cracks in the 0° plies was predicted. At about 95% of the predicted life the 90° plies failed and the 0° plies were required to carry the additional load. This quickly caused fiber failure in the 0° plies and failure of the laminate. The calculated life was most sensitive to n_2 for the 90° plies. Fig 7 shows the predicted increase in total matrix microcrack area as a function of number of applied cycles compared with electric potential drop readings taken during an SCS-6/Timetal 21S [0°/90°]$_s$ LCF test. The comparison suggests that there may be some functional relationship between the predicted matrix cracking and the PD readings.

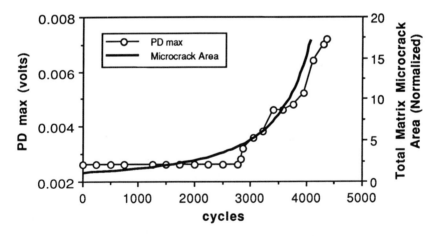

Fig 7. Measured PD and Predicted Total Matrix Microcrack Area
during SCS-6/Timetal 21S [0°/90°]s LCF at 25°C.

Crossply isothermal low cycle fatigue life was predicted for constant amplitude cycling with R=0.1 at room temperature and at 650°C. The assumptions regarding initial matrix microcrack size were the same as above. The crack area number densities, however, were taken to be linearly proportional to the matrix stress. In all cases growth of the 2-direction cracks in the 90° plies comprised the bulk of the predicted life. Failure was defined as a ninety percent reduction in laminate longitudinal modulus (meaning that the analysis was terminated when E^{ply}_L reached $0.1 * E^{ply,o}_L$). Fig 8 shows the calculated LCF life over a range of applied stress levels. Also shown are experimental results reported by Neu [9] and Ward [10] as well as those generated during the current program.

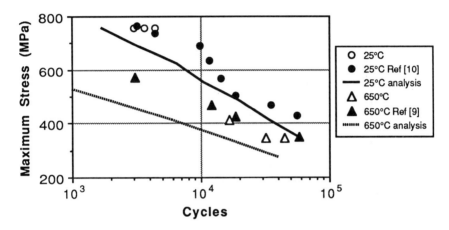

Figure 8. Calculated vs Measured Isothermal Low Cycle Fatigue Life for
SCS-6/Timetal 21S [0°/90°]$_S$ and [0°/90°]$_{2S}$, R=0.1.

SCS-6/Timetal 21S [0°/90°]$_S$ TMF

Several thermomechanical fatigue analyses were performed using the current model in a manner similar to that employed for the isothermal LCF predictions. The input loading was prescribed in terms of applied laminate resultant force and laminate front and back surface temperatures, all of which ramped linearly between reversal points. The applied temperature cycle was 316°C to 650°C in all cases.

In the first analysis in-phase TMF at an applied stress (maximum) of 414 MPa was studied. As with the LCF analysis, both 1-direction and 2-direction matrix microcracks were assumed to exist in all plies and their mean initial size was taken to be 14μm (one tenth of the fiber diameter). The calculated mechanical strain at the cycle maximum vs number of applied cycles is shown in Fig 9 where the results are compared with experimental data. The analysis predicted moderate growth of the 2-direction cracks in the 90° plies, although to a lesser extent than that predicted under isothermal conditions.

The most notable feature of the analysis was the accumulation of strain (ratcheting) which appears to have been over-predicted based on the comparison shown in Fig 9. Because the maximum applied stress occurred at the peak of the temperature cycle, the model predicted significant creep strain accumulation in the matrix causing load to be transferred to the fibers. Laminate failure was due to failure of the fibers in the 0° plies.

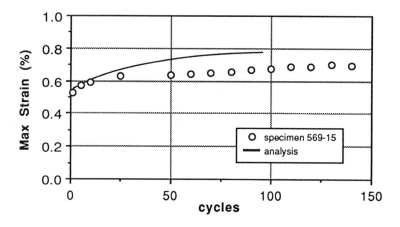

Fig 9. Calculated Maximum Mechanical Strain vs Number of Applied Cycles
for SCS-6/Timetal 21S [0°/90°]$_s$ Subjected to In Phase TMF,
Temperature Cycle: 316°C - 650°C, ν=0.011 Hz,
Stress Cycle: 41.1 MPa - 414 MPa, ν=0.011 Hz.

Finally, crossply thermomechanical fatigue life was predicted for IP and 180°OP cycling over a range of stress levels. For these calculations, the matrix microcrack area number densities were taken to be linearly proportional to the matrix stress. As was the case with the analysis presented in Fig 9, for IP TMF the model predicted moderate growth of the 2-direction cracks in the 90° plies. However, creep strain accumulation and ultimate failure of the fibers in the 0° plies were the determining factors in the predicted life. For OP TMF the model predicted very little creep strain accumulation since the maximum stress occurred at the minimum temperature. In the OP TMF case the growth of 1-direction cracks in the 0° plies and 2-direction cracks in the 90° plies comprised the bulk of the predicted life. Failure was defined as a ninety percent reduction in laminate longitudinal modulus; there was no fiber failure. Fig 10 shows the calculated IP and OP TMF life over a range of applied stress levels. Also shown are the corresponding experimental results generated during the current program.

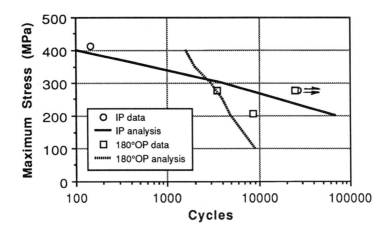

Fig 10. Calculated vs Measured In-Phase and 180° Out-of-Phase
Thermomechanical Fatigue Life for SCS-6/Timetal 21S [0°/90°]$_s$.

CONCLUSIONS

An experimental investigation of isothermal and thermomechanical fatigue of SCS-6/Timetal 21S composites was performed. Material was tested in both longitudinal [0°]$_4$ and crossply [0°/90°]$_s$ lay-ups and all specimens were unnotched. The dominant failure mechanism for the longitudinal composites (with the loading in the direction of the fibers) was crack initiation and growth at the Mo crossweave followed by fiber failure and final tensile shear of the matrix. Failure of the crossply composites was characterized by the growth of matrix cracks in the 90° plies accompanied by matrix cracking at the Mo crossweave in the 0° plies. While it was clear that the Mo crossweave consistently acted as a damage initiation site, it was not clear that this translated into shorter fatigue life. Current isothermal fatigue test results compared well with other published results which were generated using material which did not have the Mo crossweave. Comparison of inert (argon) environment isothermal fatigue test results with comparable lab air environment results indicated that oxidation was not a life limiting factor for the conditions studied.

A model for continuous fiber composite laminate thermomechanical fatigue was formulated and subjected to preliminary evaluation. The intended application of this model is the calculation of TMF life of unnotched composite laminates. Because such analyses typically require thousands or even tens of thousands of response calculations the technical approach was kept at a rudimentary level for the sake of computational efficiency. The model utilizes a micromechanics approach for lamina response in which the fiber is taken to be thermo-elastic and the matrix is taken to be elasto-thermo-viscoplastic. Three types of distributed damage are incorporated at the ply level; these are fiber breakage, fiber/matrix debonding and matrix microcracking. Potential interaction among the acting damage mechanisms is not addressed and the model is based on the assumption that the damage sites are suitably dispersed so as to provide statistical homogeneity.

Initial comparisons with experimental results showed that the model captures some aspects of cyclic TMC response and progressive failure. The model is suitable in a point stress analysis for the examination of unnotched fatigue behavior; life predictions within a factor of two are generally obtainable. It may also be used as a means for determining bulk properties and response characteristics for in a global-local analysis scheme. Finally, the model is useful as a tool for parametric and design trade studies. Significant improvements are needed in a number of areas. At a minimum, the occurrence of primary creep must be addressed. The incorporation of a unified viscoplastic model for matrix response would be preferable. In addition, the incremental energy of crack extension and its derivatives should be formulated in terms of one or more fracture parameters which are appropriate for non-linear material response.

ACKNOWLEDGMENTS: The author gratefully acknowledges Mr Jim Berling of Materials Behavior Research Corp., Cincinnati OH for the design and construction of the test apparatus and for the performance of the LCF, TF and TMF tests, and Dr. W.S. Chan of the University of Texas at Arlington for suggesting the strain energy release rate approach for matrix microcracking.

REFERENCES

[1]   Larsen, J.M., Russ, S.M. and Jones, J.W., "Possibilities and Pitfalls in Aerospace Applications of Titanium Matrix Composites," NATO AGARD Conf. on Characterization of Fiber Reinforced Titanium Metal Matrix Composites, Bordeaux France, Sept. 1993.

[2]   Castelli, M.G., Bartolotta, P.A. and Ellis, J.R., "Thermomechanical Testing of High Temperature Composites: Thermomechanical Fatigue Behavior of SiC(SCS-6)/Ti-15-3," *Composite Materials: Testing and Design (Tenth Volume), ASTM STP 1120*, G.C. Grimes, Ed., Philadelphia, 1992, pp. 70-86.

[3]   Russ, S.M., Nicholas, T., Bates, M. and Mall, S., "Thermomechanical Fatigue of SiC/Ti-24Al-11Nb Metal Matrix Composites," *Failure Mechanisms in High Temperature Composite Materials*, AMD-Vol. 122, G.K. Haritos, G. Newaz and S. Mall, Ed.s, ASME, New York, 1991, pp. 37-43.

[4]   Gambone, M.L., "Fatigue and Fracture of Titanium Aluminides," WRDC-TR-89-4145, Vol. II, Materials Laboratory, Wright Research and Development Center, Wright-Patterson AFB OH, Feb. 1990.

[5]   Johnson, W.S., "Damage Development in Titanium Metal Matrix Composites Subjected to Cyclic Loading," NASA TM-107597, April 1992.

[6]   Castelli, Michael G., "Isothermal Damage and Fatigue Behavior of SCS-6/Timetal 21S [0/90]$_s$ Composite at 650°C," NASA CR-195345, National Aeronautics and Space Administration, 1994.

[7]   Castelli, Michael G., "Characterization of Damage Progression in SCS-6/Timetal 21S [0]$_4$ Under Thermomechanical Fatigue Loadings," NASA CR-195399, National Aeronautics and Space Administration, 1994.

[8]   Castelli, Michael G., "Thermomechanical Fatigue Damage/Failure Mechanisms in SCS-6/Timetal 21S [0/90]$_s$ Composite," *Composites Engineering*, Vol. 4, No. 9, 1994, pp. 931-946.

[9]   Neu, R.W. and Nicholas, T., "Life Prediction of SCS-6/Timetal 21S Under Thermomechanical Fatigue," *Thermomechanical Behavior of Advanced Structural Materials*, AD Vol. 34, W.F. Jones, Ed., ASME, 1993, pp. 97-111.

[10]   Ward, G.T., Herrmann, D.J. and Hillberry, B.M., "Stress-Life Behavior of Unnotched SCS-6/Ti-β21S Cross-Ply Metal Matrix Composites at Room Temperature," Proceedings of the Fifth International Conference on Fatigue and Fatigue Thresholds, Fatigue 93, Vol. II, 1993, pp 1067-1072.

[11]   Naik, R.A., Pollock, W.D. and Johnson, W.S., "Effect of a High-Temperature Cycle on the Mechanical Properties of Silicon Carbide/Titanium Metal Matrix Composites," *J. Materials Science*, Vol. 26, 1991, pp.2913-2920.

[12]   Castelli, Michael G., "An Advanced Test Technique to Quantify Thermomechanical Fatigue Damage Accumulation in Composite Materials," *J. of Composites Tech. and Res.*, to appear, also NASA CR-191147, 1993.

[13]   Hashin, Zvi, "Part 6: Thermoelasticity," in "Theory of Fiber Reinforced Materials," NASA CR-1974, March 1972.

[14]   Agarwal, Bhagwan D. and Broutman, Lawrence J., Analysis and Performance of Fiber Composites, 2nd ed., John Wiley and Sons, New York, 1990.

[15]   Mendelson, Alexander, Plasticity: Theory and Application, Krieger Publishing Co., 1983.

[16]   Hinnerichs, Terry D., "Viscoplastic and Creep Crack Growth Analysis by the Finite Element Method," AFWAL-TR-80-4140, Air Force Wright Aeronautical Laboratories, Wright-Patterson AFB, OH, July 1981.

[17]   Tsai, Stephen W. and Hahn, H. Thomas, Introduction to Composite Materials, Technomic Publishing Co., Inc., 1980.

[18]   Christensen, R.M., Mechanics of Composite Materials, John Wiley & Sons, New York, 1979.

[19]   Majumdar, B.S. and Newaz, G.M., "Inelastic Deformation of Metal Matrix Composites, Part I - Plasticity and Damage Mechanisms," NASA CR-189095, March 1992.

[20]   Allen, D.H., Harris, C.E. and Groves, S.E., "A Thermomechanical Constitutive
        Theory for Elastic Composites With Distributed Damage - I. Theoretical
        Development," International Journal of Solids and Structures, Vol. 23, No. 9,
        1987, pp. 1301-1318.

[21]   Allen, D.H., Harris, C.E. and Groves, S.E., "A Thermomechanical Constitutive
        Theory for Elastic Composites With Distributed Damage - II. Application to
        Matrix Cracking in Laminated Composites," International Journal of Solids and
        Structures, Vol. 23, No. 9, 1987, pp. 1319-1338.

[22]   Kachanov, M., "Effective Elastic Properties of Cracked Solids: Critical Review of
        Some Basic Concepts," Applied Mech Review, Vol. 45, No. 8, August 1992, pp.
        304-335.

[23]   Laws, N., Dvorak, G.J. and Hejazi, M., "Stiffness Changes in Unidirectional
        Composites Causes by Crack Systems," Mechanics of Materials, Vol. 2, 1983, pp.
        123-137.

[24]   Talreja, R., "Transverse Cracking and Stiffness Reduction in Composite
        Laminates," Journal of Composite Materials, Vol.19, 1985, pp. 355-375.

[25]   van der Zwaag, Sybrand, "The Concept of Filament Strength and the Weibull
        Modulus," J Testing and Evaluation, American Society for Testing and Materials,
        1989, pp. 292-298.

[26]   Rice, J.R., "A Path Independent Integral and the Approximate Analysis of Strain
        Concentration by Notches and Cracks," Trans. of the ASME, J. of Applied
        Mechanics, Vol. 35, 1968, pp. 379-386.

[27]   Landes, J.D. and Begley, J.A., "A Fracture Mechanics Approach to Creep Crack
        Growth," Mechanics of Crack Growth, ASTM STP 590, American Society for
        Testing and Materials, 1976, pp. 128-148.

[28]   Barrett, David J. and Buesking, Kent W., "Temperature Dependent Nonlinear
        Metal Matrix Laminate Behavior," NASA CR-4016, 1986.

[29]   Wei, R.P. and Landes. J.D., "Correlation Between Sustained Load and Fatigue
        Crack Growth in High Strength Steels," Materials Research and Standards,
        MTRSA, Vol. 9, 197, JULY 1969.

APPENDIX A

Damage Index for a Single Crack in Homogeneous, Isotropic, Finite Width Strip

One evaluation of the current microcracking model may be obtained by applying it to a
problem in which the relationship between crack size and compliance is already known.
For a single, through thickness crack in an isotropic, finite width plate, equation (36)
becomes

$$\Delta w^{ce} = \frac{2\pi}{Wh} \int_a^b \frac{1}{E^m} \sigma^2 \beta^2 c \, dc \qquad (A\text{-}1)$$

where c is the half crack length, the crack area is taken to be A=2ct and β is the geometric
correction factor.

The incremental change in compliance due to crack extension is given by

$$\frac{\partial}{\partial\sigma}\left(\Delta w^{ce}\right) = \delta M \cdot \sigma \tag{A-2}$$

which upon differentiation yields

$$\delta M = \frac{4\pi}{Wh}\int_a^b M\beta^2 c\,dc \tag{A-3}$$

where the stress has been assumed to be constant. The damage index, equation (45), which for this case is a scalar becomes

$$D^b = 1 - \left[\frac{1}{1-D^a} + \frac{4\pi}{Wh}\int_a^b \frac{1}{1-D}\beta^2 c\,dc\right]^{-1} \tag{A-4}$$

Note that the unknown damage index, D, is dependent on c and therefore remains within the integral. We can obtain an approximate solution for D as a function of crack length by taking small increments in c and by assuming D is constant over the increment from a to b

$$D^b \approx 1 - \left(1-D^a\right)\left[1 + \frac{4\pi}{Wh}\overline{\beta}^2\overline{c}\Delta c\right]^{-1} \tag{A-5}$$

where the overbars indicate average values over the increment, i.e., $c=(c^b+c^a)/2$, and $\Delta c=c^b-c^a$, is the crack length increment By way of verification, results from Tada[4] for M as a function of crack length may be used to find D as

$$D = 1 - \frac{M}{M^\circ} \tag{A-6}$$

Comparisons for panel heights of 1, 2 and 3 times the panel width are shown in Fig A-1. Note that the current formulation agrees quite well with the stress analysis results, particularly for the larger RVEs. The divergence that does occur may be reduced to any extent desired by reducing the size of the crack length increment used in the current model for the calculation of D.

---

[4] Tada, H., Paris, P.C. and Irwin, G.R., The Stress Analysis of Cracks Handbook, 2nd ed., Paris Productions, Inc., St. Louis MO, 1985.

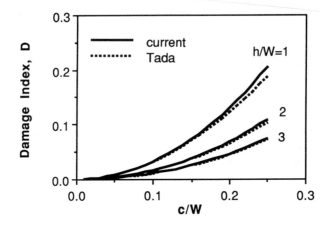

Fig A-1. Calculated Damage Index for Centered Through Thickness Crack in a Finite Width Strip. Comparison between current formulation and results from Tada et.al.

Theodore Nicholas,[1] and David A. Johnson[2]

TIME- AND CYCLE-DEPENDENT ASPECTS OF THERMAL AND MECHANICAL FATIGUE IN A TITANIUM MATRIX COMPOSITE

REFERENCE: Nicholas, T. and Johnson, D. A., **"Time- and Cycle-Dependent Aspects of Thermal and Mechanical Fatigue in a Titanium Matrix Composite,"** Thermomechanical Fatigue Behavior of Materials, ASTM STP 1263, M. J. Verrilli and M. G. Castelli, Eds., American Society for Testing and Materials, 1996.

ABSTRACT: Specimens of SCS-6/Timetal®21S in [0]₄ and [0/90]s orientations were subjected to three types of tests covering temperatures of 650, 760, and 815°C. In the first, low frequency cycles with no superimposed hold times were applied. In the second, low frequency fatigue cycles were applied with superimposed hold times of various durations at maximum loads. In the third, hold times were superimposed at minimum load. In all tests, the number of cycles to failure was recorded. The behavior of the composite in these tests was modeled by considering damage to consist of a linear sum of a cycle-dependent component and a time-dependent component. It was found that the behavior in all tests was basically time dependent. Using time at temperature and stress as a failure criterion, it is shown that data from creep, low frequency fatigue with and without hold times, and in-phase thermomechanical fatigue tests can be consolidated for all [0]₄ tests into a single curve and for all [0/90]s tests into a different curve. Both curves can be explained in terms of a single fiber bundle strength value by considering the time-dependent behavior of the matrix material. Additionally, hold times at minimum load are found to be more detrimental to fatigue life than identical tests where the hold time is applied at the maximum load.

KEYWORDS: titanium matrix composite, fatigue, hold time, time-dependent behavior, life fraction model, thermomechanical fatigue, creep

Titanium and titanium aluminides have received considerable attention over the last several years as matrix materials for composites reinforced with continuous SiC fibers for high temperature aerospace structural applications. These titanium matrix composites (TMC) must be able to withstand static as well as fatigue loading over a wide range of temperatures and cyclic frequencies. While isothermal and thermomechanical fatigue (TMF) have been addressed in several investigations which included effects of frequency, TMF cyclic phasing, material lay-up, and spectrum loading [1-10], the combined effects of fatigue cycling with superimposed hold times have received only limited attention. In an

[1] Senior scientist, Wright Laboratory Materials Directorate, Wright-Patterson Air Force Base, OH 45433.

[2] Research engineer, University of Dayton Research Institute, Dayton, OH 45469.

investigation by Revelos and Roman [11], SCS-6/Ti-24Al-11Nb was thermally cycled in air and an inert environment between 150°C and 815°C for various cycle counts. Various hold times at maximum temperature were added to assess time-dependent effects on composite integrity. Laminate orientations investigated included: [0]₄, [0]₈, [90]₄, [90]₈ and [0/90]₂S. The importance of considering total time at temperature under cyclic, and to a lesser degree, isothermal exposure conditions was demonstrated by a reduction in residual strength of the [0]₈ composite with increasing hold time at maximum temperature.

Revelos [12] extended that study to an SCS-6/β21S (Ti-15Mo-2.6Nb-3Al-0.2Si) composite. [0]₄ and [0/90]S laminates were thermally cycled in air with varying hold times at maximum temperature to determine their effects on composite integrity. Isothermal and cyclic exposure of specimens resulted in significantly decreased residual strength when longer hold times at maximum temperature were employed. The study of thermal cycling was extended by Revelos et al. [13] where the influence of thermal cycling and isothermal exposures in air on the residual ambient temperature strength of SCS-6/Timetal 21S [0]₄ and [0/90]S laminates was determined. A maximum temperature of 815 °C was used in thermal cycling and isothermal exposure. Among the variables considered, hold time at temperature was systematically varied. A reduced residual strength was noted in thermal fatigue with increasing hold time for all specimens tested in air.

To quantify degradation of isothermal fatigue life, a simple model was developed by Nicholas and Russ [1] to predict life based on cycles with superimposed hold times. The hold time model was adapted to a SCS-6/Ti-24Al-11Nb unidirectional composite under 0.003 Hz cycling with superimposed hold times up to 5000 s and produced good correlation with experimental data. To the authors' knowledge, there have been no other systematic studies of the potential interaction between fatigue and "creep," or superimposed hold times, on a high temperature metal matrix composite. Gabb and Gayda [14] evaluated the behavior of a titanium aluminide composite under cycles with superimposed hold times. They observed that the combination of high applied loads at high temperatures allowing matrix relaxation promoted fiber failure and degraded fatigue life. Similar observations on the reduction of fatigue strength due to a combination of load and time at temperature have been made by Mirdamadi and Johnson [15] and Johnson et al. [16]. To add to the experimental data base of the influence of hold times on fatigue cycling, as well as to explain the nature of the interactions so that they may be modeled, this investigation was conducted.

EXPERIMENTAL PROCEDURE

The material used in this study was a titanium matrix composite with silicon carbide (SCS-6) fibers and Timetal®21S (formerly referred to as β21S) matrix. The composite panels were manufactured by alternating layers of rolled titanium foils with Ti-Nb weaved SCS-6 fibers and then consolidated by hot isostatic pressing. Two laminate orientations were considered: [0]₄ and [0/90]S. The fibers are 141 μm in diameter and the thickness of the consolidated composite was approximately 0.9 mm for 4 plies of fibers. The plates were cut into straight-sided specimens using abrasive water jet method. The width of the [0]₄ specimens was approximately 10 mm and the width of the specimens for other orientations was approximately 12.7 mm. The specimens were heat treated in vacuum at 621°C for 8 hours.

The hold-time tests were conducted on servo-pnuematic, horizontally-loaded machines. Three minute cycles (0.0056 Hz) having a triangular wave form were used, and hold times up to 10,000 s were superimposed at either maximum or minimum load. All testing was conducted at a stress ratio (ratio of minimum to maximum load) of 0.1.

Heating was accomplished through the use of a quartz-lamp system and the temperature spatial-gradient in the gage section was typically less than 5°C. Thermocouples were used to monitor temperature, and no failure was observed to occur at thermocouple attachment-points. Loading control was very accurate, as would be expected for the low-frequency test-conditions. Strain was monitored with a high-temperature, quartz-rod extensometer. Failure was in all cases defined to be actual fracture of the specimen into two pieces. The specimens were machined from various panels, which should account for some of the variability of the results. The entire test-system is described by Hartman and Buchanan [17].

TABLE 1--Summary of test conditions and experimental results.

Temp. (°C)	σ_{max} (MPa)	Hold time (s)	Hold point	Stress ratio	Material lay-up	Cycles to fracture	Fiber volume fraction (Vf)
649	400	0	N/A	0.1	[0°/90°]s	3012	0.3587
649	400	0	N/A	0.1	[0°/90°]s	4946	0.3547
649	400	10	Max Load	0.1	[0°/90°]s	4715	0.3547
649	400	100	Max Load	0.1	[0°/90°]s	4917	0.3547
649	400	1000	Max Load	0.1	[0°/90°]s	2087	0.3547
649	800	0	N/A	0.1	[0°]4	6674	0.3843
649	800	1000	Max Load	0.1	[0°]4	2024	0.3889
760	400	0	N/A	0.1	[0°/90°]s	1475	0.3509
760	400	100	Max Load	0.1	[0°/90°]s	896	0.3547
760	400	1000	Max Load	0.1	[0°/90°]s	984	0.3889
760	400	100	Min Load	0.1	[0°/90°]s	531	0.3843
760	400	1000	Min Load	0.1	[0°/90°]s	138	0.3889
760	800	0	N/A	0.1	[0°]4	1294	0.3889
760	800	100	Max Load	0.1	[0°]4	1066	0.3889
760	800	1000	Max Load	0.1	[0°]4	659	0.3889
816	350	0	N/A	0.1	[0°/90°]s	660	0.3547
816	350	10	Max Load	0.1	[0°/90°]s	1650	0.3547
816	350	100	Max Load	0.1	[0°/90°]s	714	0.3889
816	350	1000	Max Load	0.1	[0°/90°]s	725	0.3889
816	350	10000	Max Load	0.1	[0°/90°]s	127	0.3889
816	350	100	Min Load	0.1	[0°/90°]s	429	0.3134
816	350	1000	Min Load	0.1	[0°/90°]s	335	0.3134

RESULTS AND DISCUSSION

Data obtained for a range of hold times from 10 to 10^4 s superimposed on 3 minute (0.00556 Hz) baseline cycles are summarized in Table 1. For computation purposes in modeling the data, it is assumed that the behavior is dominated by stresses in the fibers. The data are obtained from a number of different panels having different fiber volume fractions. In order to account for differences in fiber volume fraction in the various specimens, an effective stress is used in place of the applied stress. All data are normalized with respect to a reference volume fraction, 0.35. The effective stress produces the same fiber stress in the composite as would be obtained in a composite whose fiber volume fraction were 0.35, the average value for this material. From a one-dimensional analysis using a rule of mixtures approach, a relationship between applied stress and fiber stress is easily obtained as follows:

$$\sigma = \sigma_f \left[V_f + \frac{E_m}{E_f} \left(1 - V_f \right) \right] \tag{1}$$

where subscripts f and m refer to fiber and matrix, respectively, E is Young's modulus, and V_f is the volume fraction of fibers. An equivalent stress, σ_{eq}, which produces the same fiber stress as in a same with the reference volume fraction $\overline{V}_f = 0.35$, is determined from the following equations:

$$\sigma_{eq} = C \cdot \sigma \tag{2}$$

$$C = \frac{\overline{V}_f + \frac{E_m}{E_f} \left(1 - \overline{V}_f \right)}{V_f + \frac{E_m}{E_f} \left(1 - V_f \right)} \tag{3}$$

For small variations from a volume fraction of 0.35, it is easily shown that the effective stress from Eqs (2) and (3) can be accurately represented as

$$\sigma_{eff} = \sigma_{appl} \left(\frac{V_f}{0.35} \right)^{0.718} \tag{4}$$

This equation provide a simple method for representing data obtained from a number of specimens having small variations in fiber volume fraction.

In order to represent the hold-time data, a linear summation model developed by Nicholas and Russ [1]to predict life based on cycles with superimposed hold times was used to fit the experimental data:

$$\frac{N}{N_c} + \frac{N t_h}{t_f} = 1 \tag{5}$$

where N is the number of cycles to failure, t_h is the hold time per cycle, N_c is the number of cycles to failure in the absence of hold times, and t_f is the time to failure if a constant stress is applied to the material. For modeling purposes, t_f is taken in the form:

$$t_f = B \, \sigma^n \tag{6}$$

where B and n are empirical constants. To account for difference in behavior at different temperatures, a temperature-dependent constant introduced by Nicholas [18] is utilized:

$$B = \frac{B_1}{\exp\left(\frac{-B_2}{T_{abs}}\right)} \qquad (7)$$

where B_1 and B_2 are empirical constants and T is absolute temperature, $T_{abs} = T°C + 273$. The exponent n in Eq (6) is assumed to be independent of temperature, primarily for the reason of lack of sufficient data to demonstrate otherwise. In this investigation, the term N_c in Eq (5) is taken as the experimental value for cycles to failure with no hold time for each of the conditions modeled. Insufficient data were generated in order to establish a functional form for N_c as a function of temperature for the two material lay-ups studied here. To complete the modeling of hold-time data, a second version of a damage summation model was used. Nicholas and Updegraff [19] found that a non-linear representation of time-dependent damage provided a better representation of data from thermal cycling with hold times. A power law representation of the form

$$\frac{N}{N_c} + \left(\frac{N t_h}{t_f}\right)^{0.5} = 1 \qquad (8)$$

is used in addition to the linear form, Eq (5), to represent the data for cycles with hold times at maximum load in this investigation. The results of the two model representations, together with the data for tests on the [0/90]$_S$ lay-up, are presented in Fig. 1. The two models produce similar trends of decreasing life with increasing hold time, and the non linear model produces a slightly better fit. Considering the scatter in the experimental data, and the limited number of data points, the non linear (NL) model provides a reasonable representation of the data. This can be illustrated by comparing the predicted number of cycles to failure with the experimentally obtained values. Such a comparison is presented in Fig. 2, where the predictions for all of the data with hold times at maximum load for the two lay-ups are based on the NL model. It can be seen that the correlation between

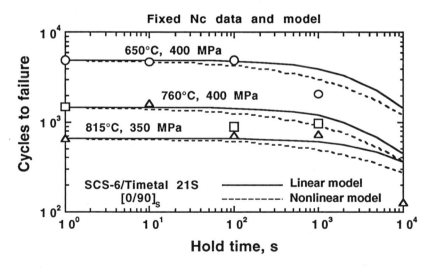

FIG. 1--Comparison of model predictions with experimental data for a fixed value of Nc for [0/90]$_S$ lay-up. Solid line is linear model, dashed line is nonlinear model.

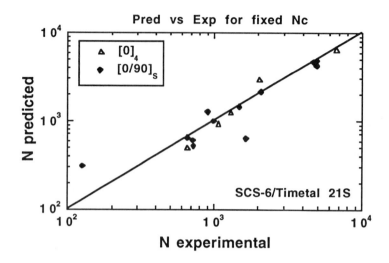

FIG. 2--Predicted life against experimental cycles to failure for a fixed value of N_c.

predictions and experiments is quite good, with perfect correlation represented by the straight line in the figure.

Since most of the test conditions involve hold times of substantial duration, it was felt that it would be instructive to compare the results of this investigation with data obtained under constant load conditions. Results of sustained load "creep" tests on the same material from the investigation of Khobaib, et al. [20] are compared with those of the present investigation in Figs. 3 and 4 for the $[0]_4$ and $[0/90]_S$ lay-ups, respectively. Shown in the plots is the total time at maximum stress to produce failure. For the "creep/fatigue" tests, only the hold time contribution is included. No contribution to the time to failure is attributed to the 3 minute cycles, since none of this time is at maximum stress. For the $[0]_4$ lay-up, Fig. 3, all of the data cluster between approximately 600 and 900 MPa, with no failures outside this range and no discernible trend of failure time as a function of applied stress. In a similar manner, all of the data for the $[0/90]_S$ lay-up seem to cluster in a stress range between approximately 300 and 450 MPa, and no trend with stress is apparent.

The apparent lack of trend of decreasing life with increasing stress would appear to demonstrate that the data are indicative of a statistical process more than a deterministic one. Further, the time-dependent nature of the tests conducted implies that failure occurs only if the applied stresses are within a narrow band and the degradation of strength with continued time of exposure is not a dominant issue. The data seem to represent the equivalent of Region 1 of a Talreja diagram [18] which is governed by the statistical failure of fibers due to fiber stresses which are sufficiently high to cause initiation and propagation of failure of a bundle of fibers. To further explore this hypothesis, the data were treated as being governed solely by a time-dependent mechanism. No contribution of cyclic processes is considered henceforth. To assess the contribution of the cycle to the overall failure process, the incremental damage attributed to time at stress from Eq (6) is integrated over a triangular wave form as in Nicholas and Russ [1], resulting in the following

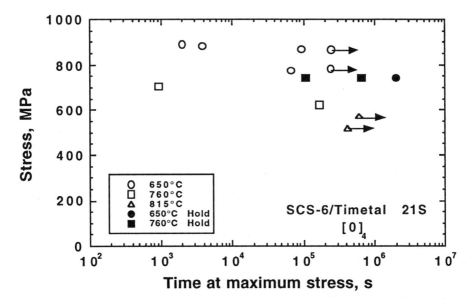

FIG. 3--Time to failure for all creep and hold time tests on [0]4 lay-up.

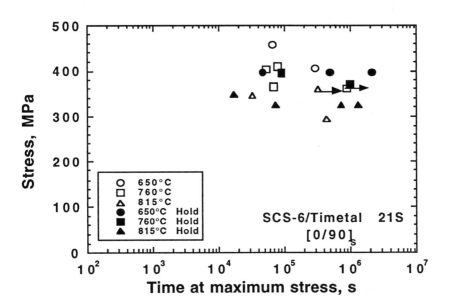

FIG. 4--Time to failure for all creep and hold time tests on [0/90]s lay-up.

equation for cycles to failure due solely to the time-dependent damage accumulation due to time at stress:

$$N_t = \frac{B\,f\,(n+1)\,(1-R)}{\sigma_{max}{}^n\left(1-R^{n+1}\right)} \tag{9}$$

where f is cyclic frequency, R is load ratio (ratio of minimum to maximum load), and B and n are the constants in Eq (6). Since the cycles in this study are at a frequency of 0.00556 Hz, it was thought desirable to include data from 0.01 Hz cycles in modeling the time-dependent behavior. Thus, all the data with hold times at maximum load from the present investigation, creep data, as well as 0.01 Hz data reported in [22] and [8] for $[0]_4$ and $[0/90]_S$ lay-ups, respectively, were used to obtain a best fit of the constants B and n in Eq (6). Note that B is further composed of B_1 and B_2 from Eq (7) to account for temperature dependence. The values for B_2 for both lay-ups were taken from prior modeling studies on the same material [18] and were found, in general, not to have a significant effect on the fitting of the data because the majority of the data were obtained at 650°C. Referring again to Figs. 3 and 4, it would be expected that a very large value of the constant n would provide a best fit to the data, since a large value of n results in an almost horizontal curve. Minimizing the least squares error between actual data and that computed using Eq (8), with N_c represented by Eq (9) for the cyclic term, no convergence could be obtained for either lay-up. In fact, the error between prediction and experiment decreased slightly as n decreased, leading to negative values of n and B_1. A value of n=5 was arbitrarily chosen to illustrate the level of correlation which could be obtained.

A measure of the correlative capability of the model is a magnification factor, M, introduced by Neu and Nicholas [22], of the form

$$M = 10^{\sqrt{\frac{1}{n}\sum_{i=1}^{n}\left(\log N^m_{f\,i} - \log N^{ex}_{f\,i}\right)^2}} \tag{10}$$

where n is the number of experiments, N_f is number of cycles to failure, and superscripts m and ex refer to the model and experiment, respectively. The factor, M, represents the standard deviation on a logarithmic scale about perfect correlation. For example, when M is 3, the standard deviation between the model and experimental lives is a factor of 3.

Using hold time, creep, and 0.01 Hz fatigue data, trial I for the two lay-ups produced the constants summarized in Table 2. To see if the 0.01 Hz fatigue cycles were contributing to the inability of the model to correlate the data, trial II was conducted without the fatigue data. As can be seen from the values in Table 2, the correlation improved slightly for the $[0]_4$ lay-up, but decreased insignificantly for the $[0/90]_S$ lay-up. Overall, the inclusion of 0.01 Hz fatigue data does not seem to change the fit of the purely time-dependent model to the data.

TABLE 2--Summary of constants for creep and hold time data fit.

TRIAL	n	B_1	B_2	R
[0] I	5	6.80 E+11	17260	2.66
[0/90] I	5	3.75 E+13	10306	3.26
[0] II	5	1.82 E+12	17260	1.92
[0/90] II	5	5.39 E+13	10306	3.34

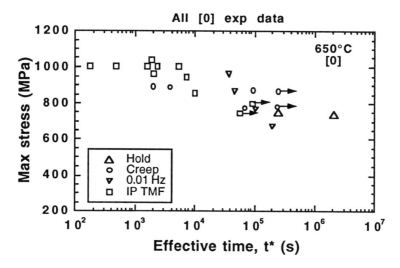

FIG. 5--Stress-effective time plot for all [0]₄ tests at 650°C.

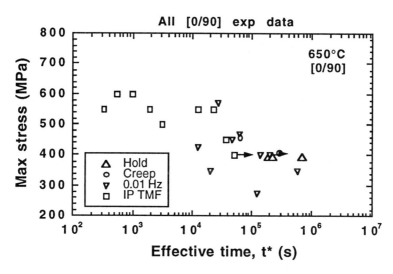

FIG. 6--Stress-effective time plot for all [0/90]ₛ tests at 650°C.

To further explore the time-dependent nature of the data, in-phase thermomechanical fatigue (TMF) data from earlier investigations [5,8,22] were also examined. To compare the time-dependent contributions to failure, an equivalent time, t*, is introduced. The parameter t* represents the equivalent time that a specimen experiences the maximum stress at a given temperature. For a hold time or during a creep test, the equivalent time is identical to the actual time. For a fatigue cycle, however, the maximum stress is reached only instantaneously in a given cycle. At other times, the stress is some fraction of the maximum. The equivalent time is the time the specimen would have to spend at maximum stress to produce the same amount of time-dependent damage over the range of stresses in the cycle. This could be computed from Eqs (6) and (9) if n were known, but as noted above, a best value of n could not be estimated for the data considered here. We arbitrarily choose the equivalent time as 0.2 of the cycle time, which would be produced by a value of approximately n=5. Using the same reasoning, and noting that during an in-phase (IP) TMF cycle the temperature as well as the stress decrease during the cycle, we choose the equivalent time to be 0.05 the cycle time. Using these arbitrary definitions, a plot of maximum applied stress against equivalent time to failure, t^*, can be produced. Such a plot is shown as Fig. 5 for all hold time, creep, 0.01 Hz, and IP TMF data with a maximum temperature of 650°C for the $[0]_4$ lay-up. A similar plot for the $[0/90]_S$ lay-up at 650°C is presented as Fig. 6. The data for the $[0]_4$ lay-up are contained in a band from 700 to 1000 MPa while the data for the $[0/90]_S$ lay-up are between 300 and 600 MPa, approximately. It is rather surprising to note that in these two plots, the data seem to follow the general trend of increasing time to failure with decreasing stress. It must be pointed out, however, that the plots are heavily influenced by the inclusion of TMF data under the hypothesis that the failure mechanism is purely time dependent.

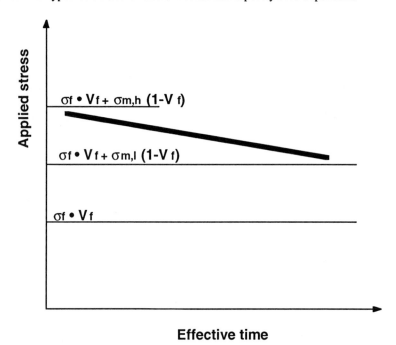

FIG. 7--Stress-effective time schematic for materials exhibiting purely time-dependent behavior.

A possible explanation for the shape of the plots for the two lay-ups at 650°C is proposed with the aid of Fig. 7. The failure process is considered to be no more than the failure of a bundle of fibers once a critical average fiber stress, $\sigma_{f,\,cr}$, is reached. It would seem that this stress is near the lower end of the distribution of fiber strengths for the

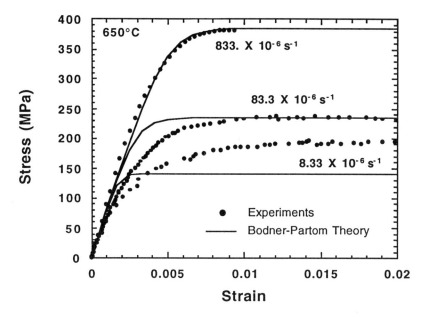

FIG. 8--Stress-strain behavior of Timetal 21S at 650°C and
constitutive model representation of data [27].

SCS-6 fibers such that once some fraction of the fibers fail, the process would lead to
higher stresses and progressive failure of the rest of the bundle. The critical average fiber
stress is denoted by σ_f in the figure, while matrix strength at low and high strain rates is
denoted by $\sigma_{m,l}$ and $\sigma_{m,h}$, respectively. The solid line in Fig. 7 represents the expected
trend in data from specimens whose time to failure is governed by fiber failure. The
reasoning is as follows. If the matrix carries no load whatsoever, then failure would occur
at a stress $\sigma_f \cdot V_f$. If the composite is loaded at a somewhat higher applied stress, then the
matrix stress will relax eventually to the strength typical of that obtained in a low strain rate
test. The lower end of the stress range for failure will then be at longer times at an applied
stress given by $\sigma_f \cdot V_f + \sigma_{m,l} (1-V_f)$. If the applied stress is even higher, then the matrix
stress will relax quickly to a stress level typical of that obtained at a higher strain rate test.
The upper end of the stress range for failure, which produces failure at short times, is given
by $\sigma_f \cdot V_f + \sigma_{m,h} (1-V_f)$. For the Timetal 21S matrix material used in this investigation, the
behavior at several different strain rates ranging from approximately 10^{-5} to 10^{-3} is shown
in Fig. 8. Additionally, at lower strain rates or under long term stress relaxation, flow
stresses under 100 MPa can occur at this temperature. It is of significance, therefore, to
note that the range of failure stresses for both lay-ups is approximately 300 MPa which is
consistent with experimental data which show a difference in flow stress of approximately
500 MPa over several decades in strain rate [23] which, when multiplied by a volume
fraction of matrix of approximately 0.65, produces approximately 300 MPa. From the

lower bound of the stress level at which failures occur, and using a low strain rate strength of 50 MPa, a value of fiber stress of approximately 2000 MPa is estimated to initiate catastrophic failure from data for both lay-ups. This value is below an estimate of fiber bundle strength of about 3000 MPa for SCS-6 fibers by Gambone [24] although assumptions in that estimate fail to include any stress concentration effects from one broken fiber to adjacent fibers. Thus, the difference between the value determined experimentally here and the calculations of Gambone is not that significant.

The observation that fatigue life is controlled by the stress range in the 0° fiber has been reported by Pollock and Johnson for SCS-6/Ti-15-3 [2]. Further, it is of interest to note that in the analysis of the same composite by Mirdamadi et al. [25], the calculated 0° fiber stress range was about 2000 MPa with R=0.07 for the fiber. Although they conclude that 0° fiber stress range controls life, they note that this is "within a given set of parameters (i.e. temperature, loading frequency, and time at temperature)." Other investigations have led to the conclusion that hold time at temperature is a very significant contributor to fatigue life under cycles with superimposed hold times [14] and under mission spectrum loading [15,16].

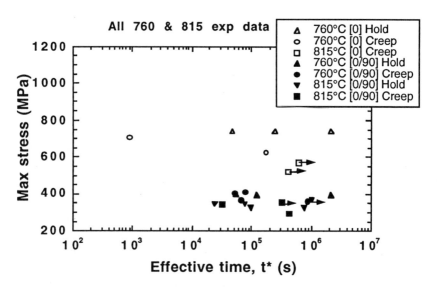

FIG. 9--Stress-effective time plot for all tests at 760°C and 815°C.

The remaining data for t* at 760 and 815°C are plotted against applied stress in Fig. 9. At these two temperatures, there are no 0.01 Hz or TMF data available, thus only hold time and creep data appear. The data are clearly in two narrow bands, one for the [0]₄ lay-up at a stress level of approximately 700 MPa, and the other for [0/90]ₛ at a stress level of approximately 350 MPa. It is significant to note that these stress levels are within the bands of data at 650°C, Figs. 5 and 6. The data at all three temperatures seem to imply that the stress for fiber failure is essentially independent of temperature over the temperature range 650 - 815°C, or that time at stress and temperature is not an important factor in the "time-dependent" failure under sustained load. The main aspect of time dependence would appear to be the time necessary for the matrix stresses to relax so that fiber stresses reach a critical level. This is true not only for sustained loading, but appears to be a significant factor in both low frequency (below 0.01 Hz) isothermal fatigue and IP TMF tests at a maximum temperature of 650°C. This observation, combined with the inherent scatter in the strength of a bundle of fibers typical of region 1 of a Talreja diagram, appears to explain the experimentally observed lives obtained in creep, hold time, 0.01 Hz, and IP TMF tests. It should be noted, however, that the temperature range investigated here is at or above the temperature where this material can be expected to be used for any extended period of time in practical structural applications [15,16]. The dominance of time at temperature on fatigue life may not hold for temperatures below 650°C or for lower stress levels where creep is less dominant and fatigue lives are longer.

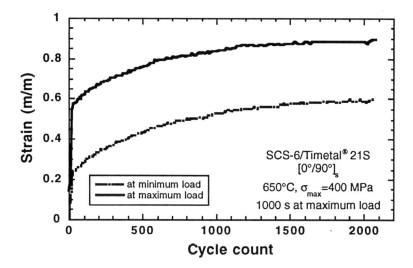

FIG. 10--Maximum and minimum strain history during 0.0056 Hz cycling of [0/90]$_S$ composite at 650°C with superimposed hold time of 1000 s at maximum, 400 MPa, load.

Another comparison of results from this investigation with behavior under in-phase TMF can be made by examining the accumulation of strain with increasing number of cycles. A typical plot of maximum and minimum strain is presented in Fig. 10 for the case of a [0/90]$_S$ lay-up at 650°C with maximum stress of 400 MPa and a hold time of 1000 s at maximum load. It can be seen that both maximum and minimum strain increase continuously, indicating that stiffness of the specimen does not decrease as would occur under extensive matrix cracking. Such behavior has been noted under in-phase TMF by Russ et al. [4] and is attributed to a combination of stress relaxation in the matrix as well as possible fiber breakage. Under either condition, composite strain accumulates without substantial reduction in stiffness.

HOLD TIMES AT MINIMUM LOAD

Only four tests were conducted with hold times at minimum load, to compare with equivalent tests with hold times at maximum load. Referring to the tabulated results in Table 1, it is clear that in all four cases the lives were significantly less with holds at minimum load than those obtained when hold times were at maximum load. The results can be seen in Fig. 11 where cycles to failure are plotted against total cycle time. Clearly,

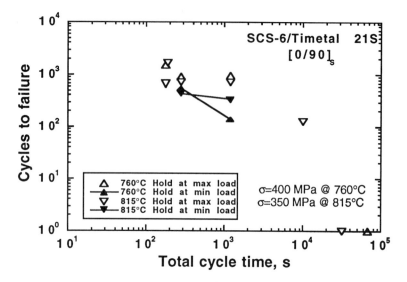

FIG. 11--Comparison of cycles to failure from tests with hold times
at minimum or maximum load.

the prior argument regarding matrix stress relaxation does not explain these results as illustrated schematically in Fig. 12. Hold times at maximum load will produce higher fiber stresses and, hence, shorter lives, contrary to the experimental results. What the schematic of Fig. 12 shows is that the matrix stress range is higher and, therefore, should produce matrix fatigue failure which, in turn, transfers the load to the fibers. Examination of the fracture surfaces of two specimens tested at 760°C with 100 s hold times, one with hold at maximum load and one with hold at minimum load, reveals some interesting differences. The two surfaces, shown in Fig. 13, show that a fatigue crack area seemed to precipitate the failure in the specimen with hold at minimum load, Fig. 13(a), while no such fatigue area was present in the specimen with hold at maximum load, Fig. 13(b). In the latter case, ductile rupture appears to have occurred everywhere in the matrix, thereby indicating that fiber failure was the dominant event. The concept that matrix cracking precipitated the failure in the test with hold at minimum load is further enhanced by observations of the surfaces of the specimens at locations away from the fracture plane. A typical crack on the surface of the specimen tested with hold at minimum load is shown in Fig. 14. While cracks were found in specimens tested with holds at both maximum and minimum load, the density of cracks at minimum load appeared to be somewhat higher than in the maximum load case.

To further evaluate the hypothesis that stress relaxation leads to higher matrix stresses in the minimum load case, a series of calculations was carried out using a uniaxial stress code described by Coker et al. [26]. The computations are based on the assumptions of thermoelastic fiber behavior and a matrix which is thermoviscoplastic. The ability of the constitutive model to represent the matrix behavior at 650°C is demonstrated in Fig. 8. The cases which were simulated involved 1000 s hold times at either minimum or maximum load at temperatures of 650, 760, and 815°C. Since the constitutive model for the matrix is based primarily on data obtained at temperatures of 650°C and lower, the ability of the

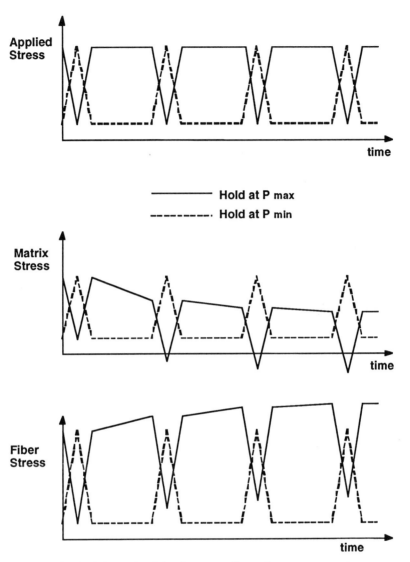

FIG. 12--Schematic of development of constituent stresses when
hold times are at minimum or maximum load.

model to represent realistic behavior at the higher temperatures is questionable. For that reason, the computations carried out at 650°C are presented, even though no hold time tests were performed at this temperature (see Fig. 11). The results, which compare the fiber and matrix stresses over 10 complete cycles, are presented in Fig. 15. It is clear that for this particular case the trends in peak stresses are the same as those depicted in the schematic of Fig. 12. One significant difference between the computed behavior and the schematic is the large amount of stress relaxation when the hold time at maximum load is applied. Other

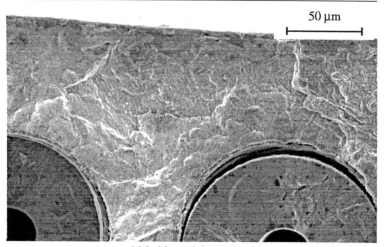

(a) hold at minimum load

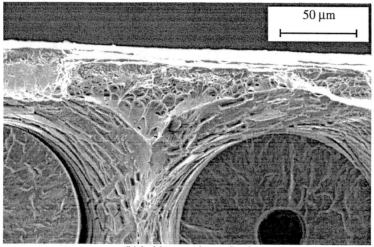

(b) hold at maximum load

FIG. 13--SEM photos of fracture surface of specimen tested at 760°C with 100 s hold.

than that, the trends are as depicted schematically in Fig. 12. The peak matrix stress, which relaxes to an almost steady state value after the first cycle, is seen to be noticeably higher in the case where hold is at minimum load, even though the peak value is attained only at the peak load of the fatigue cycle. After 10 cycles, the matrix stress is 153 MPa when the hold is at minimum load compared to 93 MPa when hold is at maximum load. Fiber stress, on the other hand, is higher when hold is at maximum load. Since the constitutive behavior of the matrix at 760°C is felt to be similar to that at 650°C, these simulations should represent the experimental conditions at 760°C in Fig. 11 reasonably well. It is not surprising, therefore, to observe matrix cracking in the minimum load hold

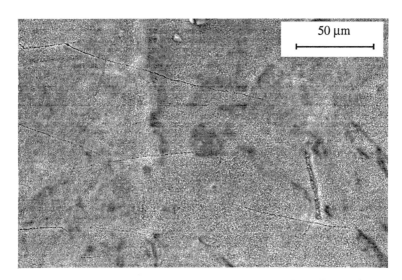

FIG. 14--SEM photograph of typical crack on the surface of a specimen
tested at 760°C with 100 s hold at mimimum load.

test and a fiber dominated failure in the test with hold at maximum load. Although the
simulations at the two higher temperatures are felt to be poorly representative of the matrix
behavior, a very soft matrix at higher temperatures produced results which showed
essentially no difference in behavior in fiber and matrix stresses depending on whether the
hold time was at maximum or minimum load. Thus, the lower the temperature, the more
sensitive the behavior is to the hold being at maximum or minimum load. The experimental
results of Fig. 11 are consistent with this speculation because they show a larger
discrepancy at 760°C than at 815°C. While the number of data points is admittedly small,
the trends in the data, the results of numerical simulations to determine constituent stresses,
and the fractography and metallography all present a consistent picture which leads to the
conclusion that hold times at minimum load can be more detrimental to fatigue life than
equivalent hold at maximum load, and the differences are greater at lower temperatures.
These results were obtained at relatively high stress levels, so the conclusions may not be
equally valid at lower stress levels.

SUMMARY AND CONCLUSIONS

The behavior of SCS-6/Timetal 21S in $[0]_4$ and $[0/90]_S$ lay-ups under creep/fatigue
conditions from 650 to 815°C is essentially purely time dependent when the cyclic
frequency is 5.6 x 10^{-3} Hz. A damage summation model which treats cycle- and time-
dependent mechanisms separately is able to consolidate data from tests with cycles with
superimposed hold times at maximum load. The cyclic contribution, however, is found to
really be a result of time-dependent phenomena. Treating the entire process as time
dependent, and introducing the concept that time to failure is related to applied stress for
each of the material lay-ups, it is shown that creep data and creep/fatigue data constitute a
single population of failure times as a function of maximum applied stress. Further,
introducing an effective time-dependent contribution for low frequency isothermal fatigue
and in-phase TMF tests, these additional data are consolidated within the scatter band of the
creep and creep/fatigue data. It is proposed that the mechanism for failure for all these test
types is the failure of a bundle of fibers when a critical fiber stress level is reached. The

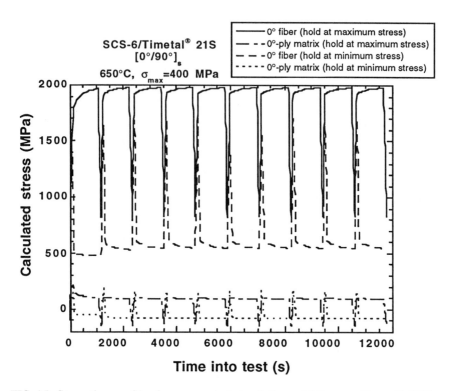

FIG. 15--Stress-time profiles from numerical simulations of fatigue cycling with 1000 s
hold times at 650°C: (a) hold at minimum load, (b) hold at maximum load.

fiber stress is influenced by the strength of the matrix which is sensitive to strain rate and
exhibits stress relaxation. It is for these reasons that the data appear to be scattered over a
narrow range of stresses and over a broad range of failure times. The process, however, is
similar to that represented in region 1 of a Talreja diagram. It is to be pointed out that the
speculated behavior indicates that under certain conditions an in-phase TMF test is not a
fatigue test at all but, rather, a complex creep test conducted under varying load and
temperature. It could also be considered a very inefficient way to test a material because of
the minimal amount of time spent per cycle at maximum load and maximum temperature
compared to a creep test, given that the only contributing mechanism is stress relaxation in
the matrix which ultimately sheds load to the fibers.

Hold times at minimum load appear to be more detrimental to fatigue life than when
the hold times are applied at maximum load under otherwise identical conditions. The
detriment in fatigue life is greater at 760°C than at 815°C. The experimental data presented
here on cycles to failure, the fractographic and metallographic evidence showing more
cracking when hold is at minimum load, and the results of micromechanical stress
calculations which show higher matrix stresses for hold at minimum load all indicate that
the mechanism of failure is different when the hold time is at minimum load. Here, rather
than a time-dependent mechanism resulting in fiber failure after matrix stress relaxation, the
mechanism appears to involve matrix cracking due to higher matrix stresses than in the

other conditions explored. In order to model this failure mode, the matrix stresses would have to be taken into account.

ACKNOWLEDGMENTS

This research was conducted at the U.S. Air Force Wright Laboratory Materials Directorate and was supported through in-house project 2302BW and Contract F33615-94-C-5200 with University of Dayton.

REFERENCES

[1]   Nicholas, T. and Russ, S.M., "Elevated Temperature Fatigue Behavior of SCS-6/Ti-24Al-11Nb," Materials Science and Engineering, Vol. A153, 1992, pp. 514-519.

[2]   Pollock, W.D. and Johnson, W.S., "Characterization of Unnotched SCS-6/Ti-15-3 Metal Matrix Composites at 650°C," Composite Materials: Testing and Design (Tenth Volume), ASTM STP 1120, Glenn C. Grimes, Ed., American Society for Testing and Materials, Philadelphia, 1992, pp. 175-191.

[3]   Castelli, M.G., Bartolotta, P.A. and Ellis, J.R., "Thermomechanical Fatigue Behavior of Sic(SCS-6)/Ti-15-3," Composite Materials: Testing and Design (Tenth Volume), ASTM STP 1120, Glenn C. Grimes, Ed., American Society for Testing and Materials, Philadelphia, 1992, pp. 70-86.

[4]   Russ, S.M., Nicholas, T., Hanson, D.G. and Mall, S., "Isothermal and Thermomechanical Fatigue of Cross-Ply SCS-6/β21-S," Science and Engineering of Composite Materials, Vol. 3, 1994, pp. 177-189.

[5]   Neu, R.W., Coker, D., and Nicholas, T., "Cyclic Behavior of Unidirectional and Cross-Ply Titanium Matrix Composites," International Journal of Plasticity, 1995 (in press).

[6]   Castelli, M.G., "Characterization of Damage Progression in SCS-6/Timetal 21S [0]4 Under Thermomechanical Fatigue Loading," Life Prediction Methodology for Titanium Matrix Composites, ASTM STP 1253, W. S. Johnson, J.M. Larsen and B.N. Cox, Eds., American Society for Testing and Materials, Philadelphia, 1995.

[7]   Castelli, M.G., "Thermomechanical Fatigue Damage/Failure Mechanisms in SCS-6/Timetal 21S [0/90]s Composite," Composites Engineering, Vol. 4, No. 9, 1994, pp. 931-946.

[8]   Nicholas, T., Russ, S.M., Neu, R.W. and Schehl, N., "Life Prediction of a [0/90] Metal Matrix Composite Under Isothermal and Thermomechanical Fatigue," Life Prediction Methodology for Titanium Matrix Composites, ASTM STP 1253, W. S. Johnson, J.M. Larsen and B.N. Cox, Eds., American Society for Testing and Materials, Philadelphia, 1995.

[9]   Neu, R.W. and Nicholas, T., "Effect of Laminate Orientation on the Thermomechanical Fatigue Behavior of a Titanium Matrix Composite," Journal of Composites Technology & Research, JCTRER, Vol. 16, 1994, pp. 214-224.

[10]   Mirdamadi, M. and Johnson, W.S., "Hypersonic Flight Simulation Testing and Analysis of SCS-6/Timetal 21S," Titanium Metal Matrix Composites II, P.R. Smith and W.C. Revelos, Eds., WL-TR-93-4105, Wright-Patterson AFB, OH, 1993, pp. 361-381.

[11]   Revelos, W.C. and Roman, I., "Laminate Orientation and Thickness Effects on an SCS-6/Ti-24Al-11Nb Composite Under Thermal Fatigue," Intermetallic Composites II, D.B. Miracle, J.A. Graves, and D.L. Anton, eds., Materials Research Society, Pittsburgh, PA, 1992, pp. 53-58.

[12]   Revelos, W.C., "The Thermal Fatigue Response of an SCS-6/Ti-15Mo-3Al-2.6Nb-0.2Si (wt.%) Metal Matrix Composite," FATIGUE 93, Volume II, J.-P. Bailon and J.I. Dickson, Eds., EMAS, 1993, pp. 963-968.

[13]   Revelos, W.C., Jones, J.W. and Dolley, E.J., "Thermal Fatigue of a SiC/Ti-15Mo-2.7Nb-3Al-0.2Si Composite. Metallurgical Transactions, 1995 (in press).

[14]   Gabb, T.P. and Gayda, J., "Isothermal and Nonisothermal Fatigue Damage/Failure Mechanisms in SiC/Ti-14Al-21Nb," Titanium Matrix Composites, P.R. Smith and W.C. Revelos, Eds., WL-TR-92-4035, Wright-Patterson AFB, OH, 1992, pp. 292-305.

[15]   Mirdamadi, M. and Johnson, W.S., "Fatigue of [0/90]$_{2S}$ SCS-6/ Ti-15-3 Composite Under Generic Hypersonic Vehicle Flight Simulation," FATIGUE 93, Volume II, J.-P. Bailon and J.I. Dickson, Eds., Chameleon Press, London, 1993, pp. 951-956.

[16]   Johnson, W.S., Mirdamadi, M. and Bakuckas, J.G., "Damage Accumulation in Titanium Matrix Composites Under Generic Hypersonic Vehicle Flight Simulation and Sustained Loads," Thermo-Mechanical Fatigue Behavior of Materials, ASTM STP 1263, M.J. Verrilli and M.G. Castelli, Eds., American Society for Testing and Materials, Philadelphia, 1995.

[17]   Hartman, G.A., and Buchanan, D.J., "Methodologies for Thermal and Mechanical Testing of TMC Materials," AGARD Report 796, Bordeaux, France, 1993.

[18]   Nicholas, T., "An Approach to Fatigue Life Modeling in Titanium Matrix Composites," Materials Science and Engineering, 1995, (in press).

[19]   Nicholas, T. and Updegraff, J.J., "Modeling Thermal Fatigue Damage in Metal Matrix Composites," Composites Engineering, Vol. 4, 1994, pp. 775-785.

[20]   Khobaib, M., John, R., Ashbaugh, N.E. and Larsen, J.M., "Sustained Load Behavior of SCS-6/TIMETAL®21S Composite," Life Prediction Methodology for Titanium Matrix Composites, ASTM STP 1253, W. S. Johnson, J.M. Larsen and B.N. Cox, Eds., American Society for Testing and Materials, Philadelphia, 1995.

[21]   Talreja, R., "Fatigue of Composite Materials: Damage Mechanisms and Fatigue Life Diagrams," Proceedings of the Royal Society of London, Vol. A378, 1981, pp. 461-475.

[22]   Neu, R.W. and Nicholas, T., "Methodologies for Predicting the Thermomechanical Fatigue Life of Unidirectional Metal Matrix Composites," Third Symposium on Advances in Fatigue Lifetime Predictive Techniques, ASTM STP, Montreal, May 1994

[23]   Kroupa, J.L., Neu, R.W., Nicholas, T., Coker, D., Robertson, D.D., and Mall, S., "A Comparison of Analysis Tools for Predicting the Inelastic Cyclic Response of Cross-Ply Titanium Matrix Composites," Life Prediction Methodology for Titanium Matrix Composites, ASTM STP 1253, W. S. Johnson, J.M. Larsen and B.N. Cox, Eds., American Society for Testing and Materials, Philadelphia, 1995.

[24]   Gambone, M.L. "The Fiber Strength Distribution and Its Effect on the Creep Behavior of SiC/Ti-1100 Composite," PhD Dissertation, University of Virginia, 1994.

[25]   Mirdamadi, M., Johnson, W.S., Bahei-El-Din, Y.A. and Castelli, M.G., Analysis of Thermomechanical Fatigue of Unidirectional Titanium Metal Matrix Composites," Composite Materials: Fatigue and Fracture, Fourth Volume, ASTM STP 1156, W.W. Stinchcomb and N.E. Ashbaugh, Eds., American Society for Testing and Materials, Philadelphia, 1993, pp. 591-607.

[26]   Coker, D., Nicholas, T. and Neu, R.W., "Analysis of the Thermoviscoplastic Behavior of [0/90] SCS-6/Timetal®21S Composites," Thermo-Mechanical Fatigue Behavior of Materials, ASTM STP 1263, M.J. Verrilli and M.G. Castelli, Eds., American Society for Testing and Materials, Philadelphia, 199#.

[27]   Kroupa, J.L., University of Dayton Research Institute, unpublished data.

Drew Blatt,[1] Theodore Nicholas,[2] and Alten F. Grandt, Jr.[3]

MODELING THE CRACK GROWTH RATES OF A TITANIUM MATRIX COMPOSITE UNDER THERMOMECHANICAL FATIGUE

REFERENCE: Blatt, D., Nicholas, T., and Grandt, Jr., A. F., **"Modeling the Crack Growth Rates of a Titanium Matrix Composite Under Thermomechanical Fatigue,"** Thermomechanical Fatigue Behavior of Materials: Second Volume, ASTM STP 1263, Michael J. Verrilli and Michael G. Castelli, Eds., American Society for Testing and Materials, 1996.

ABSTRACT: The crack growth characteristics of a 4-ply, unidirectional, titanium matrix composite, SCS-6/Ti-6Al-2Sn-4Zr-2Mo, subjected to thermomechanical fatigue were investigated. A linear summation model was developed to predict the isothermal and thermomechanical fatigue (TMF) crack growth rates of the composite. The linear summation approach assumes the total fatigue crack growth rate is a combination of a cycle-dependent and a time-dependent component. To assist the modeling effort, a series of isothermal, in-phase, and out-of-phase crack growth tests were conducted. The test temperatures ranged from 150°C to 538°C and the fastest thermal frequency was 0.0083 Hz. With the exception of the 150°C isothermal test, the model was able to correlate all the baseline fatigue crack growth test data between ΔK of 50 to 90 MPa$\sqrt{m}$. In addition, the model was able to predict the fatigue crack growth rate of a proof test which involved a continual change in temperature range and load range to produce a constant crack growth rate. The proof test began under isothermal conditions at the maximum temperature and ended under in-phase TMF conditions.

KEYWORDS: thermomechanical fatigue, fatigue crack growth, linear summation model, time-dependent, cycle-dependent, SCS-6/Ti-6Al-2Sn-4Zr-2Mo

INTRODUCTION

The behavior of metal matrix composites (MMC) over a wide range of thermal and mechanical load histories must be thoroughly understood before MMC can be confidently used by designers and engineers. The damage tolerant design concept used for military

[1] Materials research engineer, Materials Directorate, Wright Laboratory, Wright-Patterson AFB, OH 45433-7817.

[2] Senior scientist, Materials Directorate, Wright Laboratory, Wright-Patterson AFB, OH 45433-7817.

[3] Professor, School of Aeronautics and Astronautics, Purdue University, West Lafayette, IN 47907-1282.

flight vehicles requires an accurate prediction of crack growth to determine service life and inspection intervals on both airframe and engine structural components [1], for both monolithic and composite materials. To date, however, a large portion of the MMC fatigue test data base has been generated in smooth bar fatigue using a stress-life methodology. The stress-life methodology yields useful information about the fatigue life of a given composite system, but produces limited data about the fatigue crack growth rate behavior in the composite.

The crack growth rate data that are available for metal matrix composites and more specifically titanium matrix composites (TMC) were generated at either room temperature[2] or elevated temperature isothermal conditions[3-5]. There is, however, a need to determine fatigue crack growth characteristics of titanium matrix composites under thermomechanical loading. Thermomechanical fatigue (TMF) is common in engine applications and in some advanced supersonic airframe designs; severe temperature excursions coupled with cyclic mechanical loading are experienced during service. In addition to TMF crack growth rate data, models to predict crack growth rates in TMC under various load and thermal histories are needed.

This paper describes a model developed to predict the thermomechanical fatigue crack growth rates in a titanium matrix composite. The model is based on a linear summation approach and assumes no fiber bridging in the composite. An experimental program to guide and validate the modeling effort was implemented using a 4-ply, unidirectional titanium matrix composite, SCS-6/Ti-6Al-2Sn-4Zr-2Mo as the test material. The results of the model predictions and the TMF crack growth tests are presented.

LINEAR SUMMATION MODELING

The study of thermomechanical fatigue crack growth has generally been limited to monolithic materials, and the authors are unaware of TMF crack growth results for metal matrix composites in the open literature. The following discussion describes the linear summation models used to predict TMF crack growth in monolithic nickel-base superalloys and titanium-aluminide alloys. The predictive model developed for this study is based on the concepts described below, assuming that the failure mechanisms in TMC involve both cycle-dependent and time-dependent phenomenon.

The linear summation approach represents total fatigue crack growth per cycle as a combination of cycle-dependent and time-dependent phenomena. The Heil-Nicholas-Haritos Model (HNH) [1, 6] proposed a linear cumulative-damage model for TMF crack growth rates in Inconel 718 based on isothermal crack growth data in the following way:

$$\left.\frac{da}{dN}\right|_{total} = \left.\frac{da}{dN}\right|_{cycle-dependent} + \left.\frac{da}{dN}\right|_{time-dependent} \tag{1}$$

where a and N are crack length and number of cycles, respectively.

The cycle-dependent contribution in Eq 1 is determined from low temperature and/or high frequency crack growth test data, and is assumed to be independent of temperature and frequency. The time-dependent term, however, is assumed to be affected by temperature and frequency based on the experimental results and model presented by Nicholas, et al [7] who investigated the creep/fatigue interactions in Inconel 718. Initially, the time-dependent term in Eq 1 was calculated by integrating the sustained-load crack growth rate over only the loading portion of the cycle. However, to correctly predict the crack growth rates for any phase angle, the numerical integration of the da/dt term (part of the time-dependent component in Eq 1) was terminated when the sustained-load data began to decrease. Phase angle represents a time shift between the cyclic load maximum and the cyclic temperature maximum.

Mall, et al [8] proposed a similar linear summation model to predict crack growth in a titanium-aluminide alloy under sustained load and sustained load with superimposed fatigue cycles at 700°C, 750°C, and 800°C. In the study, the total crack growth rate of Ti-24Al-11Nb was successfully modeled based on a summation of cyclic damage due to fatigue cycles and due to creep (sustained load). The total crack growth rate, written in terms of crack extension per unit time, was given by

$$\left.\frac{da}{dt}\right|_{total} = \frac{F_1(T_H)}{6T}\left.\frac{da}{dt}\right|_{fatigue} + \frac{F_2(T_H)T_H}{T}\left.\frac{da}{dt}\right|_{creep} \tag{2}$$

where T is total cycle time, T_H is the duration of the hold time, $F_1(T_H)$ is the correction factor to account for retardation of fatigue crack growth which develops during the hold time and $F_2(T_H)$ is the correction factor to account for acceleration of the creep crack growth due to crack sharpening by the fatigue cycle. Unlike the observations made in nickel-base superalloys [1], a hold time in titanium-aluminides appears to retard the fatigue contribution of the creep-fatigue cycle, and pure mechanical fatigue tends to accelerate the contribution of the creep portion of the cycle.

In a follow-on study, Nicholas and Mall [9] attempted to extend the model presented in [8]. Their model represented three components: purely cyclic fatigue; environmentally enhanced fatigue; and sustained load growth. They also included an interactive term as Mall, et al [8] did. Thus, their model looked like the following:

$$\left.\frac{da}{dN}\right|_{total} = F_1(T_H)\left.\frac{da}{dN}\right|_{fatigue} + F_2(T_H)T^{\gamma-1}\left[\int_{cycle}\frac{da}{dt}dt + T_H\frac{da}{dt}\right] \tag{3}$$

Here γ is an empirical constant and $F_1(T_H)$ and $F_2(T_H)$ are correction factors similar to those in Eq 2. The model captures several features of fatigue crack growth in titanium-aluminide alloys. For instance, hold times tend to blunt the crack tip, reducing the crack growth rate, while environmental exposure over time accelerates crack growth. A unique feature of this model is its ability to predict a decrease in growth rate due to the addition of hold times. This is not the case with other linear summation models no matter what correlating parameter is used.

The most recent model to predict fatigue crack growth behavior under elevated temperature as well as TMF conditions in Ti-24Al-11Nb was by Pernot [10] and Pernot, et al [11]. While other linear summation models [8, 9] use retardation coefficients to account for crack-tip blunting, these retardation coefficients are limited to a particular temperature. The model proposed by Pernot [10] incorporates retardation coefficients that vary continuously during a TMF cycle, resulting in effective cycle-dependent and time-dependent contributions. This model has the form:

$$\left.\frac{da}{dN}\right|_{tot} = \int_0^{t_{ul}} \beta(t)\left(\frac{2}{\tau}\right)\left(\frac{da}{dN}\right)_{ur\ cd}dt + \int_0^{t_{nd}} \beta(t)\left(\frac{da}{dt}\right)_{ur\ td}dt \tag{4}$$

where t_{ul} equals the uploading time, t_{nd} equals the sum of t_{ul} and t_{hld} (t_{hld}=hold time), $\beta(t)$ is the retardation coefficient, τ is the cycle period, $(da/dN)_{ur\ cd}$ is the unretarded cycle-dependent crack growth rate and $(da/dt)_{ur\ td}$ is the unretarded time-dependent crack growth. The $d\beta/dt$ term is expressed as the combination of an increasing (crack-tip sharpening) term and a decreasing (crack-tip blunting) term. Predictions over a wide range of test conditions correlated test data well. Test conditions included isothermal crack growth (T = 649°C) for 0.01, 0.1 and 5.0 Hz, isothermal crack growth at 315°C, 482°C, and 649°C at 0.01 Hz, in-phase and out-of-phase TMF tests at 0.01 Hz, and tests with hold times at various times in the cycle.

The linear summation model developed for this study employs a methodology similar to the linear summation models described above. A complete description of the newly proposed model is given in a later section.

EXPERIMENTAL METHODS

Material

The material used in this study was a 4-ply, unidirectional titanium based metal matrix composite reinforced with silicon-carbide fibers. The nominal composition (weight %) of the matrix alloy was Ti-6Al-2Sn-4Zr-2Mo-0.1Si (Ti-6242). The fiber volume fraction of the composite was approximately 36%. The material was manufactured by the induction plasma spray technique and was tested in the as-fabricated condition. The fabrication quality of each composite panel was evaluated using radiography, reflector plate ultrasonic scanning, as well as metallography. A complete description of the pedigree of the SCS-6/Ti-6242 composite is found in Ref [12].

Specimen Description

Both single-edge notched, SE(T), and center notched, M(T), geometries were used for the crack growth tests. Each had nominal dimensions of 1 by 25 by 136 mm. Electric discharge machining was used to cut the edge notch (a/W=0.3) in the SE(T) and center notch (2a/W=0.25) in the M(T) specimens perpendicular to the fiber and loading direction. Approximately 18 mm of each specimen end was gripped with a rigid grip system yielding an unclamped height-to-width ratio, H/W, of 4.0. It should be noted that the height-to-width ratio of a rigidly clamped single-edge notched specimen significantly affects the stress intensity factor solution[13]. The stress intensity factor and compliance solutions for the SE(T) specimens with clamped ends used in this study are described in detail in Ref [14]. The rigid grip constraint does not affect the stress intensity factor solution for the M(T) geometry for height-to-width ratios greater than 1.0 [15].

Test Apparatus

A 22 kN closed-loop servo-hydraulic test frame applied the mechanical loads. The load train was oriented horizontally, not vertically, as conventional fatigue test frames commonly are. The horizontal load frame has two primary advantages over the vertical frame for TMF tests: 1) the quartz rod high-temperature extensometer is more readily mounted in the horizontal orientation, and 2) there is little chimney effect from the heating lamps that vertical systems experience. A four-zone, quartz-lamp, radiant heating system coupled with forced air jets for "power-on" cooling provided a controlled and uniform temperature profile over the specimen. Crack length was monitored by optical inspection of both sides of the specimen as well as direct-current electric potential (DCEP). The crack length measured by DCEP typically correlated well with the optically measured crack length for both the isothermal and TMF conditions and both geometries as illustrated in Figs. 1 and 2. A triangular waveform for both thermal and mechanical cycling was used for all fatigue crack growth tests with no hold times. In addition, some tests were run under isothermal conditions. To automate the test procedure, a micro-computer and the control

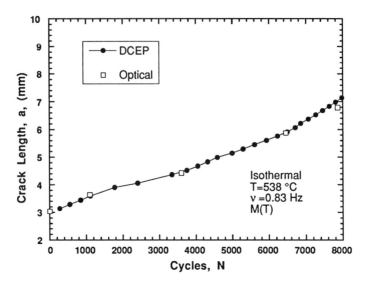

FIG. 1--Crack length from DCEP and optical measurements as a function of applied
cycles.

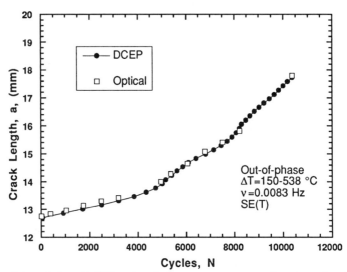

FIG. 2--Crack length from DCEP and optical measurements as a function of applied
cycles.

software MATE [16] were used in conjunction with the servo-hydraulic test frame electronics and lamp controllers. A complete description of the automated TMF test system is given in Ref [17]. All fatigue crack growth tests were performed under constant maximum load, P_{max}, conditions and at a load ratio of 0.1. The fastest cycle time achieved for the TMF tests was 120 seconds or 0.0083 Hz for a temperature range of 150°C to 538°C. The load levels and corresponding ΔK's were chosen based on previous experiences on crack growth testing in similar materials at the low frequencies used in this study.

EXPERIMENTAL RESULTS AND DISCUSSION

Seven baseline isothermal and TMF tests were conducted initially. The test matrix describing the conditions tested is given in Table 1. A complete description of the test results is found in Ref [12].

TABLE 1--Summary of baseline crack growth test conditions for the SCS-6/Ti-6Al-2Sn-4Zr-2Mo.

Test Type	Temperature (°C)	σ_{max} (MPa)	Stress Ratio	Period (sec/cyc)
Isothermal	538	348	0.1	120
Isothermal	538	560	0.1	1.20
Isothermal	150	560	0.1	120
In-phase	150-538	348	0.1	120
In-phase	150-538	560	0.1	1200
Out-of-phase	150-538	348	0.1	120
Out-of-phase	150-538	560	0.1	1200

Isothermal Fatigue Crack Growth

Three baseline isothermal tests were conducted. Isothermal data were generated at T = 538°C and T = 150°C, both at 0.0083 Hz, and T = 538°C at 0.83 Hz as shown in Fig. 3. From these data the influences of temperature and frequency on fatigue crack growth rates were illustrated. The crack growth rate data for the isothermal test at T = 538°C and 0.0083 Hz are approximately 2 orders of magnitude higher than the rates generated from the 150°C test. This behavior is expected since the environmental (time-dependent) contribution to the total crack growth rate for the 150°C test is almost negligible compared to the test at 538°C. Moreover, the time dependent contribution to the crack growth rate at 150°C is about two orders of magnitude less than the cycle dependent contribution. The 150°C crack growth data also suggest that some fiber bridging occurred throughout the test. The crack growth rate remained relatively constant although the applied stress intensity factor range increased; this behavior is commonly observed in a composite in which fiber bridging is a dominate mechanism. Similar crack growth behavior was reported by Larsen, et al [4] who observed full-scale fiber bridging at room temperature in SCS-6/Timetal 21S.

The 538°C, 0.83 Hz isothermal crack growth rates were about one order of magnitude lower than the 538°C rates generated at 0.0083 Hz. This test gave insight into how the isothermal crack growth rate changed as a function of frequency. Typically, isothermal fatigue crack growth rates decrease as frequency increases [10, 18-21]. In fact, for truly time-dependent behavior, an order of magnitude increase in frequency results in an order of magnitude decrease in crack growth rate. Note that higher frequencies elapse less time per cycle and, therefore, less time per cycle is available for the environment to contribute to the overall fatigue crack growth rate. The data suggest, however, that the crack growth rates are not fully time dependent. The frequency was increased by a factor of 100, but the crack growth rates only differ by about a factor of 10.

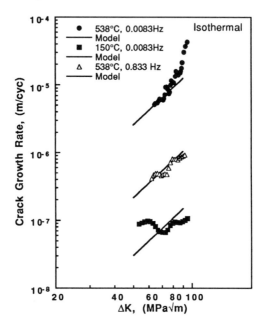

FIG. 3--Correlation between model and SCS-6/Ti-6Al-2Sn-4Zr-2Mo fatigue crack growth data from the isothermal tests at 150°C and 538°C.

Thermomechanical Fatigue Crack Growth Data

Four baseline TMF crack growth tests were run to observe the general behavior of the composite and to verify the TMF crack growth model. All TMF tests were cycled between 150°C and 538°C at frequencies of 0.0083 and 0.00083 Hz. The phase angles were limited to 0° and 180°, or in-phase and out-of-phase, respectively. The in-phase and out-of-phase data are given in Figs. 4 and 5, respectively.

The crack growth rates from the in-phase and out-of-phase tests at 0.0083 Hz were nearly identical. TMF data of monolithic Ti-24Al-11Nb [10] and Inconel 718 [6] showed that typically the in-phase fatigue crack growth rates were higher than the out-of-phase rates. In contrast, the 0.00083 Hz in-phase and out-of-phase growth rates were noticeably different. This suggests that the in-phase condition is more frequency dependent than the out-of-phase. Neu saw a similar response in the TMF behavior of unnotched SCS-6/TIMETAL®21S [22]. Micromechanical modeling of the fiber and matrix stresses during in-phase and out-of-phase conditions revealed that the fiber stresses in the in-phase condition rose more rapidly with cycle period than the out-of-phase condition. During in-

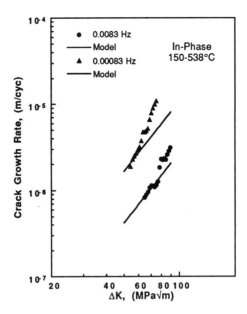

FIG. 4--Correlation between model and SCS-6/Ti-6Al-2Sn-4Zr-2Mo fatigue crack growth data from the in-phase tests at 0.0083 Hz and 0.00083 Hz.

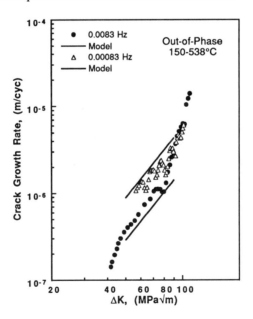

FIG. 5--Correlation between model and SCS-6/Ti-6Al-2Sn-4Zr-2Mo fatigue crack growth data from the out-of-phase tests at 0.0083 Hz and 0.00083 Hz.

phase loading over longer cycle periods, the matrix relaxes and the load is transferred to the fiber. Over shorter cycle periods the load transferred to the fiber from the matrix is lower.

Correlation of Growth Rates with ΔK_{app}

The crack growth rate data generated by both the isothermal and TMF tests as well as high magnification optical microscopy indicated that large scale fiber bridging did not occur. While not explicitly determined, visual inspection of crack opening displacements during each crack growth tests suggested fiber bridging was limited to 2-3 fibers near the advancing crack tip under all test conditions, except the isothermal test at 150 °C. Since large scale bridging was not present, the crack growth rates were correlated with the applied stress intensity factor range instead of an effective stress intensity factor range, where $\Delta K_{eff} = \Delta K_{app} - \Delta K_{brdg}$. The ΔK_{brdg} component was negligible for each test as evidenced by the equivalent crack growth rates produced by different specimen geometries, M(T) and SE(T), with different initial conditions (i.e., crack length and applied load) over a range of applied stress intensity factor range [8]. Since no large scale bridging occurred ($\Delta K_{brdg} << \Delta K_{app}$), it is, therefore, acceptable to correlate all the crack growth rates with the applied stress intensity factor range.

MODELING TMF CRACK GROWTH RATES

The modeling effort was based on concepts embedded in the previously mentioned linear summation models [1, 6-11] as well as by the experimental data generated during this study. The basic assumption for linear summation models for TMF fatigue crack growth behavior is that damage for any given cycle is a combination of a cycle-dependent and time-dependent damage component [1, 6-11] as expressed mathematically in Eq 1. The challenge is choosing the correct form of the cycle- and time-dependent terms that represent the physical phenomena experienced by the material. The expressions that define $(da/dN)_{cyc-dep}$ and $(da/dN)_{time-dep}$ vary widely depending on the type of behavior they attempt to model. The process by which the cycle- and time-dependent term were determined for the current linear summation model is discussed below.

Cycle-Dependent Term

The cycle-dependent term represents contribution to crack growth from pure mechanical fatigue experienced during a given cycle of the composite material. In general, the cycle-dependent term in a monolithic material is a function of ΔK_{app}, the stress ratio, and the temperature[6, 11]. Because purely cycle-dependent fatigue crack growth data for this composite was unavailable as a function of temperature and stress ratio, the cycle-dependent term was assumed to be a function of ΔK_{app} only. Vesier and Antolovich [23] reported that for monolithic Ti-6242, stress ratio affected fatigue crack growth rates for a given temperature, whereas temperature only slightly affected the fatigue crack growth rates for temperatures in the range used in this study. Since only one stress ratio (R = 0.1) is used during all fatigue crack growth tests for this study, and temperature does not influence the matrix crack growth rate markedly, the assumption that the cycle-dependent term is only a function of ΔK_{app} is considered sufficient to describe the results generated in this study.

In the TMF modeling efforts by Heil et al. [6] and Pernot et al. [11] a modified sigmodial equation (MSE) was used to describe the cycle-dependent behavior. Heil et al. [6] made the cycle-dependent MSE a function of ΔK_{app} and stress ratio, whereas Pernot et al. [11] made the cycle-dependent MSE a function of ΔK_{app} and temperature. A modified sigmodial equation has six independent empirical parameters, typically determined from experimental data. In the present case, since insufficient data were available to determine these constants, a simple Paris Law was used for the cycle-dependent term. Since the Paris Law is valid only in Region II, only the crack growth data in this region are modeled in this study. Predictions of crack growth rates in Regions I and III were not attempted because insufficient crack growth data were generated in those regions to determine an expression which would represent all the conditions used in this study.

The Paris Law has the form:

$$\frac{da}{dN} = C(\Delta K)^n \tag{5}$$

where C and n are empirical constants. The constant n was determined from the slope of the isothermal and TMF crack growth data generated for this study between ΔK of 50 to 90 MPa $\sqrt{m}$, with the exception of the 150°C isothermal data, where n was nearly zero due to fiber bridging (see Fig. 3). In general, the slope of the linear portion of each data set was similar enough to be considered a constant, and not a function of temperature, frequency or phase angle. For convenience in modeling, the exponent n was set equal to 2.7. The coefficient C was chosen so that, along with n = 2.7, the purely cycle-dependent crack growth rates were slightly less than the isothermal, 150°C, 0.0083 Hz fatigue crack growth data. Since the Paris Law does not accurately represented the slope of 150°C data because of fiber bridging, the fit to the 150°C data represents an idealization of the behavior as if there were no bridging. With this understanding, C was set equal to 7.9 x 10^-13 (where ΔK is in units of MPa $\sqrt{m}$ and da/dN is in m/cycle). The purely cycle-dependent term is now:

$$\frac{da}{dN} = 7.9 \times 10^{-13}(\Delta K)^{2.7} \tag{6}$$

which represents the crack growth data ranging from ΔK_{app} of 50 to 90 MPa $\sqrt{m}$.

Time-Dependent Term

The isothermal condition at 538°C and 0.0083 Hz, produced the highest crack growth rate of all the conditions tested in this study, whereas the isothermal condition at 150°C and 0.0083 Hz produced fatigue crack growth rates which were approximately two orders of magnitude slower (Fig. 3). The in-phase and out-of-phase tests cycled between 150°C and 538°C at 0.0083 Hz and had crack growth rates approximately a half an order of magnitude slower than the isothermal condition at 538°C. This behavior suggests that the time-dependent contribution to the total crack growth rate is not a linear function in temperature, since as the temperature approaches T_{max}, the time-dependent contribution greatly dominates the total crack growth rate.

An expression similar to the fiber damage term in Neu's model [24] for TMF in titanium matrix composite was used to model the time-dependent crack growth behavior. Neu suggested an Arrhenius-type expression to represent the weakening of a fiber because of time, temperature, and fiber stress. In general, the fatigue and fatigue crack growth behavior of titanium based composites reinforced with silicon-carbide fibers are greatly dependent upon the fibers' behavior [4]. Even the creep behavior of TMC with SCS-6 fibers is associated with stress-assisted environmental degradation of the carbon-rich layers of the SCS-6 fibers, followed by subsequent weakening and fracture of the fibers [25]. Therefore, the time-dependent component was chosen in a form similar to Neu's Arrhenius expression, except the fiber-stress component was replaced with an applied ΔK dependence. Because of the strong dependence on temperature and a weaker dependence on K, the ΔK^n term is taken outside the integral. This insures all curves representing time-dependent behavior have the same slope, m. The time-dependent term had the form:

$$\frac{da}{dN} = \int_{cycle} \frac{da}{dt} dt = (\Delta K)^m \int_0^{t_{nd}} C_1 e^{\frac{-C_2}{T_{abs}}} dt \qquad (7)$$

where C_1 is an empirical constant, T_{abs} is the absolute temperature (T °C + 273 K) at any instant during the cycle, and m is the exponent of ΔK. Based on the experimental data from a variety of tests involving combinations of cycle- and time-dependent behavior, m is taken to be the same as n because all the da/dN-ΔK curves are essentially parallel over the range of ΔK's between 50-90 MPa $\sqrt{m}$. The constant C_2 represents the ratio Q_{fib}/R, where Q_{fib} is the apparent activation energy (kJ/mol) for environmental attack of the fiber and R is the gas constant (kJ/mol/deg). The time, t_{nd}, represents the sum of the times during which load is increasing and held at P_{max}. This is the same upper limit of integration used by Pernot et al. [11] on the time-dependent term. The TMF crack growth model of Pernot et al. considered the effects of hold times; they found that the hold times at max load had to be accounted for in the time-dependent term for the predictions to match the experimental data. Since no hold times are included in this study, t_{nd} is actually equivalent to t_{inc}, the time the load is increasing.

To determine C_2, two of the TMF crack growth tests were used: the isothermal, 538°C, 0.0083 Hz condition and the in-phase 150-538°C, 0.0083 Hz condition. A value of C_1 was also found during this process, but it was later modified when frequency effects were taken into account.

C_1 and C_2 were initially estimated, but through trial and error, both C_1 and C_2 were adjusted such that the total crack growth rates for the isothermal and in-phase conditions matched the experimental data based on a best-eye fit. The best values of C_1 and C_2 were estimated to be 1.70E-5 and 12000, respectively. The relatively large value of C_2 suggests that the majority of the damage occurs when the temperature is at or near the maximum temperature. In fact, the time-dependent contribution calculated for the isothermal condition at 150°C is about two orders of magnitude smaller that the cycle-dependent term. In contrast, the time-dependent term is about an order of magnitude greater than the cycle-dependent contribution for the isothermal 538°C condition.

Effects of Frequency

Until now frequency effects were not explicitly considered in either Eq 5 or Eq 7. In general, the cycle-dependent term is independent of frequency, but since the time-dependent term in Eq 7 is implicit in time (T_{abs} depends on time during a TMF cycle), it is a function of frequency. While the data reveal that the crack growth behavior is affected by a change in frequency, the data never exhibit a completely time-dependent behavior for the conditions tested.

To account for crack growth that is not truly time-dependent, Nicholas and Mall [9] modified the time-dependent term in their linear summation model (Eq 3) by a coefficient of the form $T^{\gamma-1}$, where T is the total cycle time (or period). They used $\gamma = 0.5$, which yielded a time-dependent term that would vary with frequency as $\sqrt{T}$. A similar modification was used in the current study to account for frequency effects, $t_c^{\gamma-1}$. The value of γ for each type of test was determined by comparing the total crack growth rates for isothermal, in-phase, and out-of-phase conditions at two different frequencies. The value of γ for the in-phase, isothermal, and out-of-phase conditions was 0.62, 0.57, and 0.52, respectively.

The γ–term is not a constant because the frequency affects the magnitude of the fiber stress range at or near the crack tip over the life of the test according to the test type. Micromechanical modeling by Neu [22] of SCS-6/TIMETAL®21S $[0]_4$ under thermomechanical loading investigated the effect of frequency on the stress in the matrix and fiber. That study showed that during in-phase loading the fiber stress range became noticeably higher at a longer cycle period (30 min/cyc) than at the shorter cycle period (3 min/cyc). The fiber stress range increase was due to relaxation of stresses in the matrix at the longer cycle periods requiring the fiber to carry a greater percentage of the applied load. The out-of-phase condition showed very little change in fiber stress range between the long and short cycle periods. In the out-of-phase loading the maximum temperature occurs at the minimum load; a condition that does not lend itself to matrix stress relaxation.

The larger value of γ for the in-phase condition and lower value for the out-of-phase condition is expected since the influence of frequency on the in-phase condition on the fiber stress range is greater than the out-of-phase condition. The small difference, however, between the γs indicates that while the frequency affects the fiber stress range depending on the test conditions, the effect is not as pronounced as the micromechanical modeling of the SCS-6/TIMETAL®21S showed [22]. The change in fiber stress range due to a frequency change was not as pronounced during the fatigue crack growth tests because the applied loads and temperatures were not conducive to large scale matrix plastic deformation and stress relaxation away from the crack tip region.

Proposed TMF Linear Summation Model

The linear summation model proposed to predict the fatigue crack growth rates under isothermal and thermomechanical fatigue is written as a combination of Eqs 1, 5 and 7 and a frequency term.

$$\left.\frac{da}{dN}\right|_{total} = C(\Delta K)^n + t_c^{\gamma-1}(\Delta K)^m \int_0^{t_{tol}} C_1 e^{\frac{-C_2}{T_{abs}}} dt \tag{8}$$

The model in this form has five parameters, C, n, m, C_1 and C_2, which are fixed for all conditions, and the variable, γ, which is test-type dependent. The terms t_c and T_{abs} are the total cycle time and the absolute instantaneous temperature, respectively. Substituting the values for each of the five parameters (where m=n), Eq 8 becomes

$$\frac{da}{dN} = 7.9E-13(\Delta K)^{2.7} + t_c^{\gamma-1}(\Delta K)^{2.7} \int_0^{t_{tol}} 1.70E-5e^{-12000/(T+273)} dt \tag{9}$$

where γ equals 0.62, 0.57 and 0.52 for the in-phase, isothermal and out-of-phase conditions, respectively. Recall, that since the cycle-dependent term is valid for ΔK's between 50 and 90 MPa $\sqrt{m}$, Eq 9 is accurate only within this ΔK range.

Based on Eq 9, crack growth rates were calculated for each of the seven different baseline conditions for ΔK's between 50 and 90 MPa $\sqrt{m}$. The calculated growth rates were then compared to the experimental data to verify the model's ability to correlate the baseline data. The correlation between the three isothermal data sets and results from the model (Eq 9) is shown in Fig. 3. As expected, the crack growth rates calculated from the model for each of the isothermal conditions correlate well with the experimental data in Region II. The correlation between the two in-phase and two out-of-phase data sets and results from the model (Eq 9) is shown in Figs. 4 and 5. Again, the computed crack growth rates correlate well with the experimental data in Region II.

PROOF TEST

A proof test was conducted to check the predictive capability of the linear summation model. The goal of the proof test was to use the predictive model to develop a complex TMF history which would yield a constant crack growth rate over a wide range of ΔK. The proposed proof test begins under isothermal ($T_{max}=T_{min}=538°C$) conditions, continues with T_{min} decreasing throughout the test and finishes under in-phase conditions with $T_{min}=150°C$. Decreasing T_{min} according to a predefined profile will produce a constant crack growth rate for the life of the specimen. A constant crack growth rate is possible since two competing damage mechanisms are acting. As the matrix crack grows through the composite (assuming no fiber bridging) the stress intensity factor range increases which leads to an increase in the crack growth rate. If during this same test, the minimum temperature is decreased the crack growth rate will tend to decrease since growth rate is dependent on time-at-temperature. Appropriately combining these two opposing mechanisms in one test should produce a constant growth rate.

The critical factor in achieving a constant crack growth rate test is determining the minimum temperature profile as a function of elapsed cycles. To determine the exact T_{min} profile for the proof test, the linear summation model (Eq 9) was used. The proof test, using an M(T) geometry, was to start at T=538°C with a constant cyclic force, ΔP, of 10 kN. The desired constant crack growth rate was 1.41×10^{-6} m/cycle. Knowing the beginning and ending ΔK_{app}, 42 and 80 MPa $\sqrt{m}$, respectively, the number of cycles the test would run was found to be 2750. The minimum temperature profile needed to produce a near constant crack growth rate for the given initial conditions is described by the equation

$$T_{min} = 150 + 2.0839(2750 - N)^{0.66} (°C) \qquad (10)$$

Here N is the cycle count, and at N=2750 the minimum temperature equals 150°C as the proof test required.

Experimental Results of Proof Test

To conduct the proof test with the desired T_{min} profile, the test system control software was modified so that the minimum temperature setpoint would update according to Eq 10 each cycle. The minimum and maximum temperature followed the desired profile successfully as is illustrated in Fig. 6. Because of an output hardware error, however, the test stopped at cycle 2500 and not at the desired 2750. Nonetheless, a significant portion of the test was completed and the data generated were similar to what was predicted during the simulation. The DCEP crack lengths were similar to those calculated using Eq 9 as shown in Fig. 7.

The crack growth rate is plotted as a function of ΔK_{app} and compared with the growth rate from the simulation. Fig. 8 shows that the simulation predicted a slightly decreasing growth rate over the life of the specimen. Nevertheless, the crack growth rates generated during the proof test followed the general trend predicted by the simulation. More specifically, the experimental crack growth rate never deviated from the expected growth rate by more than a factor of 1.5. This deviation from the expected crack growth rate is quite small considering the amount of variability that is commonly observed in the crack growth rates [26] of monolithic materials. Depending on the data reduction technique [27]

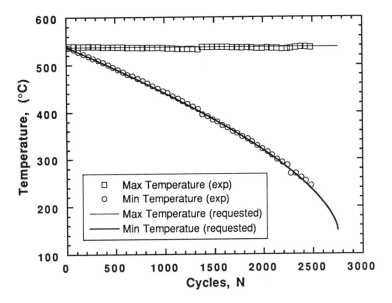

FIG. 6--Experimental data verifying the minimum and maximum temperatures during the proof test followed the requested profiles.

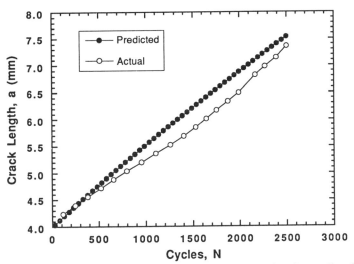

FIG. 7--Experimental crack lengths from the proof test compared to the predicted values as a function of applied cycles.

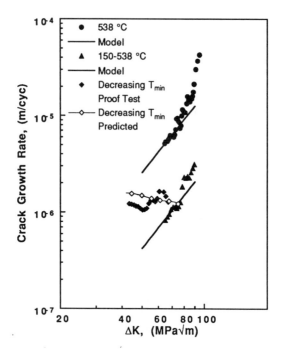

FIG. 8--Experimental crack growth rates from the proof test compared to the predicted values as a function of ΔK_{app}.

and the homogeneity of the material, fatigue growth rates can vary up to a factor of 3 [26] for replicate crack growth tests of the same material.

The results of the proof test indicate that the proposed linear summation model successfully captured the primary features which control the crack growth behavior of the SCS-6/Ti-6242. Namely, the time-at-temperature significantly influences the crack growth behavior. While a single proof test does not offer absolute evidence that the model is entirely accurate, it does strongly support the fundamental notions built into the model.

PARAMETRIC STUDY OF MODEL

Having provided a limited experimental evaluation of the numerical model, it was decided to use it as a tool to further investigate TMF in SCS-6/Ti-6242. This section describes the results of two parametric studies. One study investigated the effect of load hold times during isothermal cycling. The other study determined the effect of decreasing the minimum temperature during in-phase TMF conditions. All the studies were conducted at a ΔK_{app} of 60 MPa $\sqrt{m}$ since that represented the approximate middle of Region II of the experimental data.

The results of the load-hold-times study indicated, as expected, that as the load hold times increase under isothermal conditions, the crack growth rates increase as illustrated in Fig. 9. Since the upper integration limit of the time-dependent component includes hold times, the model provides that the crack growth rate should increase. For the slower frequency (0.00083 Hz) an increase in hold time increases the growth rate only slightly, but at the higher frequency (8.3 Hz) the shorter hold times yielded the greatest change in crack growth rates. As the hold times increased the resulting change in crack growth rates was less and less since, at the slower frequencies, the hold times contribute proportionately less to the time-dependent component than the loading portion of the cycle.

The in-phase study illustrated how increasing the minimum temperature of an in-phase cycle increases the crack growth rate toward that of the isothermal condition at the maximum temperature, in this study, 538°C. The magnitude to which the growth rate increased as the minimum temperature was increased to a completely isothermal condition was directly influenced by the frequency. This is clearly shown in Fig. 10 in which the growth rate, normalized with respect to the growth rate for $T_{min} = 150$°C, is plotted as function of T_{min} for frequencies ranging from 0.00083 Hz to 8.3 Hz. For the fast frequency, the growth rate only increased about a factor of two when T_{min} changed from 150°C to 538°C or T_{max}. For the slowest frequency, the growth rate differed by a factor of eight between $T_{min} = 150$°C and $T_{min} = T_{max} = 538$°C.

SUMMARY AND CONCLUSIONS

The thermomechanical fatigue crack growth behavior of 4-ply, unidirectional SCS-6/Ti-6Al-2Sn-4Zr-2Mo was evaluated. TMF experiments, using a fully automated TMF test frame, generated baseline isothermal and non-isothermal fatigue crack growth data. A linear summation model was developed to predict the fatigue crack growth rates of titanium matrix composites in the absence of large scale fiber bridging under isothermal as well as thermomechanical fatigue conditions.

None of the crack growth tests, except for the isothermal test at 150°C, produced evidence that full scale fiber bridging occurred. It is believed that fiber bridging was limited to approximately 2-3 fibers behind the crack tip. Based on this observation, the crack growth in this composite was modeled essentially as a monolithic material using a linear summation approach, assuming that fiber bridging only occurs on a small scale. The linear summation approach assumed that the fatigue crack growth rate could be decomposed into a cycle dependent component and a time dependent component. The cycle dependent term is dependent upon the applied stress intensity factor range, whereas the time-dependent term also depends on the cycle period and the temperature profile.

The experimental fatigue crack growth data suggested several conclusions that the model incorporates. One dominant factor that influenced the fatigue crack growth behavior was the time that the composite was cycled at elevated temperatures, especially when temperatures approached 538°C. Accordingly, the crack growth rates under the thermomechanical conditions were lower than those generated isothermally at the maximum temperature of the TMF test at the same frequency and load ratio. This behavior is attributed to the extended exposure to the elevated temperatures which degraded the integrity and strength of the fiber at and behind the matrix crack tip. The degradation of the mechanical properties of the fiber led to fiber fracture and little fiber bridging.

The TMF crack growth data indicated that the test frequency influenced the growth rates depending on the phase angle. The in-phase condition tended to have higher growth rates at the slower frequency than the out-of-phase condition. This was attributed to more matrix relaxation in the in-phase condition than the out-of-phase condition which lead to higher fiber stresses. The higher fiber stresses lead to accelerated fiber fracture, and in turn to faster crack growth rates.

The linear summation model was able to predict the fatigue crack growth rate of a proof test which involved a continual change in temperature range and load range to produce a constant crack growth rate. The proof test began under isothermal conditions at the maximum temperature and ended under in-phase conditions.

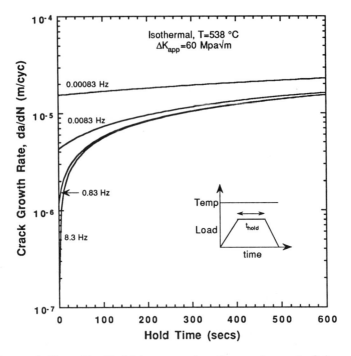

FIG. 9--Proposed effect of load hold times at various frequencies on the fatigue crack growth rate of [0]₄, SCS-6/Ti-6Al-2Sn-4Zr-2Mo.

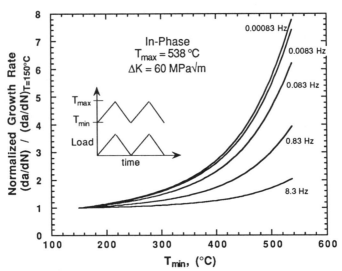

FIG. 10--Proposed effect of varying T_{min} at various frequencies on the in-phase fatigue crack growth rate of [0]₄, SCS-6/Ti-6Al-2Sn-4Zr-2Mo.

ACKNOWLEDGMENTS

The first author was supported by the US Air Force Palace/Senior Knight Program. The technical support of the University of Dayton Research Institute under contract number F33615-91-C-5606 is gratefully acknowledged.

REFERENCES

[1]   Nicholas, T., Heil, M. L., and Haritos, G. K., "Predicting Crack Growth Under Thermo-Mechanical Cycling," *International Journal of Fracture*, Vol. 41, No. 3, 1989, pp. 157-176.

[2]   Davidson, D. L., "The Micromechanics of Fatigue Crack Growth at 25 °C in Ti-6Al-4V Reinforced with SCS-6 Fibers," *Metall. Trans. A*, Vol. 23A, 1992, pp. 865-879.

[3]   Ghosn, L. J., Telesman, J., and Kantzos, P., "Fatigue Crack Growth in Unidirectional Metal Matrix Composite," Prepared for the International Fatigue Series (Fatigue 90), NASA-TM-103102, July, 1990.

[4]   Larsen, J. M., Jira, J. R., John, R., and Ashbaugh, N. E. "Crack Bridging Effects in Notch Fatigue of SCS-6/Timetal 21S Composite Laminates," *Submitted for publication in Life Prediction Methodology for Titanium Matrix Composites*, ASTM Special Technical Publication, eds. W.S. Johnson, J.M. Larsen, and B.N. Cox, American Society for Testing and Materials, Philadelphia, PA, 1994.

[5]   McMeeking, R. M. and Evans, A. G., "Matrix Fatigue Cracking in Fiber Composites," *Mech. Mater.*, Vol. 9, 1990, pp. 217-227.

[6]   Heil, M. L., Nicholas, T., and Haritos, G. K., "Crack Growth in Alloy 718 Under Thermal-Mechanical Cycling," *Thermal Stress, Materials Deformation and Thermo-Mechanical Fatigue,* H. Sehitoglu and S.Y. Zamrik, Editor, American Society of Mechanical Engineers, New York, 1987, pp. 23-29.

[7]   Nicholas, T., Weerasooriya, T., and Ashbaugh, N. E. "A Model for Creep/Fatigue Interactions in Alloy 718," *Fracture Mechanics: Sixteenth Symposium*, ASTM STP 868, ed. M.F. Kanninen and A.T. Hopper, American Society for Testing and Materials, 1985, pp. 167-180

[8]   Mall, S., Staubs, E. A., and Nicholas, T., "Investigation of Creep/Fatigue Interaction on Crack Growth in a Titanium Aluminide Alloy," *Journal of Engineering Materials and Technology*, Vol. 112, 1990, pp. 435-441.

[9]   Nicholas, T. and Mall, S. "Elevated Temperature Crack Growth in Aircraft Engine Materials," *Advances in Fatigue Lifetime Predictive Techniques*, Vol. ed. M.R. Mitchell and R.W. Landgraf, American Society of Testing and Materials, 1992, pp. 143-157

[10]   Pernot, J. J., "Crack Growth Rate Modeling of a Titanium-Aluminide Alloy Under Thermal-Mechanical Cycling," Air Force Institute of Technology, Wright-Patterson AFB, Ohio, Ph.D. Thesis, 1991.

[11]   Pernot, J. J., Nicholas, T., and Mall, S., "Modeling Thermomechanical Fatigue Crack Growth Rates in Ti-24Al-11Nb," *International Journal of Fatigue*, Vol. 16, No. 2, 1994, pp. 111-122.

[12]   Blatt, D., "Fatigue Crack Growth Behavior of a Titanium Matrix Composite Under Thermomechanical Loading," Purdue University, Ph.D. Thesis, 1993.

[13]   Dao, T. X. and Mettu, S. R., "Analysis of an Edge-Cracked Specimen Subjected to Rotationally-Constrained End Displacements," NASA - Johnson Space Center, Technical Memorandum, NASA JSC 32171, August, 1991.

[14]   Blatt, D., John, R., and Coker, D., "Stress Intensity Factor and Compliance Solutions for a Single Edge Notched Specimen with Clamped Ends," *Engineering Fracture Mechanics*, Vol. 47, No: 4, 1994, pp. 521-532.

[15] Tada, H., Paris, P. C., and Irwin, G. R., "The Stress Analysis of Cracks Handbook," St. Louis, Del Research Corporation, 1985,

[16] Hartman, G. A., "MATE and MATE Modules — Version 2.22A, Crack Growth Analysis and Test Environments," University of Dayton Research Institute, Technical Report, UNR-TR-88-138, 1988.

[17] Blatt, D. and Hartman, G. A., "A Methodology for Thermomechanical Fatigue Crack Growth Testing of Metal Matrix Composites," *(submitted) Experimental Mechanics*, 1994,

[18] Balsone, S. J., Maxwell, D. C., Khobaib, M., and Nicholas, T. "Frequency, Temperature, and Environmental Effects on Fatigue Crack Growth in Ti3Al," *Fatigue 90*, Vol. II, ed. H. Kitagawa and T. Tanaka, Materials and Components Engineering Publications LTD, Birmingham UK, 1990, pp. 1173-1178

[19] DeLuca, D. P., Cowles, B. A., Haake, F. K., and Holland, K. P., "Fatigue and Fracture of Titanium Aluminides," Wright-Patterson AFB, OH, WRDC-TR-4136, Feb, 1990.

[20] Parida, B. K. and Nicholas, T., "Frequency and Hold-Time Effects on Crack Growth of Ti-24Al-11Nb at High Temperature," *Material Science Engineering*, Vol. A153, 1992, pp. 493-498.

[21] Weerasooriya, T., "Effect of Frequency on Fatigue Crack Growth Rate of Inconel 718 at High Temperature," *Fracture Mechanics: Nineteenth Symposium*, American Society of Testing and Materials, 1988, pp. 907-923.

[22] Neu, R. W., Coker, D., and Nicholas, T., "Cyclic Behavior of Unidirectional and Cross-ply Titanium Matrix Composites," *(submitted) International Journal of Plasticity*, 1994,

[23] Vesier, L. S. and Antolovich, S. D., "Fatigue Crack Propagation in Ti-6242 as a Function of Temperature and Waveform," *Engineering Fracture Mechanics*, Vol. 37, No. 4, 1990, pp. 753-775.

[24] Neu, R. W., "A Mechanistic-based Thermomechanical Fatigue Life Prediction Model for Metal Matrix Composites," *Fatigue and Fracture of Engineering Materials and Structures*, Vol. 16, No. 8, 1993, pp. 811-828.

[25] Khobaib, M., "Damage Evolution in Creep of SCS-6/Ti-24Al-11Nb Metal Matrix Composites," *Journal of Reinforced Plastics and Composites*, Vol. 12, 1993, pp. 296-310.

[26] Virkler, D. A., Hillberry, B. M., and Goel, P. K., "The Statistical Nature of Fatigue Crack Propagation," *Journal of Engineering Materials and Technology, Transactions of the American Society of Mechanical Engineers*, Vol. 101, 1979, pp. 148-153.

[27] Ostergaard, D. F., Thomas, J. R., and Hillberry, B. M., "Effect of Δa-Increment on Calculating da/dN from a versus N Data," *Fatigue Crack Growth Measurement and Data Analysis,* J. Hudak S.J. and R.J. Bucci, Editor, American Society for Testing and Materials, 1981, pp. 194-204.

Author Index

Subject Index